Ingenious Yankees

INGENIOUS YANKEES

The Rise of
the American System
of Manufactures
in the Private
Sector

DONALD R. HOKE

Columbia University Press
NEW YORK

Publication of this book was supported by a gift from David Bendel Hertz, Ph.D. Columbia University 1949, and charter member of the Friends of the Columbia University Press.

Columbia University Press
New York Oxford

Library of Congress Cataloging-in-Publication Data

Hoke, Donald.
Ingenious Yankees : the rise of the American system of manufactures
in the private sector / Donald R. Hoke.
p. cm.
Originally issued as the author's thesis (Ph.D.)—
University of Wisconsin—Madison, 1984.
Bibliography: p.
Includes index.
ISBN 0-231-06756-9
1. United States—Manufactures—History—Case studies.
2. Clocks and watches—United States—History.
3. Axe industry—United States—History.
4. Typewriter industry—United States—History.
5. Manufacturing processes—United States—History.
I. Title.
HD9725.H65 1989
338.4'767'0973—dc19
89-965
CIP

Casebound editions of Columbia University Press books are Smyth-sewn and printed on permanent and durable acid-free paper

Book design by Claire B. Counihan

Printed in the United States of America

Dedication

EARLY in a scholar's career, he is influenced by and often greatly indebted to a particular professor. In my case, that professor was Dr. Stephen Salsbury, now of the University of Sydney. In 1975 as Chairman of the History Department at the University of Delaware, Salsbury looked beneath the antiquarian crust of my intellect and detected an undeveloped enthusiasm for history. Salsbury (with the assistance of Reed Geiger) undertook the job that any good teacher would do, to teach me what history was and how historians did it. This was a tedious, often tiring and frustrating experience for both of us, but it was also exciting and stimulating. In 1978, following his move to the University of Sydney in Australia, Professor Salsbury wrote the letter of recommendation that gained my entrance to the University of Wisconsin at Madison, where I took my Ph.D. The success of *Ingenious Yankees,* first as the 1985 Nevins prizewinning dissertation and now as a monograph, reflects as much as anything else the influence of Steve Salsbury. He, more than anyone else, should bask in its reflected glory. Its shortcomings merely manifest my inability to absorb all that Salsbury had to offer. I hope he is not disappointed.

To Stephen Salsbury
with deepest gratitude.

Introduction to the 2007 Reprint

First a very special thanks to Chairman Dave Gorrell not merely for the invitation to deliver the James Arthur keynote lecture at the 2007 Ward Francillon Seminar, but also for partnering to reprint *Ingenious Yankees* as a part of the Seminar. The Francillon Seminar is sponsored by the National Association of Watch & Clock Collectors Chapter 1, Philadelphia, PA. This annual horological seminar is named in memory of Ward Francillon, who was an expert in wooden movement clocks. Having hosted the 1980 annual meeting of the Society for the History of Technology in Milwaukee, I fully appreciate the time and effort Dave has invested in the Francillon Seminar.

Ingenious Yankees is dedicated to Stephen Salsbury who successfully salvaged my young intellectual career when others would not or could not, first at the University of Delaware and then later from the University of Sydney in Australia. Twenty-nine years ago Steve's letter to Mort Rothstein guaranteed my entry into graduate history program at the University of Wisconsin at Madison. I never saw Steve's letter, but when I interviewed with Mort during the application process in 1978, Mort simply said "I've read Steve's letter and you're in if you want to come." Six years later, I completed my Ph.D. under Mort Rothstein's and Diane Lindstrom's direction at Madison.

Sadly, Steve Salsbury died in March 1998, apparently while working at his desk in Sydney. Steve wrote to me on March 28, 1990 after receiving the book. "I was thrilled to receive a copy of your book. I have never had a book dedicated to me before and it is a real pleasure. I will treasure it for many years to come."

Twenty-nine years later I retain a strong emotional attachment to Steve and an undying gratitude for what he did for me. I will always remember sitting in Steve's office while he was trying to teach me how to write and how to grow from an antiquarian into a historian. He had a habit of sliding down in his easy chair so that his back was on the seat of the chair and his head on the back. Steve continues to inspire me to work with my younger museum staff members.

Returning to *Ingenious Yankees* to prepare the James Arthur Lecture was an interesting intellectual exercise, as my career has taken me into museum administration, which is primarily fund raising. I was privileged to work for nine years in Appleton, WI, helping Doug and Myrt Ogilvie see their dream of a professional museum come to fruition at the Outagamie County Historical Society. I was honored to be the first director to hold the Niels C. & Elisabeth Miller Executive Director position. Three Outagamie Museum staff hold positions endowed by Doug and Myrt, a permanent tribute to their dedication to history.

My interest in the History of Technology remains strong. Having passed through my "watch and clock" phase and my "typewriter phase," I have suffered a relapse of "steam car disease," from which I suffered in high school. Today, I find myself restoring a 1925 Stanley Steamer Model SV 252G, writing a book on the death throws of the Stanley Steamer, and creating a virtual steam car museum on the web. www.stanleysteameronline.com

There is an error in the Wooden Movement Clock chapter on page 55. Figures 2.4A and 2.4B illustrate an Eli Terry clock from the pre-Porter Contact period. It is not one of the first 1000 clocks produced under the Porter Contract. Ward Francillon and Chris Bailey brought this error to my attention at the Houston Seminar on October 26, 1990. I thank them for their correction.

Donald R. Hoke, Ph.D.
Dallas, TX

Dr. Stephen Salsbury
October 12, 1931 - March 1, 1998

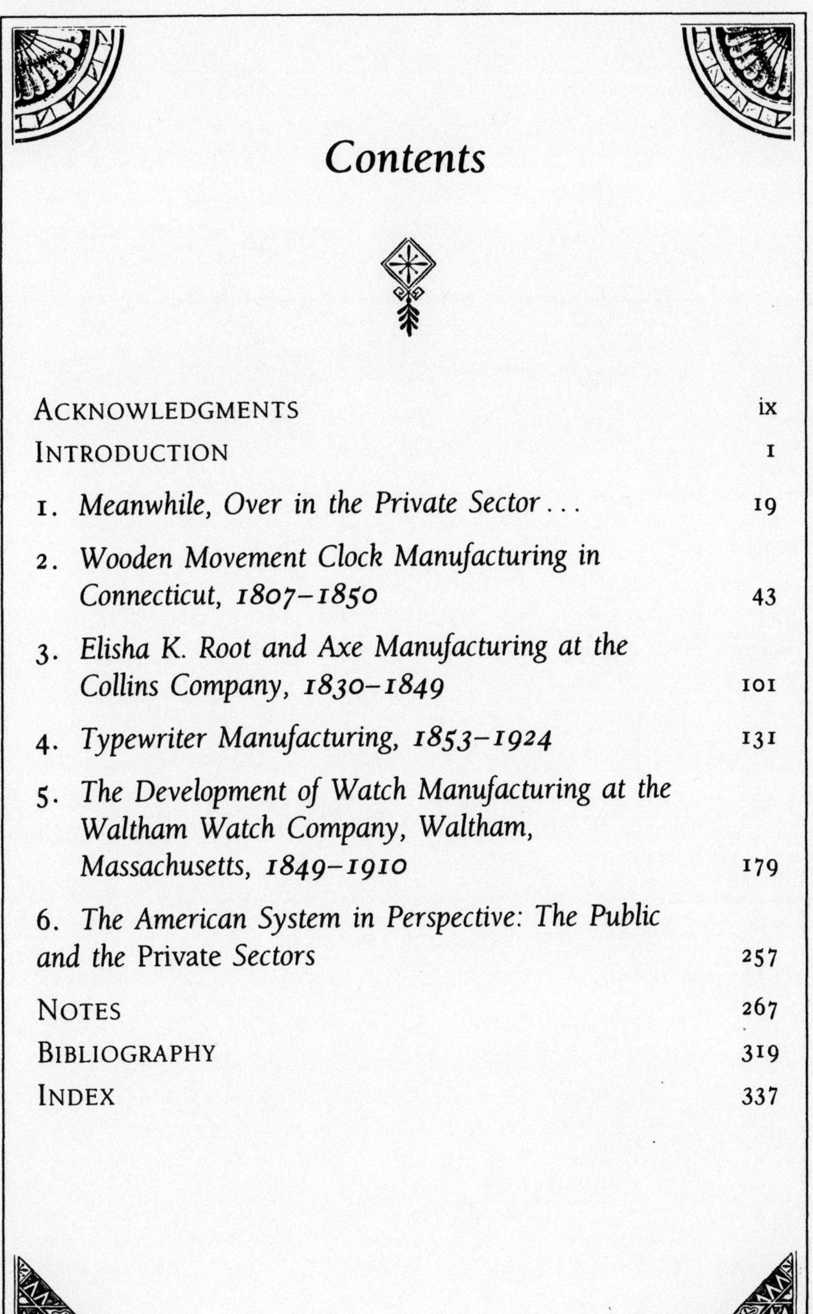

Contents

The 2007 Ward Francillon Time Symposium

October 24-26, 2007

The Impact of Mass Production on the American Clock Industry

at The Yorktowne Hotel, York, PA

Sponsored by NAWCC Chapter 1 Philadelphia, PA

Wednesday, October 24, 7:00 p.m. —
"The Beginning of Mass Production, Eli Terry, and the Porter Contract" by James Arthur Lecturer Donald R. Hoke, Ph.D., author of *Ingenious Yankees*

October 25 Presentations

"Significant Contributions of the Chauncey Jerome, E. C. Brewster, and Elias Ingraham Companies from 1835 to 1850"
by Tom Grimshaw

"Joseph Ives—The Man and His Clocks" by Wayne Laning

"New England's Revolution in Precision Manufacturing, 1847"
by Dr. Kenneth Aiken

"Atkins and the Early Entrepreneurs" by Phil Gregory

"The Ansonia Clock Co. and the Movements They Used" by Allen Stevenson

"Seth Thomas, A Unique Way of Doing Business" by Ralph Pokluda

October 25 Evening Activities

A visit to the National Watch and Clock Museum, Columbia, PA, to view the Lewis B. Winton Collection of Seth Thomas' original jigs, fixtures, and gauges and to celebrate the 30th Anniversary of the National Watch and Clock Museum

October 26 Presentations

"Workers and the Twentieth-Century Workplace in Bristol, Connecticut's Clock and Watch Industry" by Dr. Philip Samponaro

"Methods of Modern Clock Movement Production" by Mark Butterworth

October 26 Evening Activities

Symposium Banquet—The Yorktowne Hotel, York, PA

Acknowledgments

IN writing *Ingenious Yankees,* I received the help of many people. First, I must thank the Smithsonian Institution and the Smithsonian's pre-doctoral screening committee at the Museum of American History for choosing my proposal for support. I simply could not have conducted the research without the Smithsonian's pre-doctoral fellowship. I trust the committee finds its choice a wise one. The Smithsonian's Office of Fellowships and Grants was cooperative and supportive in every way during my pre-doctoral fellowship. This was my second opportunity to work at the Smithsonian through that office. I sincerely appreciated its efforts in 1983 as I did in 1972 as a National Science Foundation undergraduate research scholar.

Otto Mayr (now the Deutches Museum director), Carlene Stephens, David Todd, and Kay Youngflesh of the Division of Mechanisms at the Museum of American History put up with many requests and inquiries. Pete Daniels, Brooke Hindle, and Debbie Warner read an early wooden movement clock chapter draft. Barb Janssen provided Remington data from the Division of Textiles.

Raylene Leavitt saved me hours of typing time by providing me with a floppy diskette of my work from the Wang word processor at MAH. The Smithsonian's photo lab supplied many photographs, many of which are found in the pages that follow. The keepers of the Warsaw trade literature collection put up with my many requests for photographs and Xerox copies.

Dr. Rudolph Dornemann, my section head at the Milwaukee Public Museum, and Dr. Kenneth Starr, most generously allowed me the time to accept my Smithsonian fellowship. Without their cooperation and support, my work would have been long delayed. Every curator should have a supervisor as patient and understanding as Rudy.

Anita BaergVatndal, of the Milwaukee Public Museum History Section, was of immense help in finding objects, running down stray catalogue and negative numbers, and assisting in numerous details. Mel Scherbarth, the Milwaukee Public Museum photographer, shot and printed several specimens, producing excellent views and exposures. Susan Otto, the Milwaukee Public Museum Photo Collections Manager, contributed in no small way by helping find and print negatives and providing negative numbers. Howard Madaus, the Curator of the Nunnemacher Arms Collection, generously shared his expertise and bibliographic sources and helped procure several photographs on rather short notice. It was a pleasure to have worked for eight years at one of the world's great museums and with such dedicated staff. I hope my affiliation with the Milwaukee Public Museum will continue.

Several members of the National Association of Watch & Clock Collectors gave generously of their time and expertise. Ward Francillon and especially George Bruno come instantly to mind, as do those collectors who contributed to the *Cog Counter's Journal* and the *Bulletin of the N.A.W.C.C., Inc.* Chris Bailey of the American Watch & Clock Museum in Bristol, Connecticut, kindly allowed me to work with the Hopkins & Alfred wooden movement clock material as well as other specimens. Dana Blackwell also contributed to my understanding of horology, as he does every time I speak with him.

Philip Dunbar of the Connecticut Historical Society was most patient with me during my thorough examination of the Seth Thomas wooden movement clock material. He opened his files and shared information which was most helpful.

Florence Bartosheky, Margorie Kierstead, and George Pano of the Harvard Business School's Baker Library kindly helped me with the Waltham Watch Papers. Margorie Kierstead was especially sympathetic during a lost week of work after my automobile was robbed in Cambridge. Clyde Helfter and Patrick Gabor of the Buffalo & Erie County Historical Society kindly gave me access to the Dawes Typewriter, and the late Violet Hosler, Secretary-Treasurer of the Onondaga County Historical Association in Syracuse, New York, graciously allowed me to use her superb local history research files. Every local historical society should have a director as dedicated as Ms. Hosler.

Charles Molloy of the Customer Affairs Department of SCM, Inc., in DeWitt, New York, and his assistants Virginia McElroy and Linda Dann cheerfully allowed me to work in the SCM historical files and subsequently loaned me several items which were instrumental in writing the typewriter chapter. Edson Mosher, a long time SCM employee, was primarily responsible for compiling this data.

Sunny Addison, reference librarian at the Public Library of Anniston and Calhoun County, Alabama, and Eugenia Rankin of the State of Alabama's Department of Archives and History kindly sent me material on John Pratt. Carol Emrich of the Local History Room at the Rochester Public Library took my search for information on John Jones as a personal crusade and found several interesting bits of data. Larry Goodstall, curator of the Remington Gun Museum at Ilion, New York, allowed me to use the Remington Sewing Machine Company's board of directors' minutes which we found in a corner of the factory. Jane Spellman, director of the Herkimer County Historical Society, provided me with some information I had not found elsewhere. Charles Wilt of the Franklin Institute Library helped me find material on John Cooper.

Patty Atwood, the director of The Time Museum in Rockford, Illinois, was particularly helpful in securing photographs, especially from Marsh's book on automatic machinery. In addition, she graciously granted permission to publish chapter five, some sections of which will appear in greater detail as the major historical essay in The Time Museum's catalogue of American watches. Steve Pitkin took many of The Time Museum's photographs.

Judy Cochran, who works in the History Department at the University of Wisconsin, has probably helped hundreds of graduate students survive the process of completing a Ph.D. at Madison, but she was especially helpful in my case. During my fellowship at Smithsonian, she handled a number of important details, conveyed information to my advisers, and saw that I met all the necessary requirements. Her advice was perfect in every instance, and while she may find the process of repeating the same advice to each graduate student year after year a bit routine, she should know that in my case her help was indispensable. I am deeply appreciative of her help.

Finally, I owe an immense debt to Professors Rothstein and Lindstrom, my principal dissertation advisers. Mort Rothstein is unequivocally the most sensible, level-headed academic I have ever known. Diane Lindstrom seemed to know exactly when to praise and when to criticize. And she does know how to criticize! Together, these two scholars helped me plan a strategy to succeed. They advised and prodded, questioned and encouraged, and pushed and drove me throughout my seemingly endless part-time career at Madison. No graduate student was ever better guided than I was by Diane Lindstrom, and no graduate student was ever better advised than I was by Mort Rothstein.

I have read many sensitive dedications and dozens of amusing acknowledgments. I could choose a shelf in the Madison library at random and find a good example in any one of its books. I always thought that I too would write something clever and witty when the time came. The time is here and my creativity has failed me. I can think of nothing to say to Mort and Diane to express my sincere gratitude except a heartfelt thank you.

Ingenious Yankees

Introduction

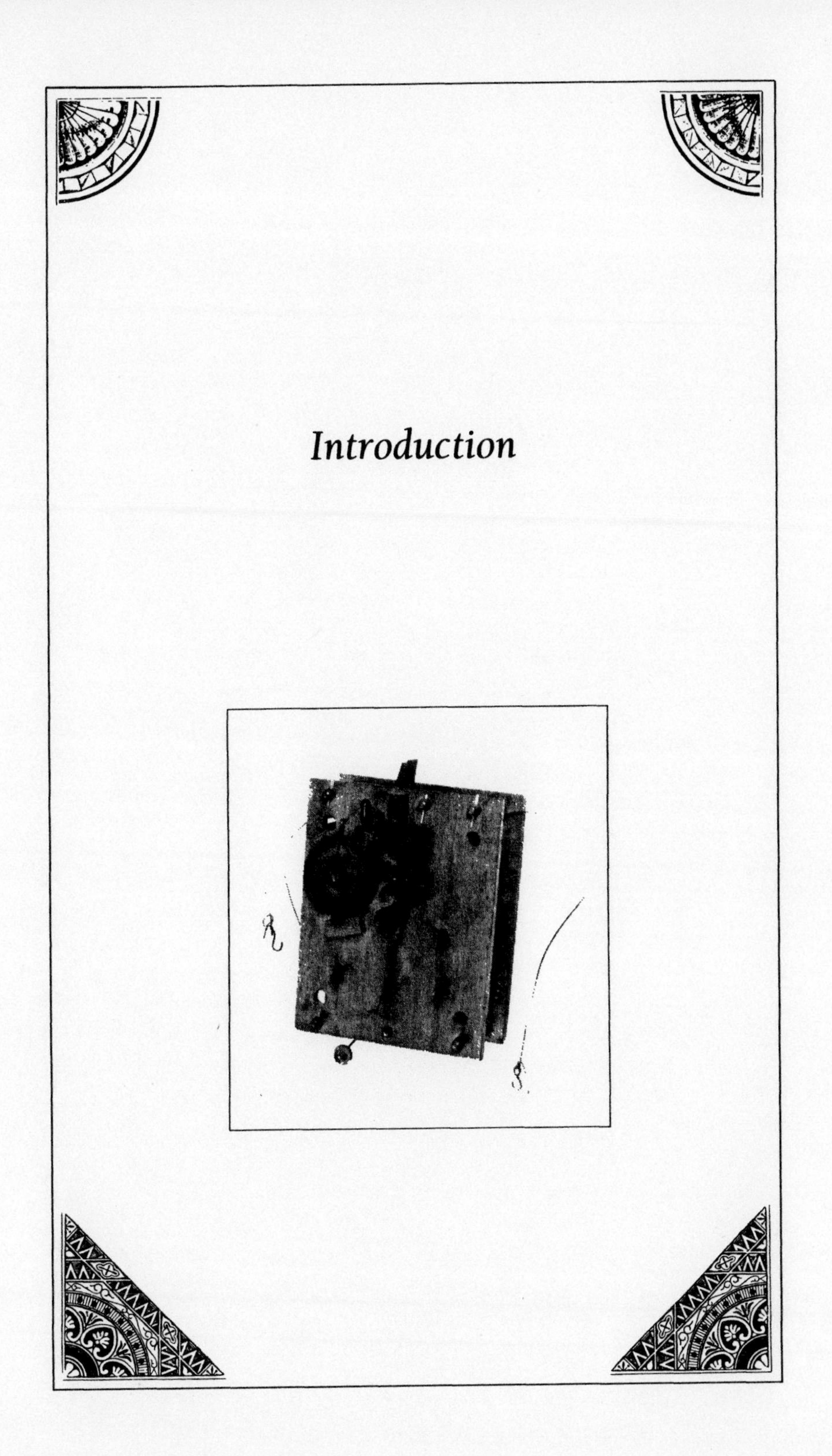

As is well known, the uniformity system developed in the armories became the basis of the so-called American system of manufactures, characterized by special machinery, precise gauges, and interchangeability of parts. Men left the arms business to set up the machine tool industry and went on from there to carry the principle of uniformity into the manufacture of railroad equipment, sewing machines, pocket watches, typewriters, agricultural implements, bicycles, and so on. The rest, as they say, is history, the history of progress.

But there was another side to this story, which we have not heard much about.

DAVID F. NOBLE
"Command Performance"[1]

THERE certainly is "another side!"

Ingenious Yankees is the progeny of the marriage between my life long interest in old mechanical things and the inspiration of Stephen Salsbury that I become more than a mere antiquarian. In exploring the origins of the American System of Manufactures, my methodology and my conclusions differ strikingly from the works on which they build, particularly Merritt Roe Smith's and David Hounshell's. This book interprets the evidence drawn from objects to reach conclusions about nineteenth-century manufacturing. Most studies of nineteenth-century technology begin with documents and use objects to illustrate the written evidence found in them. The documents used in this study serve primarily to corroborate the physical evidence found in the objects.[2] Readers will not be surprised to learn that I spent several summers working with Edwin Battison at the Smithsonian's National Museum of History and Technology.

The use of objects in writing the history of technology is critically important. As my early mentor Ed Battison so eloquently illustrated in his analysis of Whitney muskets, the objects often speak for themselves.[3]

Prior to Battison's work, many historians had written about Eli Whitney but none had "read" a Whitney musket. Perhaps they didn't know how

The objects under study in *Ingenious Yankees*—wooden movement clocks, axes, typewriters, and watches—were manufactured and sold by the private sector mechanics and entrepreneurs who originated the American System of Manufactures. In many instances, and particularly in the case studies of wooden movement clocks and typewriters, surviving objects represent the primary source and most significant body of evidence on which to base any interpretation. This research is largely based on "reading the objects."

"Reading an object" implies the careful analysis of a specimen and drawing conclusions from the surviving physical evidence. Certainly "reading an object" is not perfect. Many objects have long histories of alteration and change as well as damage and repair, not to mention deterioration. Museum curators and clock collectors, for example, spend much of their time trying to determine how much of a particular timepiece is "original" and how much is "restoration." "Was this seventeenth century lantern clock converted from a crown wheel escapement to an anchor escapement in the eighteenth century and "restored" in the late nineteenth century?" There is always the problem of fakes and forgeries.

Such pitfalls not withstanding, historians may generate solid data from the examination of objects and may interpret these data as legitimately as they may interpret the data found in documents. The question for historians of American technology and the American System is simple. How much can we learn from objects? Lots—if we look.

Such new ideas as the importance of product design in American System products, varying degrees of interchangeability based on the nature of the product, the concept of designing adjustability into the product and adjusting the product as an integral part of its manufacturing process are not to be found in paper documents. They are only to be found by reading three dimensional documents, the objects themselves.

Ingenious Yankees builds upon the efforts of several exceptional historians, notably Merritt Roe Smith and David Hounshell.[4] It is a substantially better work for having followed Smith's and Hounshell's research and analysis. In his prize-winning book, *From the American System to Mass Production, 1800–1932*, Hounshell set the terms of the debate on the origins of the American System for the first time. Historians finally have something to debate. So let's debate.

There are many areas of contention; three critical issues bear special attention:

1. The origin of the American System (government armories vs. the private sector),
2. why the American System appeared (idealistic army officers vs. profit maximizing entrepreneurs), and
3. the interpretation of the surviving data.

Both Smith and Hounshell believe the American System originated in the federal armories, supported by federal funds and promoted by a military bureaucracy. "[I]n the mid-1840s . . . the national armories and private contractors began to produce the . . . first fully interchangeable firearms to be made in large numbers anywhere, one of the great technological achievements of the modern era."[5] "This system had come about only after a forty- to fifty-year period of relentless effort on the part of the United States government to realize in practice a technical-military ideal that was born in Enlightenment France."[6] The American System then spread at an uneven rate to the private sector via the migration of armory trained mechanics and engineers. "In due course industry accepted, even welcomed, peculiarly military forms of organizing work through hierarchy, discipline, and the interchangeability of men as well as machines."[7] Armory Practice was adopted in fits and starts in the private sector, notably the sewing machine and bicycle industries. Armory Practice was the product of government research and development, which the private sector adopted and adapted. It did not spring from the entrepreneurs and mechanics outside government.

Yet the historical record indicates that it took the federal armories some forty years to develop the "great technological achievement of the modern era." By the mid-1840s, Chauncey Jerome was exporting ship loads of cheap brass mass-produced interchangeable clocks to England. By the mid-1840s the axe industry, typified by the progress at Collins & Co., had a broad range of highly specialized machinery with which it produced hundreds of thousands of tools per year. By the mid-1840s the entire wooden movement clock industry had come and gone, leaving hundreds of thousands of wooden movement clocks reliably ticking away in American homes. To what extent, one wonders, did the presence of Eli Terry's Pillar and Scroll Top shelf clock in so many New England homes influence the New Englanders working in the federal armories?

The Armory Practice interpretation has a curious outlook on the economics of the American System. Was it cheaper to produce things using this technology? Were these people driven by economic forces? Did the new technology lower costs, expand markets, and increase profits? "The benefits of the system, clear to the military, were not so clear to many manufacturers, given the high cost, uncertainties, and inescapable industrial conflict it engendered."[8] "It is naive to think that interchangeability lowered costs," argues Hounshell.[9] "The unit cost of Springfield small arms with interchangeable parts almost certainly was significantly higher than that of arms produced by more traditional methods."[10] Yet, in his discussion of sewing machines, Hounshell found that Singer's labor costs were cut in half using interchangeable manufacture.[11] The falling price of Henry Ford's Model T certainly suggests that his costs were also falling, despite the introduction of the $5 day. The prices of wooden and brass movement clocks, axes, and watches also fell as manufacturers found the new technology of the American System to be highly cost effective.

This economic issue is confused by comparing production costs of guns at government armories with consumer goods made by private manufacturers. The antebellum federal armories, where Armory Practice appeared and matured, were isolated from financial concerns. "Uniform parts manufacture . . . would be worth almost any price."[12] In contrast, private sector arms makers were driven by financial concerns. The firearms industry was effectively composed of two sectors, the federal armories which sought complete interchangeability at any price vs. the private sector armories which made their guns only as interchangeable as necessary. The consequences of these vastly different economic atmospheres and hence the attitudes of their mechanics and financiers were very different.

The entire historiography of the American System and interchangeability has been skewed toward the arms makers and particularly the federal arms makers. The interchangeability issue has always turned on whether machine-made parts could be chosen at random and assembled without fitting by a skilled workman.[13]

In the federal armories, where cost was of little concern and a large military bureaucracy pursued the ideal of complete interchangeability, the cost of producing guns was predictably quite high. It was apparently higher than producing non-interchangeable guns either by craft or machine methods in the private sector. This is, however, the wrong criterion against which to judge the economic efficiency of the American System. Comparing federally produced firearms with private sector civilian goods is comparing the proverbial apples and oranges.

These perfectly interchangeable federal guns were supposed to be repairable in battle by salvaging parts of damaged guns to repair other damaged guns. Producing complex objects of iron and steel to such precise standards of interchangeability was certainly more expensive than producing guns nearly (but not quite) interchangeably by craft or machine methods. Historians err when they extrapolate this particular case study of costs and a particularly restrictive definition of "interchangeable" to the private sector in general. They assume that if it is more expensive to produce guns by perfectly interchangeable manufacture then it must necessarily be more expensive to produce clocks, axes, typewriters, sewing machines, bicycles, and watches by the same methods.

This is a false assumption, based on the nature of guns, the particulars of government standards, and the nature of non-military products made in the private sector. Neither the products nor the methods are comparable. It may be a plausible theory, but only in the absence of private sector data. *Ingenious Yankees* provides the antebellum private sector cost data to balance the picture of the origins of the American System.

The problem facing historians is to interpret technological change in *competitive markets.* What is happening in the antebellum private sector, the second critical issue—why the American System appeared at all. First, however, a word about interchangeability.

The term "interchangeable" almost certainly meant different things to different nineteenth-century manufacturers, based on the particular object each was producing and the quality of that object. In the federal armories, interchangeable meant perfectly interchangeable. However, in the private sector, where no such restrictions applied, interchangeable meant something very different. Each private sector manufacturer chose to produce his product only as interchangeably as needed.

For example, wooden movement clocks meet the criteria for perfect interchangeability. They were assembled with parts chosen at random without fitting. No historian would suggest that a wooden movement clock matched the precision or finish of a federally manufactured fire arm or the clocks being hand-produced by Boston and Philadelphia craftsmen such as the Willards. Wooden movement clock escapements were designed to be adjusted as an integral part of the manufacturing process (a feature also designed into typewriters, watches, bicycles, reapers, sewing machines, and automobiles) so that parts could be manufactured within certain tolerances and still remain perfectly interchangeable. It is well worth noting that the only firearm produced with adjustable parts was the breach loading rifle of John Hall, who designed his breach loading mechanism long before he had any connection with the federal armory at Harpers Ferry.

Hounshell's Armory Practice theory postulates that the American System was the product of idealistic army officers in a bureaucratic setting that permitted them to develop technical ideas without regard to cost or price. Once developed, these ideas were carried to the private sector via the migration of skilled mechanics. Implicitly, it was evidently not clear at all to the private sector entrepreneurs that these ideas and techniques were profitable, otherwise they would have been developed in the private sector. The implication seems to be that the private sector could not develop these ideas independently (they were too expensive?) and that only the government had the resources to invest in such an enterprise. The enthusiastic engineers were responsible for spreading the American System to the private sector, but not on the basis of cost. What was the basis for this diffusion? On what data—the third critical issue—is such an implication based?

The most disappointing aspect of Hounshell's very important book, *From the American System to Mass Production,* was his heavy reliance on documentary evidence and his failure to employ a critically important data set in his study—objects.[14] Hounshell chose objects to illustrate his ideas, but in several cases chose the wrong objects. For example, his analysis of wooden movement clocks is clearly *out of beat* and perhaps best illustrates our different approaches to the use of objects in the history of technology.[15]

The first point focuses on Hounshell's choice of a movement to illustrate wooden movement clock manufacturing. On page 53, figure 1.10, he illustrates an Eli Terry Pillar and Scroll Top wooden movement shelf clock. However he chose the wrong clock (figures 1 and 2). This is indeed a Pillar and Scroll Top shelf clock, but it is an example of Terry's "outside escapement" clock, apparently, according to Chris Bailey, Terry's first production model. It features a distinctly unsuccessful design based on a bad engineering idea (although with the thought of cutting costs) that was in production for only a very short period of time, ca. 1818–1821. Terry hung the movement of this clock on the back of the dial and ran the escape wheel arbor through the dial to an adjustable cock mounted on the dial. The pendulum also hung from the dial. This design was quickly succeeded in 1822 by the five-arbor movement with an "inside-outside escapement" as it is known by collectors.

Today this "outside escapement" clock is a very rare and highly prized collector's item often selling for $5,000 to $8,000. The more common but much more important Eli Terry Pillar and Scroll Top shelf clock has the "inside-outside escapement" movement. It too is highly collectible, but it is relatively common and for only the very best examples commands a price of $1,500 to $2,000. Hounshell simply does not understand enough

FIGURE 1. The "outside escapement" model wooden movement clock found in Eli Terry Pillar and Scroll Top shelf clocks, ca. 1818–1821. COURTESY: American Clock & Watch Museum, Bristol, Conn.

FIGURE 2. This is the five-arbor, thirty-hour, "inside-outside escapement" movement. It appeared about 1821–1822 and was widely copied by numerous other makers. Compare it with the "outside escapement" model in figure 1. This particular example is found in an early E. Terry & Sons Pillar and Scroll Top Clock. Author's collection.

about the design of wooden movement clocks, about the different movement variations in the Pillar and Scroll Top shelf clocks, and more fundamentally between Terry's earlier "pull-up" clocks (1807–1810) and the later shelf clocks (1814+).[16] This is not merely an antiquarian's quibble.

Failing to recognize that Terry was producing unadjustable "pull-up" clocks between 1807 and 1810 and adjustable shelf clocks after 1814 leads Hounshell to group Terry with "other manufacturers such as Gideon Roberts and James Harrison."[17] Roberts and Harrison were neither "manufacturing" the same clock as Terry nor using the same techniques. They were craftsmen making "hang-up" clocks before Terry went into production in 1807. Although they may have adopted some of Terry's designs and methods, neither was a manufacturer.

Hounshell's hang up with Armory Practice leads him to misinterpret completely the surviving Seth Thomas tools and parts in the Winton Collection at the Connecticut Historical Society. In describing the Seth Thomas marking and checking gauges (page 55, figure 1.11), he finds them "roughly constructed—sheet iron hastily [?] cut and filed—and so would give only a very rough accuracy."[18] Yet these tools were used to produce tens if not hundreds of thousands of fully interchangeable clocks. One need only examine the hundreds of identical wheels and parts that survive in the Winton Collection and in the Hopkins & Alfred Collection at the American Clock & Watch Museum in Bristol, Connecticut. The very existence of the Seth Thomas plate drilling jig (page 57, figure 1.13, in Hounshell) proves that the wheels and pinions for this particular eight-day clock were *perfectly interchangeable*. Were they not perfectly interchangeable, then clockmakers would have had to depth them individually and the plate drilling jig would have been of no use.

Most startling is Hounshell's assertion that "the various devices used to aid the production of Seth Thomas clocks were by no means uniquely American."[19] There is nothing in the history of either European or American horology with which to compare them. To denigrate them by describing them as "an attempt to achieve . . . workmanship of certainty" defined by David Pye and to say "although very ingenious, these devices have little in common with the jigs, fixtures, and gauges used at the Springfield Armory,"[20] is to overlook the revolutionary importance of these tools. They were used to produce perfectly interchangeable clock parts. They were the basis of a highly sophisticated putting out system that insured that individual contractors produced parts within tolerances fine enough to enable clock assemblers to combine the parts of various contractors to produce finished clock movements.

Hounshell's conclusion—that the difference between the Seth Thomas and Armory Practice jigs, fixtures, and gauges "in number, precision, refinement, and use is of such a degree as to be different in kind"—exposes a gross misunderstanding about product design and manufacturing technology.[21]

The concept of product design is almost totally lacking in Hounshell's analysis of the American System, yet it is crucial to understanding the development of the American System outside the federal armories.[22] For example, Hounshell's chapter 6, "The Ford Motor Company and the Rise of Mass Production in America," is clearly the best description of Henry

Ford's technical achievements available. Yet despite exceptional details on manufacturing, only once is there a discussion of the design of the Model T. How did its design change from 1908–1915–1927? The 1908 Model T Ford is vastly different than its 1927 descendent. How? Why? Similarly, Singer's first sewing machine was a "cast iron device weighing over 125 pounds with three large spur gears" (Hounshell's figure 2.8), that was succeeded by the Model A Family Sewing Machine in 1858 (Hounshell's figure 2.13) and later by the New Family Sewing Machine (Hounshell's figure 2.14) and the Improved Family Sewing Machine (Hounshell's figure 2.25).[23] What is the importance of these design changes for Singer's manufacturing process? Are these new designs introduced primarily for marketing reasons or are they easier and cheaper to manufacture?

To stress the importance of design, let us return to a short comparison of wooden movement clocks and Armory Practice guns in the antebellum period. It should not surprise historians to learn that cheap clocks were the first consumer durables produced in large quantities. They are comparatively simple mechanisms whose parts are very simple shapes—round wheels. These shapes are relatively easy to produce on a lathe and lend themselves to being easily made in quantities by stacking many gear blanks on a single arbor and cutting the teeth in dozens of wheels simultaneously. The wooden movement clock has no complex or oddly shaped parts, as does a musket, and its mechanism can function quite satisfactorily with only a single precision part, its escapement. The clock's gears revolve in a simple action, and what few oddly shaped parts it has are made of soft iron wire. Only the adjustable escapement is complex, but even it is simple compared with the lock of a musket.

Virtually all of a musket's parts are oddly shaped. It is made of iron and steel that was annealed, worked soft, then hardened during its manufacturing process. No such involved process is required for wooden movement clocks.

In essence, the manufacturing process for each object—clock and musket—was strictly constrained by the very nature of the product; its design, its materials, and the methods required to work those materials. Each manufacturer faced a *technological imperative* that sharply limited his choice of techniques. Gun makers had primarily to deal with iron and steel, while clock makers had primarily to deal with wood. Each developed techniques to solve particular problems and achieve particular objectives. Given the nature of wooden movement clocks and muskets, these techniques were predictably quite different, yet highly comparable.

Ingenious Yankees differs basically and fundamentally from the thesis in Hounshell's *From the American System to Mass Production* on the origin of the American System, the forces that caused it, and the interpretation of significant data. The American System is primarily a private sector phenomenon, the end result of mechanics and entrepreneurs joining together and taking risks. It is not merely the result of federal government military research and development that spread from the armories to the private sector.

When considering the development of technology in the nineteenth century in general vis-à-vis the private and public sectors, the private sector held the technological lead from 1807 to the mid 1820s. Between the mid 1820s and the late 1840s the federal armories shared the forefront of technological change with the private sector. In the 1850s, however, the federal armories stagnated after achieving an acceptable degree of mass production and interchangeability. By the 1860s and the 1870s innovation in the private sector had again surged past the public sector, especially in such industries as watch and typewriter manufacturing which pioneered new techniques such as precision press work, precision automatic machining, precision gauging, and vulcanizing. By the early twentieth century the private sector had left the public sector far behind.

These shifts in American technological leadership over the nineteenth century centered on the incentive to innovate. In the closed environment of the federal armories, technology stagnated as change occurred in gun design but not in manufacturing methods or new materials. In the private sector, enthusiastic mechanics found expression for their ideas through economic incentives provided by the entrepreneurs who hired them. This seems to have been the case in several of Hounshell's case studies, notably sewing machines, where one finds such mechanic/entrepreneur pairs as Willcox & Gibbs, Singer & Clark, and Wheeler & Wilson.[24]

The phenomenon of migrating mechanics—from the armories to the private sector—raises intriguing questions. Why were these men leaving the gun factories for the consumer durables factories? Were there greater economic opportunities? Was there more freedom to develop new methods and work with new materials? Had the arms makers reached some kind of technological dead end? Had the federal armories deteriorated into a kind of trade school for basic machine shop skills? Was the "technological action" to be found only in the private sector?

In differing so sharply with Hounshell's work, this book draws very precise distinctions between my ideas and his, often in rather stark language. I have never believed in ambiguity or euphemistic language. I have always subscribed to the words of that famous philosopher/blacksmith and former Hagley Fellow Roger Moore, who said, "The only way to do it is to do it."[25] However, the reader should infer no animosity toward David Hounshell or other historians with whose interpretations I differ. I have known David since 1973. During the later years of his graduate career at the University of Delaware, he lived in Washington, D.C., and commuted to Newark. He often stayed at my apartment, as we were both Hagley Fellows.

David is as fine a historian as I know and I greatly respect his work and his ideas.[26] We simply interpret the same data in very different ways. But then that's why there's chocolate and vanilla. That makes life (and history) interesting and fun.

Our different perceptions lead us to interpret the same data in very different ways. For example, David's outstanding bicycle chapter provides historians with the first detailed history of bicycle manufacturing. "Clearly," says Hounshell, "the bicycle industry as a *staging ground for the diffusion of armory practice cannot be overemphasized*" (italics added).[27] But what one finds instead is a private sector industry that initially used some Armory Practice but which then quickly departed radically from it to develop and adopt new technology. Such techniques as electric resistance welding for wheels, the manufacture and use of seamless steel tubing for bicycle frames, and the development of presswork are all outside the Armory Practice experience.[28] Furthermore, the bicycle's frame and wheels and chain were specifically *designed to be adjusted as an integral part of their manufacturing process* (as were wooden movement clocks, typewriters, watches, sewing machines, reapers, and automobiles). Even such assembly tools as Pope's wheel truing stand were described as "adjustable all over."[29] One could argue convincingly that Pope's manufacturing process parallelled other private sector manufacturers, particularly in responding to the technological imperative of the bicycle. Pope was forced for technical reasons related directly to the nature of the bicycle itself to develop and adopt particular (non-Armory Practice) technologies such as frame and wheel truing. So were watchmakers, typewriter manufacturers, and wooden movement clockmakers.

Following Pope's creation of the bicycle industry in 1878 and the expiration of his patent protection in 1886, competing firms rapidly entered the bicycle market. Predictably, many chose to produce cycles that were, by Pope's definition, "awfully cheap looking."[30] Yet these new cycle manufacturers with their innovative press work and stamping techniques (Seth Thomas was stamping escape wheels in 1814) captured a greater share of the cycle market. Notice particularly that press work and stamping are not Armory Practice. They were new techniques developed in the private sector that enabled their users to compete very successfully with the Armory Practice producers.

The firms like the Western Wheel Works not only brought new techniques to the industry, but broadened the bicycle's public appeal by producing cycles over a range of quality, precisely what Pope had sneered at. Thus the bicycle industry represents not so much a transfer of Armory Practice into the private sector as it is the appearance of a new industry with some—*but very little*—Armory Practice, noted particularly for developing new materials, technologies, and designs, and attracting new innovative firms who broadened the range of available quality. This is precisely what happened in the watch and typewriter industries, both of which used some—*but very little*—Armory Practice, but neither of which relied heavily on it.

The most rewarding aspect of writing history is the joy of insight, the flash of cerebral electricity when one recognizes something that may be important. Several moments stand out in my mind while writing *Ingenious Yankees*. One occurred while sitting at the word processor in the Milwaukee Public Museum front office at 11:30 p.m. and suddenly realizing the synthesis of Ferguson's enthusiastic engineers and the businessman's obsession with profits.[31] However one particular moment stands out, the concept of adjustability and the importance of design.

I recall cataloguing a Corona No. 3 folding portable typewriter (manufactured in Groton, New York, ca. 1924, by the Corona Typewriter Company), in the basement ballistics range of the Milwaukee Public Museum (figures 3 and 4).[32] These are particularly common machines, which survive in attics and garages in surprisingly large numbers. This example was part of a typewriter collection containing some very rare and valuable machines. As I looked for the serial number (539,897), stamped on the bottom in the back, I noticed an adjusting screw for a particular part of the mechanism.[33] I thought it strange that a machine made in the 1920s needed an adjusting screw—after all, by that time everything was perfectly interchangeable, right? Wrong.

I suddenly realized that the Corona No. 3 folding portable typewriter was *designed to be adjusted as a part of its manufacturing process*. I then looked at wooden movement clocks, other earlier typewriters, and watches and discovered that they too were designed to be adjusted as a part of the manufacturing process. So were Hounshell's bicycles and sewing machines and reapers and automobiles. Gun locks, however, are not adjustable."[34] Perhaps this is due to the function of a small arm—to contain

FIGURE 3. Corona No. 3 folding typewriter front view in typing position with carriage folded down. COURTESY: Milwaukee Public Museum, Carl P. Dietz Typewriter Collection (Mrs. Wilson E. Mayer, donor). Negative no. H-627-19-J-2. MPM H 43403/26819.

and channel an explosion routinely. Perhaps it is due to the design of the gun lock. Almost certainly it is due to some technological imperative that sets fire arms apart from wooden movement clocks, watches, sewing machines, bicycles, typewriters, and automobiles.

The American System appears to rise from two separate and independent sources prior to the Civil War. Clearly important work was being done in the federally financed armories. Yet more significant work was being done outside the armories in the private sector. Each strain of the American System was driven by a series of incentives which differ radically in purpose. Following the Civil War, that strain of the American System developed in the armories stagnates, while a few of its key ideas were adopted at varying rates in the private sector. To the extent that manufacturers started with a model and manufactured duplicates of this ideal with some form of metalworking, most nineteenth-century factories had some form of Armory Practice. Yet many of the metalworking tech-

FIGURE 4. The serial number location on the Corona No. 3 folding portable typewriter. Note the adjusting screws. COURTESY: Milwaukee Public Museum, Carl P. Dietz Typewriter Collection (Mrs. Wilson E. Mayer, donor). Negative no. H-672-K-9. MPM H 43403/26819.

niques employed in the armories were applicable—indeed indispensible—in other industries.

One of the issues now facing historians of the American System is to decide just how much Armory Practice was adopted by the private sector and how the private sector chose that technology. For example, the Singer Sewing Machine Company developed the go/nogo gauge system to an art in the 1890s when it had some 15,000 different gauges.[35] However this technology was totally inappropriate for the Waltham Watch Company that discarded the go/nogo gauge system in favor of a system of measurements. These differing degrees of Armory Practice adoption reflect the nature of sewing machines and watches and the techniques required to manufacture them. To see what was new and different and innovative in the light manufacturing sector of the American economy in the nineteenth century, one must look beyond the federal armories to the private sector manufacturers.

In conclusion, I recognize the hazard of drawing such sharp distinctions between my work and David Hounshell's. Time marches on and as certainly as Eli Terry manufactured fully interchangeable clock parts in 1814, there is some hot shot graduate student out there ready to challenge my thesis and clean my clock. When he does, I hope he'll ream my pivot holes and polish my pivots with zest. I hope he'll adjust my rate to "six positions, temperature, and isochronism" with the same exhilaration and pleasure as I had in adjusting David's escapement. As Lynn White, Jr., wrote, "The most important thing that can be said about any scholarly pursuit is that it is fun. The history of technology is, emphatically, fun."[36] Let the fun begin!

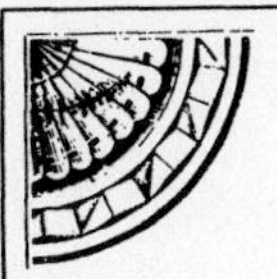

CHAPTER

1

Meanwhile, Over in the Private Sector . . .

A bald definition of the word "History" would be—"a narrative of events",—"an account of that which is known to have occurred." But as ordinarily used, the word is understood to include more than a simple statement of facts, in that there is an attempt to discern contemporary as well as previous conditions, as affecting the events which are narrated. For the real aim of History should be, not simply to gratify curiosity, but to inform the reader of actions, occurrences or events which have taken place, in connection with, or in consequence of, certain conditions, and thereby to serve as a means of instruction, as well as of information to the reader.

EDWARD A. MARSH
Master Mechanic, Waltham Watch Company
History[1]

THIS is a study of the interaction between technology and the market place and the mechanism through which that interaction occurred in the nineteenth century. It concerns the economic and cultural/technological matrix through which mechanics and entrepreneurs created the private sector of the American System of Manufactures. The American System of Manufactures is generally characterized by the mass production of interchangeable parts on specialized machinery arranged in sequential operation.[2]

The examination of four nineteenth-century industries—wooden movement clocks, axes, watches, and typewriters—traces the development of the American System without government subsidy in any form, either in capital, in transfer of technology, large orders for goods, or tax incentives. The data, many of which are derived from new and somewhat unusual sources, are especially interesting since they illuminate such important interpretive areas of American economic and technological history as the economic reasoning of nineteenth-century entrepreneurs, economic and technical forces behind new product design, the interchangeability debate, new materials, the economics of mass production, the existence of a "technological imperative," and market response. Finally, these case studies complete the study of the classic American System industries first enumerated by Joseph W. Roe in 1916 and so superbly developed by David Hounshell in 1984.[3]

Nineteenth-century mechanics played the single most important role in the rise of the American System. As individuals, they invented the new machinery of mass production and designed (or redesigned) the products to be mass produced. Their ingenuity inspired capitalists to invest. Eli Terry developed techniques to produce large quantities of wooden movement clocks; then, reacting to an assembly problem, he redesigned the clock and its escapement to solve that problem. David Hinman and Elisha K. Root invented axe manufacturing machinery and changed the process and the product dramatically. Aaron L. Dennison and a host of brilliant mechanics who followed him at the Waltham Watch Company invented the most sophisticated automatic machine tools in the nineteenth century, introduced the metric system to American manufacturing, developed an entirely new gauging concept and system (now standard practice worldwide), and designed new watches in order to achieve mass production. William K. Jenne and Jefferson Clough redesigned the crude Sholes & Glidden typewriter and eventually designed and built the Remington typewriter factory which mass produced the writing machine.

In each case history, the particular mechanics and engineers stand out as the agents of technological change. They were the individuals having the bright ideas about product design and machine design. They were not alone in their work, but functioned in a particular environment, the environment of the nineteenth-century economy. This economy was populated by the nineteenth-century businessman who viewed his economic world in his own particular way, and who was inspired by the ingenuity of the mechanics with whom he worked.

This work advances the historical view of the rise of the American System in several ways. First, it provides private sector data to balance the excellent studies of the public sector (i.e. the arms industry). Second, this study provides the data to synthesize the two opposing historical schools of thought on the rise of the American System, the economic and the noneconomic. This synthesis more accurately interprets events in American manufacturing and more plausibly explains the nature of technological change in the American economy.[4] Third, this study joins the historical debate with new ideas about the origin and development of the American System.

The four case studies span the entire nineteenth century and the early twentieth century, from 1807 to 1924, and each deals with a different product which provides insights into the rise of the American System. Eli Terry's inventions illustrate how advances in production technology led to design innovation in wooden movement clocks. Together, the new production technology and product design drastically reduced the cost of clocks, creating new demand and greatly expanding their market. The development of axe manufacturing technology at Collins & Co. took place in an already strong market. Collins learned to mass produce axes (1832–1849), cut costs dramatically with highly efficient machinery which allowed him to compete very effectively in a crowded market. The successful manufacture of watches at Waltham (1849+) was an entirely new industry and its mechanics and entrepreneurs faced the problem of developing a new and exceedingly precise technology. They responded by inventing and building the most sophisticated automatic machinery in the nineteenth century. The typewriter industry faced the problem of producing the most complex consumer durable good manufactured in the nineteenth and early twentieth centuries. These manufacturers developed new production technologies and assembly techniques, notably "exercising" machines, and fully adjustable typewriter designs.

One of the most exciting discoveries of this study is the economic importance of product design. Wooden movement clock design was an integral part of the manufacturing technique. Eli Terry first produced clocks which were difficult to assemble. In 1816 his new clock featured an adjustable escapement, which lowered its price through cheaper manufacturing. This invention remained a standard feature of American clocks for well over a century (figures 1.1 and 1.2). In the watch industry, American designers discarded the complicated power transmission system of the English watch (fusee and chain) and substituted a simpler design, the "going barrel," eliminating hundreds of parts and a very difficult manufacturing process (figures 1.3, 5.4a, and 5.4b). American mechanics designed their products to reduce manufacturing costs by easing assembly through adjustment during assembly and increasing tolerances during fabrication.[5]

There is a second aspect of product design in the private sector. Product design not only made things cheaper and often better, but also provided a range of goods across a broad economic spectrum. In 1890 typewriters ranged from the very expensive to the very cheap, from the Remington No. 6 ($125) to the World ($10), manufactured by the Pope Manufacturing Company (figures 1.4, 1.5a, and 1.5b).[6] Waltham watches

FIGURE 1.1. E. Terry & Sons Pillar and Scroll Top wooden movement shelf clock, ca. 1821–1822, showing the adjustable "inside-outside escapement." Author's collection.

ranged from $75 to $3 (figures 1.6 and 1.7). The spectrum of products included a broad range of price, quality, reliability, speed, and flexibility based primarily on design. The same phenomenon was found in the cycle industry, the clock industry, the sewing machine industry, and the automobile industry. Henry Ford may be properly noted for having devel-

FIGURE 1.2. This Sessions shelf (mantle) clock movement, ca. 1920, shows the same concept of an adjustable escapement, although in somewhat different form, found on the 1821–1822 E. Terry & Sons Pillar and Scroll Top wooden movement in figure 1.1. While the escape wheel is between the plates, the verge is clearly designed to be adjusted as a part of the manufacturing process. Carolyn Nichols Hoke collection.

oped the assembly line, but, in a different light, he was simply another in a long line of entrepreneurs producing cheap goods, including Waterbury watches, Champion typewriters, and Iver Johnson firearms.

Historians—perhaps more than the nineteenth-century capitalists and engineers they study—are concerned with interchangeability. In the federal armories, interchangeability was a technical ideal to be pursued.[7] In the private sector, historians now believe, interchangeability was merely an advertising device, never achieved in practice.[8] This is a start-

FIGURE 1.3. Although this watch movement is engraved "J. R. Brown and Sharpe, Providence, R.I.," it is, in fact, a low-grade 18-size Liverpool movement. It is a typical English watch movement with an American name engraved on it, ca. 1860. The gilt, full plate, key wind back, key set front movement features the typical English fusee and chain with maintaining power. It has only three jewels in settings with a diamond cap stone and engraved balance cock. The escapement and balance feature a right angle lever with a brass, typically English ratchet tooth escape wheel and a cut, bimetallic, compensation balance with brass screws regulated by a blued steel, volute hairspring. The movement is inscribed "J. R. Brown and Sharpe No. 12002"—on the top plate, "Providence, R.I."—on the barrel bridge, "Patent F S"—on the balance cock. COURTESY: The Time Museum, Rockford, Ill., Inventory no. 1770.

ling conclusion, given the attention paid by nineteenth-century observers. The four case studies in this work provide a new definition of interchangeability and new interpretation of the debate. Before entering that debate, historians need an accurate and feasible definition of interchangeability.

Historians have defined interchangeability as some absolute degree of

FIGURE 1.4. Remington No. 6 typewriter, serial number 31,405, the standard office-quality American typewriter in the late nineteenth and early twentieth century. COURTESY: Milwaukee Public Museum, Carl P. Dietz Typewriter Collection. Negative no. H-627-4-D-4. MPM E 40659/11418.

sameness which allows the worker assembling a product to select the parts at random and put them together without any fitting.[9] However nineteenth-century manufacturers never thought of interchangeability as an absolute and interchangeability meant different things to different manufacturers.[10] This spectrum of concern was a function of the product itself. At one end, for example, Collins & Co. didn't attempt to manufacture interchangeable axes but concerned itself with the general shape

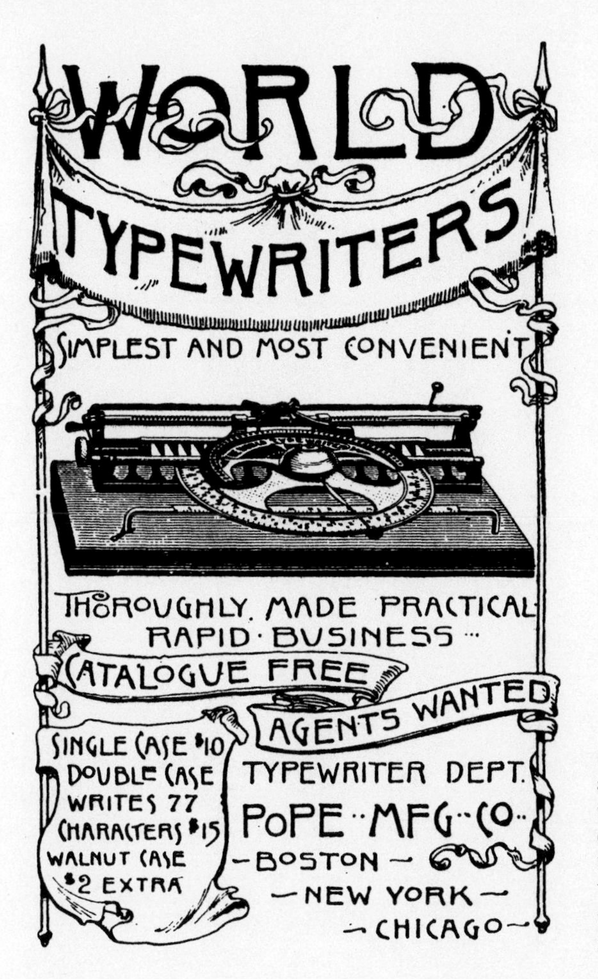

FIGURE 1.5A. Magazine advertisement for the World typewriter, ca. 1889. COURTESY: Milwaukee Public Museum, Carl P. Dietz Typewriter Collection. Negative no. H-657-J-4. MPM H 32844.370/12045.

only. Axe manufacturers were simply unconcerned about small variations in size and shape.

Further along the spectrum, wooden movement clock manufacturers were concerned with interchangeability and did make fully interchangeable parts. But they made their parts only as interchangeable as necessary. They designed their clocks to be easily adjusted as a part of the assembly process to compensate for this rough degree of interchangeability.

Near the opposite end of the spectrum, at the Waltham Watch Company, interchangeability reached its nineteenth-century zenith in parts produced on automatic machinery. But it still fell short of some absolute criteria. Despite virtually total interchangeability of screws, wheels, pivot sizes, and depthing—all produced on automatic and semi-automatic machinery—there still remained much hand work in assembling and ad-

FIGURE 1.5B. The World typewriter, serial number 13,338, ca. 1890. This model is the higher grade machine, featuring 77 characters. COURTESY: Milwaukee Public Museum, Carl P. Dietz Typewriter Collection. Negative no. H-627-13-A-2. MPM H 48319/14181.

justing despite the fact that such parts as hairsprings and balance wheels were produced with such machinery. Higher grade watches all required special attention; notably adjustment to both temperature and position was important. The most interchangeable product of the American System was not completely and absolutely interchangeable.

Typewriter manufacturers faced the most severe production problems with the most complex product manufactured on the "interchangeable system." The typewriter required many aligners, assemblers, and adjusters to finish its assembly. Many manufacturers struggled to solve the problem of "perfect alignment" by designing adjustability into their machines and relying on skilled labor to do the adjusting and aligning. As late as 1953 alignment and final adjustment were critically important aspects of typewriter manufacturing at the Royal Typewriter Company, often requiring two to three hours per machine in the final stage alone.[11]

Thus, each manufacturer faced a different criterion of interchangeability based on the product he was manufacturing and the quality of the product.

FIGURE 1.6. Waltham's Model 1892, 18-size *Crescent St.* grade railroad watch made about 1899. It features a nickel, split full plate with recessed balance in top plate. It is a stem wind, lever set, open face movement with twenty-one jewels in gold settings, nicely damaskeened with an engraved balance cock. The escapement and balance feature a straight line lever with a steel club-tooth escape wheel and a cut, bimetallic, compensation balance with gold screws regulated by a blued steel, Breguet hairspring with a micrometer regulator. The movement is inscribed "Crescent St. Waltham, Mass. 21 Jewels. Adjusted."—on the top plate, "10567241"—on the barrel bridge, "F S"—on the balance cock, black filled lettering. Author's collection, The Time Museum photograph.

FIGURE 1.7. This is the lowest grade of Waltham's famous *Model 1857*, its *Home Watch Co.* It was made by the American Watch Company about 1869 and cased in a plain silver hunting case. It was such a low grade that the company chose not to put its name on the movement. It is a gilt, full plate, key wind back, key set front, seven-jeweled movement with an engraved balance cock. The escapement and balance feature a right angle lever with a brass club-tooth escape wheel and a plain polished steel three-arm balance regulated by a blued steel, volute hairspring. The movement is inscribed "Home Watch Co. F S"—on the top plate, "425954 Boston. Mass."—on the barrel bridge. COURTESY: The Time Museum, Rockford, Ill. Inventory no. 4131.

Interchangeable manufacture was itself only the first step, and interchangeability in and of itself was of no great concern to a manufacturer. It was the assembly and adjustment stage that concerned him most. Hence there were different degrees of interchangeability, depending on the assembly and adjustment needs of a particular manufacturer and his product. Even perfect interchangeability could still require adjusting and assembly. Consider the assembly of bicycle wheels and spokes.[12] Here cycle assemblers had fully interchangeable spokes—the threads on the ends were the same, the lengths were the same, the diameters were the same, they could be chosen at random for assembly—yet the wheel still required the careful adjustment of the spokes in order to run true. If the wheel were improperly trued, the cycle was virtually unridable.[13] Assembly and adjustment of cycle wheels simply had to do with the nature of the product.

The principal is the same for other products as well, including watches and typewriters. In the manufacture of watches and typewriters, entrepreneurs and mechanics confronted the same kinds of assembly and adjustment problems as the cycle manufacturers, differing only in the requirements of their different products. The more complicated the mechanism (watches and typewriters) the more adjusting was necessary in the final stages of production. The number of parts were vastly larger in typewriters and watches compared to wooden movement clocks, thus requiring more adjusting simply because more parts had to fit together.

The kinds of adjustments necessarily differed as a function of the product. Watch manufacturers matched escapements, poised balances, and adjusted timing screws. Typewriter manufacturers aligned type and adjusted the tension, shift, etc. Manufacturers found it impossible to manufacture perfectly interchangeable parts for increasingly complex products. Logically, if all parts were the same, there would be no fitting or assembling or adjusting. But in practice there was (and still is).[14] Manufacturing—even given an extremely high degree of interchangeability—necessarily required the adjustment of fine mechanisms.[15]

Nineteenth-century engineers understood the concept of manufacturing interchangeable parts that were not perfectly interchangeable as well as the need to adjust mechanisms. W. F. Durfee wrote concerning "the Art of Interchangeable Construction in Mechanisms" that "super-refinement of accuracy of outline and general proportions is not always necessary or even desirable. There is a recognized roughness of interchangeability."[16]

This "roughness of interchangeability" found expression everywhere in the American economy, including the production of many cheap and simple goods. Simpler designs and fewer working parts meant fewer gauges and manufacturing steps. Thus it was easier to hold to fewer (and less strict) tolerances.

Many of the materials used in the cheaper goods also made manufacturing easier. Compare the paper dials of Waterbury watches to the enamel dials of Waltham watches (figure 1.8). Again, the nature of the product

FIGURE 1.8. Note the paper dial on this Waterbury Rotary watch made by the Benedict and Burnham Manufacturing Company of Waterbury, Conn., ca. 1879. The 16-size "long wind" movement is a gilt, skeletonized full plate, stem wind, hand set, open face movement. The tourbillion escapement and balance feature a duplex, brass escape wheel and a solid brass balance regulated by a blued steel, volute hairspring. The movement is inscribed "The Waterbury Watch Patented in the United States, Great Britain, Canada, France, Germany, Austria, Russia, Spain, Sweden, Denmark, Belgium. Benedict & Burnham Mf'g Co. Manufacturers Waterbury Conn. U.S.A."—on the movement cover, "Patented May 21, 1878" on the top of crown. COURTESY: The Time Museum, Rockford, Ill. Inventory no. 1118.

was critical. Conversely, interchangeability is relatively difficult to achieve in high quality, expensive, precision goods such as railroad watches and typebar typewriters. These goods had many more parts, which implied more manufacturing steps and hence more gauging and checking. With more parts fitting together there were far more opportunities for problems in assembly. The statistical probability of needing adjustment is simply far greater in more complex mechanisms, a kind of technological imperative.

The private sector manufacturers of American System products faced a very real "technological imperative." This is especially true in the following sense. When a manufacturer decided to make a product of a particular quality, he was forced to employ given techniques and certain general designs—he had no alternatives. This is not to imply that manufacturers did not develop new technology and new designs, but many adopted and adapted existing technology. The mix of existing and new technology was determined by the mechanics in charge of production, the agents of technological change.

Nineteenth-century mechanics worked closely and harmoniously with nineteenth-century entrepreneurs to develop the new technology and implement their new designs. Eli Terry was both inventor and entrepreneur, but relied heavily on local merchants to provide the capital and sell the output of his embryonic factory. Several mechanics are known to have worked with him. David Hinman and especially Elisha K. Root worked quite closely with Samuel W. Collins to develop axe manufacturing technology. Aaron L. Dennison worked closely with Edward Howard in the early years of his watch manufacturing enterprise. William K. Jenne and Jefferson Clough worked with the Remingtons for years in apparent harmony.[17]

Together, mechanics and entrepreneurs developed an economical technology to mass produce their particular item. They were able to discover what techniques they needed and to develop them. They understood the technological imperative of the goods they were making and the production techniques required. That is what all the private sector American System industries have in common, the concept held by the entrepreneurs and implemented by the mechanics of developing the technology to produce goods economically for the particular market they were in. Each industry had its specific manufacturing needs and each entrepreneur understood that his economic success depended directly on technical success which was in turn based on specialized tools. This characteristic feature of the American System—the development of specialized tools—

was not simply for production of parts, but for the subsequent modification of those parts, their assembly, and testing.

Entrepreneurs understood the concept of lowering unit costs through the use of specialized machinery for mass production. Those who failed to comprehend these ideas did not survive. They were simply marking time before their inevitable failure.

Nineteenth-century mechanics were fully aware of the threefold economic implications of the technological changes they brought about: a) the economics of mass production, b) the economic reasoning of the entrepreneurs with whom and for whom they worked, and c) the need to respond to the market. These mechanics understood the economics of machine production. They realized that the use of specialized machinery could sharply improve productivity and hence lower costs. They knew that spreading manufacturing costs over a large number of units lowered unit costs. They understood that a properly designed product could speed assembly, thus saving time and cutting costs. They understood the structure of nineteenth-century demand, especially the demand for lower-priced goods. Contrary to twentieth-century economists who see technology as simply "another factor of production," nineteenth-century mechanics and entrepreneurs comprehended technology as the most important factor of production.

Technology changes the proportionate use of all other factors of production, particularly raw materials, labor, and capital. Technology is more than simply another economic factor of production. It is the critical factor of production because it shapes the use of labor, the use of raw materials, and the use and rate of return on capital. As the case studies that follow will demonstrate, entrepreneurs did not explicitly consider the cost of labor or the cost of capital or the cost of raw materials in deciding to innovate. They considered the technical production needs of their products (the technological imperative) as well as the need to produce larger quantities, and then responded to the productivity of their machinery. At Collins & Co., for example, David Hinman's first die forging machine increased the productivity of a single skilled striker (with an assistant) from 12 to 300 axe heads per day. The cost of labor and capital were virtually inconsequential in the face of such an improvement in output. The labor cost per axe poll dropped so dramatically that Collins had primarily to worry about finding the capital, not what the machine cost.

The psychological impact that such returns on investment (returns typical of the first stages of technological innovation in manufacturing) had on the outlook of the typical nineteenth-century entrepreneur was substantial. Confronted with the reality of sharply falling costs, entrepreneurs were surely willing to invest in such an increase in productivity and the mechanics who promoted such machinery.

Both economic and technological historians have missed the most important aspect of technological change and economic development in the nineteenth century, the relationship between the mechanic and the entrepreneur. Economic historians seem to have been too preoccupied with finding a general theory of economic development and technological change, while historians of technology have simply ignored economics and economic historians altogether.[18] Both groups have tended to ignore the businessman whose needs are quite practical and immediate—how to produce a better product more cheaply and how to expand the market and meet the competition. Only by focusing on the entrepreneur and the mechanic working together can historians begin to understand how and why technological change took place in general and how the American System in particular became a dominant feature of the American economy.

The interaction between the mechanic and the entrepreneur is the critical relationship in nineteenth-century technological change and hence economic growth. It is the mechanism though which changes in technology found expression in practical applications and transformed the light manufacturing sector of the American economy from a craft tradition to a series of major industries. The mechanic and entrepreneur generally knew each other well, often having a close personal relationship as well as a business (employer/employee) relationship.[19]

Perhaps the most pervasive, economy-wide impact of the private sector American System was the entrepreneur's incredible flexibility and his response to changes in the market. In virtually all of the private sector American System industries there was a pattern of competition along the following lines. First, inventors and entrepreneurs struggled to develop and manufacture a new product or an American variation of an existing product, usually as a fairly high-grade item. Subsequently, the market was flooded with competitors, both at the same quality level and at a lower, or cheaper, level. A period of savage competition ensued during which most firms and products disappeared from the market.[20] Finally, the enduring firms emerged, each with its share of a particular market.

This pattern emerges in clockmaking, watchmaking, and typewriter manufacturing. In the watch industry, for example, the Waltham Watch Company pioneered the manufacturing of watches. Shortly after its success was insured, about 1864, a host of competitors entered the market, hiring away Waltham mechanics and operatives to build and run their

FIGURE 1.9. This is the highest grade of Waltham's *Model 1857*, its *Appleton, Tracy & Co.* grade. It was made by the American Watch Company of Waltham, Mass., ca. 1866. The 18-size gilt, full plate, key wind back, key set front movement has fifteen jewels in settings with an engraved balance cock. The escapement and balance feature a right angle lever with a brass club tooth escape wheel and a cut, bimetallic, compensation balance with gold screws regulated by a blued steel, volute hairspring. The movement is inscribed "Appleton, Tracy & Co. F S 279405"—on the top plate, "Waltham, Mass."—on the barrel bridge. The hand painted, seconds sunk dial is inscribed "Penna Rail Road." COURTESY: The Time Museum, Rockford, Ill. Inventory no. 2323.

new factories. The watches of these first competitors were strikingly similar to the Waltham *Model 1857* (figures 1.9 and 1.10). By the early 1880s competition appeared in the form of cheap (later known as "Dollar") watches (figure 1.11). By the early 1890s the Waterbury Watch Company had created and secured a market in cheap watches that Waltham was unable

FIGURE 1.10. This is the second watch (serial number 102) completed on April 1, 1867, by the National Watch Company of Elgin, Ill. (later known as the Elgin Watch Company). Compare the general design of this watch with Waltham's *Model 1857* in figure 1.9. Although the balance cock has been moved to the opposite side of the barrel bridge, the general train layout and design are strikingly similar. The same is true of other firms, including Newark and Tremont. This 18-size, first model (Grade 69) movement is a full plate, gilt, key wind back, key set front, cased hunting, with fifteen jewels in brass settings, and an engraved balance cock. The escapement and balance feature a straight line lever with a brass club tooth escape wheel and a cut, bimetallic, compensation balance with gold screws regulated by a blued steel, volute hairspring. The movement is inscribed "B. W. Raymond Burt's Patent F S"—on the top plate, "Elgin. Ills. No. 102"—on the barrel bridge. The gilding is rather poor, suggesting that in the early days of its production, Elgin had yet to master the art. COURTESY: The Time Museum, Rockford, Ill. Inventory no. 1053.

FIGURE 1.11. In the mid 1870s the Auburndale Watch Company of Auburndale, Mass., introduced this "Rotary" model watch that started the cheap or "dollar watch" industry. Although the Auburndale Rotary failed, some of its key features were picked up by the Benedict & Burnham Manufacturing Company (later the Waterbury Watch Company). The Waterbury Watch's price soon fell to $1.00, hence the term "Dollar Watch."
This example dates from about 1878. Its 18-size, rotary, stem wind, lever set, open face, nickel movement has only four jewels. The escapement and balance feature a straight line lever with a brass escape wheel and a plain, polished steel, three arm balance wheel, regulated by a blued steel, volute hairspring. The movement is inscribed "Auburndale Rotary Mass. 447 Patents Mar. 30 1875. July 20, 1875. June 20, 1876. Jan. 30 1877."—on the pillar plate, "F S"—on the balance cock. COURTESY: The Time Museum, Rockford, Ill. Inventory no. 328.

or unwilling to enter. By the late 1890s most of the jeweled watch companies that competed with Waltham were dead or dying; only three major firms remained. The same phenomenon occurred in the typewriter industry and the clock industry as well. The promise of mass production through the new technology of the American System made these significant changes in the market possible. Without these new techniques, there could not have been and would not have been such a swift market response by private sector entrepreneurs and such a wide variety of quality.[21]

One aspect of this rapid market response and the wide range of quality was the introduction of new materials in the private sector industries.[22] In the private sector there were no technical constraints on either the materials used or their quality as there were on the federal armories. Manufacturers used wire, wood in many forms, paper, enamel, paint, nickel plating, fine steel springs, gold, silver, cast iron, and plastic (celluloid) (figure 1.12). There was not only a much greater need to innovate in the private sector, but also a much greater freedom to do so. These manufacturers were unrestricted in product design and material use in developing new products.

Finally, this study synthesizes two historical schools of thought on the rise of the American System and the nature of technological change: the economic and the noneconomic. Economist and economic historians have offered theoretical arguments to explain the rise of the American System, but none has studied the technology itself. H. J. Habakkuk, whose work started the debate and whose concept of labor scarcity still dominates the literature, cast his arguments on two theoretical grounds. First, the relatively scarce supply of labor in America compared to Britain was important. Second, the interest rate influenced the rate of technological change.[23] Most economists and economic historians consider technology as no more than another input such as labor or raw material or capital, simply responding to economic pressure.[24] Historians of technology have objected to this myopic approach but have never rejected the argument on theoretical grounds.

Technological historians have also developed a curiously blind perspective on technological change, a view focused on the intellectual character of the engineer termed "technological enthusiasm."[25] The theory of "technological enthusiasm" explains technological change as the result of engineers promoting new ideas with no economic basis.

The concept of "technological enthusiasm" is not inconsistent with the economic forces at work. Indeed, they are fully compatible. Engineers and mechanics were not blind to the economic needs of their firms and often had strong economic evidence on which to base further work. As the following chapters on wooden movement clocks, axes, and watches will demonstrate, the economic rewards for technological innovation in manufacturing were both substantial and entirely obvious. The price of Eli Terry's wooden movement clocks fell quite fast as the new technology spread. At Collins & Co., Elisha K. Root and Samuel W. Collins saw

FIGURES 1.12A AND 1.12B. Note the double sunk celluloid dial of this 18-size pocket watch, manufactured by the Keystone Standard Watch Company of Lancaster, Pa., ca. 1886–1890. This high-grade watch features a 3/4 plate, nickel, stem wind, lever set, hunting movement with fifteen jewels in settings and damaskeened plates. The escapement and balance feature a straight line lever with a brass club-tooth escape wheel and a cut, bimetallic, compensation balance with gold screws regulated by a blued steel, volute hairspring with a micrometer regulator. The movement is inscribed "F S Pat. Reg."—on the balance cock, "Keystone Watch Co. Adjusted Pat. Dust Proof. Safety Pinion Pat. S.W. 351319"—on the top plate, with black filled lettering. COURTESY: The Time Museum, Rockford, Ill. Inventory no. 3011.

worker productivity skyrocket with the introduction of David Hinman's axe forging machine in 1832. After the Civil War, Royal Robbins, the treasurer and major stock holder of the Waltham Watch Company, repeatedly justified the company's investment in automatic machinery by insisting that the new technology reduced watch prices. Simultaneously, the mechanics at Waltham reveled in the technology they developed in response to noneconomic forces as well as the economic pressure of competition. Clearly, nineteenth-century mechanics and entrepreneurs understood the concept of spreading their unit costs over a large volume by producing more units with machinery.

None of these economic incentives is inconsistent with the theory of "technological enthusiasm." Indeed, the economic rewards (or the possibility of such rewards) for innovation could be interpreted as the mechanism through which the mechanics gave free reign to their imaginations.

Their enthusiasm for invention and design found an outlet in the market place, but the market place provided only a part of the incentive, the other part being the engineer's excitement and enthusiasm for his project and its production machinery.

Neither the market alone nor the engineer's enthusiasm was enough to bring technological change to fruition; both were required in some undefinable and ever changing proportion. Without the market, the engineer's new idea remained just another new idea. Through the market, it became reality. However, the market alone was unable to create new ideas. The mere existence of an economic demand does not imply that a successful product will appear to meet it. The two forces (the economic and the noneconomic) must coincide, must work simultaneously, in order to result in a technically and economically successful product or process.

The study of the interrelationships between mechanics and entrepreneurs illustrates that both the economic and noneconomic approaches to technological change are incomplete. In the absence of an economic system through which to express themselves, these nineteenth-century mechanics would have failed to transform their ideas into reality. Yet the economic system by itself was (and still is) incapable of producing new technology. It was through the combination of a viable economic system capable of transforming the engineering ideas produced by enthusiastic mechanics into reality that the American economy in general and the American System of Manufactures in particular grew and flourished in the nineteenth and early twentieth centuries.

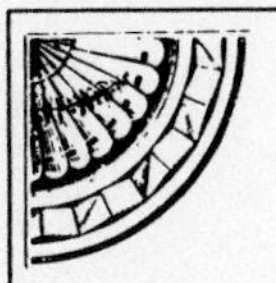

CHAPTER

Wooden Movement Clock Manufacturing in Connecticut, 1807–1850

John J. Murphy, "Establishment of the American Clock Industry: A Study in Entrepreneurial History, (Ph.D. dissertation, Yale University, 1961), . . . sees the early clock industry as the first link in the chain that was to become the American System of Manufactures. [He] describes quantity production of clocks before 1814. It is surprising that nobody has built on Murphy's solid work.

EUGENE S. FERGUSON,
"History and Historiography," in *Yankee Enterprise*[1]

THE American System as it came to be practiced in the nineteenth century originated in the wooden movement clock factories of Connecticut between 1807 and 1825. Wooden movement clock manufacturers mass produced interchangeable parts on specialized machinery, redesigned the clock to ease assembly problems and relax manufacturing tolerances, and adopted an industry structure to exploit the new technology within the agricultural cycle of life in the early antebellum economy. The mass production of wooden movement clocks took place in the private sector free from any form of government subsidy. The first surge of clock production materialized through the joint efforts of a mechanical genius, Eli Terry, and two local merchants and entrepreneurs, Levi and Edward Porter. This is the first instance of this essential relationship in the American System industries, a relationship repeated in all the private sector American System industries. The manufacture of wooden movement clocks dramatically lowered their price and greatly expanded their market.

Wooden movement clock manufacturing never became a major part of the American economy. In 1807, and even at its height in the late 1820s, the industry was virtually unmeasurable, a minuscule part of the GNP, capital invested, the work force, etc. However, it was important beyond any measure compared with its tiny size. It was the first industry to pioneer successful mass production technology—however crude—and in doing so demonstrated the economic returns to technological change.

Eli Terry initiated the manufacture of the wooden movement hang-up clock in 1807. This clock became the first mass-produced, interchangeable parts, consumer durable good on the American market. Terry's work was solidly based in the craft tradition of producing similarly designed wooden movement clocks. He retired in 1810, only to return to shock the market again in 1814 with his wooden movement shelf clock. In the late 1830s and early 1840s the brass shelf clock—basically identical in design to Terry's wooden movement shelf clock—became the first mass-produced, interchangeable parts, consumer-durable good on the world market.

Clockmakers achieved precision mass production far earlier than the gun makers, who relied heavily on federal subsidies and purchases. Their economic success was based on several technical factors: 1) new clock design, 2) choice of materials—primarily wood with some brass and iron, and 3) new production technology. Clockmakers employed the new technology in a form of the putting-out system, which fit well into the seasonal nature of work in the early antebellum economy.

The scholarly study of wooden movement clocks and the work of America's most important clockmaker, Eli Terry, began and ended with a dissertation by John Joseph Murphy.[2] Murphy focused on the entrepreneurial aspects of the clock industry, emphasizing marketing, industrial structure, and entrepreneurship. From the size and extent of the industry he describes, it is clear that wooden movement clocks were produced in vast quantities. The number of surviving clocks also supports this hypothesis. Murphy had little to say about manufacturing techniques. "There is very little evidence to indicate precisely what new machinery was put into use in . . . Terry's early factories."[3]

Later historians reached the same conclusion. "How Terry actually produced his clocks is unknown." "The historical record in clock manufacture is so inadequate—for example, the lack of any significant description of machinery or the survival of it—that the historian risks peril with any kind of generalization."[4] Most economic and technological historians mention the manufacture of wooden movement clocks, but none has explored the field in depth, the reason given being the presumed scarcity of sources.[5]

Contrary to these assertions, much new material has surfaced since the pioneering work of Murphy.[6] No historian has collected or analyzed this new data on wooden movement clock manufacture. The data offer evidence of an industry continually jolted by technological change.

From about 1750 to 1807 wooden movement clockmakers used the traditional designs and methods to make hang-up or wag-on-the-wall clocks. These clocks were cheap wooden versions of expensive brass clocks. In 1807 Eli Terry radically changed the technology of clock production with the use of circular saws, although he retained the basic wooden movement clock design features. After 1810 his techniques spread quickly, leaving the traditional wooden movement makers unable to compete. They either adopted the new technology or went out of business.

Terry shocked the clock industry again in 1814–1816 with the thirty-hour, wooden movement shelf clock. This design spread faster and further than his earlier hang-up clock, and by 1825 the new machine-made shelf clock had driven the machine-made hang-up clock from the market, just as it had destroyed the earlier, hand-made hang-up clock market. The wooden movement shelf clock industry flourished for two decades until the cheap, rolled brass clock began to displace it about 1837. By 1850 the era of the wooden movement clock had passed, itself the victim of technological change.

The new data provide enough information to reconstruct the operation of a "typical" wooden movement factory, ca. 1825–1830. From surviving tools, unused material, literary accounts, and the clocks themselves much detail about the methods for wheel making, pinion making, plate making, wire work, and assembly emerge. In short, we have a fairly clear picture of the manufacturing technology.

The study of wooden movement clocks is particularly important because these clocks were the first American mass-produced product. Wooden movement clock mass production co-dates Eli Whitney's so-called "interchangeable" guns and much of the legitimate work on interchangeable manufacture at the Federal armories.[7] For example, in 1824, when John Hall was tooling up at Harpers Ferry to produce his breech loading musket, tens, perhaps hundreds of thousands, of wooden movement clocks had already been produced and sold.[8] The manufacture of these clocks represents the first break with the craft tradition of producing consumer (nonmilitary) goods financed by the private sector. Hence, it is in the manufacture of wooden movement clocks in Connecticut, that one finds the origin of mass production in America.

Eli Terry, who began the revolution in clock manufacturing, was trained in the craft tradition of clockmaking. To appreciate the magnitude of the changes Terry brought to the industry, one must briefly examine the traditions in which he apprenticed.

CLOCKMAKING IN AMERICA BEFORE ELI TERRY

> The manufacture of clocks has been properly assigned a special place in the rise of machine technology. Early, in the handicraft period, a standard of measured precision and of machine production became a feature of clockmaking; it was not attained in other crafts until much later.
>
> BROOKE HINDLE, *Technology in Early America*[9]

Clockmaking in America began in the seventeenth century with the immigration of London-trained clockmakers. They passed on the tradition of the London Guild in making brass clocks. American clocks at this time "closely reflected the styles and techniques of Europe."[10] The eighteenth-century clockmaker produced his clocks individually, or perhaps in small batches, but never in great quantity. In this process, the clockmaker filed his plates from hammered cast brass, cut each wheel individually on a wheel-cutting engine, depthed the wheels and pinions individually (insuring that the gears meshed properly) in laying out the train, and finished the entire clock by hand in a very attractive manner.[11] His tools were strictly traditional clockmaker's tools, such as files, gravers, depthing tool, wheel cutting engine, saw, and bow driven lathe.[12] Each clock became as much a work of art as it was a functional timepiece.[13]

Many clockmakers had specialized in certain aspects of the craft by the late eighteenth century, as had horological tool makers who were active as early as 1750.[14] Parts makers supplied hands, springs, dials, etc. to clockmakers who used them together with their own parts. The clockmaker who made the entire clock was a rare and exceptional craftsman, and most clockmakers bought parts.[15] The sale and use of parts produced by specialists spread throughout England and into America by the end of the eighteenth century.[16] Thus, an early Federal period clockmaker usually produced his own plates, pivots, pinions, wheels, and striking work, but often bought his hands and dials from Birmingham.

A second tradition of clockmaking flourished simultaneously in America, a "native school of clockmaking," using wood instead of brass.[17] American wooden movement clocks may have been made as early as 1712 in Connecticut, although the first date of manufacture and the first maker remain unknown. In 1761 wooden movement clocks were said to be common in East Hartford, Connecticut, implying that the trade was then well established. Clockmakers outside Connecticut, notably in Massachusetts, also made these clocks or clocks generally similar to them.[18] These clock movements, commonly called "Cheney movements" after the family of Connecticut clockmakers who made many of them, found a market as late as 1812 (figures 2.1a, 2.1b, and 2.1c).

The manufacture of these clocks, "characterized by heavy construction, rough workmanship, and large lenticular wheels with hand-cut teeth," has been romanticized by late nineteenth- and early twentieth-century authors. Their errors have been repeated uncritically by modern authors to draw a sharper contrast between the craft tradition of clockmaking and the factory methods introduced by Terry.[19] Samples of this mythology abound. The tools used to make these clocks were supposedly those of the cabinetmaker, according to one scholar, "a handsaw and plane, bow drill or hand brace, file, knife, compass, and foot lathe."[20] Another "expert," Chauncey Jerome, claimed that before Terry "the wheels and teeth [were] cut by hand; first marked out with square and compasses, and then sawed with a fine saw, a slow and tedious process."[21] Eli Terry, still another author believed, was "using no machinery, but cutting the wheels with a saw and jack-knife."[22] These clockmakers were all, the myth is repeated, "making their wheels with a saw and knife." These are wonderfully romantic descriptions, but they have no basis in fact. Such statements could only have been written by authors who had never examined these clocks.

Two sets of data provide a more realistic view of pre-Terry wooden movement clock manufacturing technology: a) surviving clocks and b) Henry Terry's assessment. The Cheney-type clock movements are—especially when compared to Terry's later hang-up clocks—rather clumsy looking, but they are "gross in size only."[23] Antiquarians who have examined dozens of these clocks find no teeth cut with hand saws, no knife whittling marks, in short, no evidence to support the romantic view of the eighteenth-century wooden movement clockmakers.

Henry Terry, Eli Terry's son, commented accurately on the early technology:

> He [Eli Terry] made clocks both of wood and brass in the then ordinary way, having a hand engine for cutting the teeth or cogs of the wheels and pinions, and using a foot lathe for doing the turning. It is probable that he used a knife, as well as many other tools then in use, in doing some part of the work, but that the different parts of the clock "were cut out with a penknife," is a tale of many years' growth, having no foundation,

FIGURE 2.1A, 2.1B, AND 2.1C. A typical eighteenth-century American wooden movement clock by Benjamin Cheney of Hartford, Conn., with whom Eli Terry apprenticed. Most clocks of this style had plate pillars protruding through the back plate of the clock which provided clearance for the pendulum (not shown) if the clock were simply hung on the wall without a case. The short pillars on this clock suggest that it was made to be cased. COURTESY: Smithsonian Institution, Museum of American History. Catalogue no. 317,030, negative nos. 74018, 74019, and 74020.

and ought not to be stereotyped as part of the history of clock making in this country.[24]

Gideon Roberts and John Rich (two of Terry's contemporaries) are both reported to have used wheel cutting engines before 1812. Unquestionably, the wooden movement makers before (and including) Terry, employed the same techniques as their brass clockmaking brethren, modified only to suit the medium in which they worked, wood rather than brass. They cut their wheels individually on hand-operated wheel cutting engines, just as did the brass clockmakers. One of these early wheel cutting engines survives at the Connecticut Historical Society (figure 2.2).[25]

While these movements do show a degree of similarity suggesting to collectors the use of templates or patterns of some sort, no hard data survives to substantiate this impression.[26] There is only the slightest data to suggest that a single maker, Gideon Roberts, made these clocks in any great quantity.

These wooden movement clocks were sold complete with face, hands, and weights and were designed either to hang on the wall as sold (hence the term wag-on-the-wall), or to be cased in a tall case. All these makers produced thirty-hour, pull-up (a method of winding) movements.[27] Clock collectors have identified a dozen makers through surviving examples of their work (see table 2.1).[28]

A strong tradition of clockmaking flourished in eighteenth century America consisting of two branches, the traditional brass clockmakers and the wooden movement clockmakers. None of these clockmakers, and especially the wooden movement makers, produced large quantities of clocks. Each of them produced a clock primarily in response to a customer's order, but some produced a few clocks in advance of sales and then set

TABLE 2.1
Hang-Up or Wag-on-the-Wall Clock Makers

Maker	*Location*	*Dates*
James & Lemuel Harrison	Waterbury, Conn.	1790–1830
Gideon Roberts	Bristol, Conn.	1780–1815?
John Rich	Bristol, Conn.	....–1820
Levi Lewis	Bristol, Conn.	....–1810
Eli Terry	Plymouth, Conn.	1792–1833
Nathaniel Hamlen	Oxford, Maine	1800–....
A. & C. Edwards	Ashby, Mass.	1780–1830
Benjamin Cheney	E. Hartford, Conn.	1740–1800
Timothy Cheney	E. Hartford, Conn.	1750–1790
Alex Willard	Ashby, Mass.	1796–1800

FIGURES 2.2A AND 2.2B. Early wheel cutting engine for wooden movement clocks. This engine is missing its indexing wheel and the wheel (probably similar to a spinning wheel) held in the two arms at the left that turned the cutters. COURTESY: The Connecticut Historical Society, Hartford, gift of Kenneth D. Roberts.

about to sell them. All the wooden movement makers produced essentially the same thirty-hour, pull-up, wooden movement clock using brass clockmaker's technology modified to suit their medium, wood. Among these makers Eli Terry made both brass and wooden clocks (which was not uncommon) from 1793 to 1807. Between 1807 and 1822 he produced the new technology and new clock designs that destroyed the world of the clockmaker/craftsman in which he had been raised.[29]

ELI TERRY: A BIOGRAPHICAL SKETCH

Eli Terry was the single most important clockmaker in American history. Through his inventions in clock design and manufacturing technology, he transformed a small craft activity into an industry. Terry's importance lies in his technical accomplishments based on his conceptual leap from individual clockmaking to mass production. He developed production techniques to produce his parts, but equally significant, he redesigned the clock to ease both production and assembly constraints. Through the new technology of mass production, Terry "turned a luxury into a necessity."[30] Terry's venture was far outside the norm for a clockmaker of his time. When Terry began his contract with the Porters, he was ridiculed. Few people believed that the Porters could sell 4,000 clocks, and hardly anyone believed that Terry could make them.[31]

No other clockmaker could conceive of making 4,000 clocks in a lifetime, much less in three years. But Terry could, because, unlike other clockmakers, he had only to worry about manufacturing the clocks. The Porters guaranteed their sale.

Eli Terry was born on April 13, 1772, in East Windsor, Connecticut, and died on February 24, 1852, in Terryville, Connecticut. He served an apprenticeship with two clockmakers, Daniel Burnap and a "Mr. Cheeney."[32] Burnap made brass clocks in the tradition of the London Guild.[33] Cheney made wooden movement clocks."[34] There is a consensus that Terry learned an important lesson from each of these very different craftsmen. From Burnap he learned the concept of production in advance of sales. From Cheney he learned about wooden movement clocks.[35]

In 1793, following his apprenticeships, Terry moved to Plymouth in west central Connecticut. Here Terry produced both brass and wooden movement tall clocks until about 1800, when he ceased making brass clocks and made only wooden movement clocks. Between 1793 and 1806

Terry never standardized his production but continually experimented with variations of the wooden movement clock (figure 2.3).[36]

In 1807 Terry contracted with Levi and Edward Porter, Waterbury merchants, to produce 4,000 wooden movement clocks in three years. Terry produced the traditional wooden movement hang-up clock, including the movement, dial, hands, weights, and pendulum (figure 2.4). Terry's relationship with the Porters proved critical, since they provided the stock (capital) and guaranteed the sale. Terry neither worked in a vacuum nor monopolized clock invention, but he most certainly made the most important design developments and many of the manufacturing inventions. Relieved of the concern for sales, which reportedly took much of his time prior to the Porter contract, Terry focused all his energy on developing the machinery to make clocks and improving movement design. Before the contract, Terry was just another clockmaker. Without it, Terry would probably have remained just another wooden movement clockmaker. After it, Eli Terry became *THE* great American clockmaker. Because of it, American clockmaking and American industry were forever altered, and Terry's problems were reduced to technical problems, making clocks cheaply, not disposing of them.

Terry completed his contract within the three years specified, according to his son, by spending the first year constructing the machinery, the second year producing the first 1,000 clocks, and the third year producing the last 3,000 clocks. The first 1,000 clocks were made with milled teeth (i.e., cut with the traditional wheel cutting engine), the last 3,000 were made with machine sawn teeth (i.e., the new technology of wooden movement clockmaking) (figure 2.5). This is strong evidence of significant technological change and innovation during the contract.

Terry's production of 3,000 wooden movement clocks in a single year was a truly remarkable accomplishment. Indeed, it must have seemed almost incredible to contemporary clockmakers who could hardly dream of producing 3,000 clocks in a lifetime, much less in a single year. Terry employed (among others) two joiners, Seth Thomas and Silas Hoadley. Both later became important clock manufacturers using Terry's technology. In 1810 Terry sold his factory to Thomas and Hoadley for $6,000 and retired a wealthy man.[37]

Terry did not long remain retired. In 1812 he acquired land which included a series of mills on the Naugatuck River, and there he continued to design clocks. Terry completed the initial phase of his experimental work in 1814 when he applied for a patent, granted in 1816, covering his new clock. This clock was a four-arbor (gear and shaft), thirty-hour,

FIGURES 2.3A AND 2.3B. Tall clock movement by Eli Terry made prior to his contract with the Porters, pre-1807. Notice the vastly improved design over the same kind of clock produced by his master, Benjamin Cheney (figures 2.1a, 2.1b, and 2.1c). The movement clearly illustrates the shape of the wheel teeth. These teeth have been cut with the traditional wheel cutting engine and are referred to as "milled" teeth. COURTESY: Smithsonian Institution, Museum of American History. Catalogue no. SI 2479, JA 1C173P, negative nos. 78-15094 and 77-14619.

FIGURES 2.4A AND 2.4B. Eli Terry manufactured this tall clock (or hang-up) clock movement as one of the first 1,000 of his historic "Porter Contract" in 1807. This clock has several important features, particularly the shape of its gear teeth. These teeth have been cut with a conventional wheel cutting engine, and are commonly called "milled" teeth. Compare their shape with those in figures 2.3 and 2.5 COURTESY: American Clock & Watch Museum, Bristol, Conn.

wooden movement shelf clock (figure 2.6). The Patent Office granted six other patents on August 22, 1814, for the manufacture of such clocks to individuals with technical connections to Terry.

Terry manufactured this four-arbor, thirty-hour, wooden movement shelf clock as early as 1814 in a simple box case.[38] Seth Thomas also began manufacturing the shelf clock, perhaps as early as 1814, but certainly by 1816.[39] The classic case style, known as the "Pillar and Scroll Top," appeared as early as 1816, but certainly by 1818. Its design has been attributed to Heman Clark, but no conclusive evidence supports this attribution (figure 2.7). The Pillar and Scroll Top case is nothing more than the original box clock design surrounded with some molding and painted glass.

The new shelf clock differed radically from every clock that preceded it. It was cased in a simple box with numerals painted directly on the glass door which served as the dial. Straps of wood formed the front plate of the clock movement while the back plate formed part of the back of the case. The design of this clock indicates clearly that Terry was trying

FIGURES 2.5A AND 2.5B. Eli Terry also manufactured this tall clock (or hang-up) clock movement under his contract with the Porters. This clock dates from 1808–1810 and is significantly different from the 1807 version. Note the shape of its gear teeth. They have been cut with circular saws, not milled. Compare their shape with those in figures 2.3 and 2.4. Note also the cheaper design. The back plate is now attached to the seat board with wooden dowels and pins rather than let in and screwed as with the earlier design. Again, compare with figure 2.4a. COURTESY: American Clock & Watch Museum, Bristol, Conn.

to produce a *very cheap* clock. There is no decoration or ornamentation, nothing extra to sell the clock other than its timekeeping and striking features. This was a strictly utilitarian clock designed to be sold in the lower end of the American market where clocks had never been sold before.

This new shelf clock possessed many advantages over its hang-up clock predecessor. It was smaller, easier to move about the house, and easier to set up and put in beat after such a move. It could be placed anywhere in the house, upon the mantle, on a table, hung on the wall, or placed on a shelf attached to the wall. Its movement was protected from dust, it ran the same length of time, was easier to transport from maker to customer, and came complete with a case. It may have been better adapted than hang-up or tall clocks to the smaller houses seemingly being built in America.[40] Most important, it was cheaper than the hang-up clock and hence available to a vastly wider market. The hang-up clock sold for about $20 plus $20 for a case (optional) in 1807.[41] By 1810 its price had

FIGURE 2.6. Terry's first shelf clock movement, designed ca. 1814. Note the adjustable verge pin to the left of the escapement wheel. This movement was manufactured in small numbers by Eli Terry and Seth Thomas and cased in a Pillar and Scroll Top case. It was eventually superceded by the final form of wooden movement shelf clock. This example was made by Seth Thomas. COURTESY: Smithsonian Institution, Museum of American History. Catalogue no. 335,722, negative no. 77-3688.

FIGURE 2.7. A typical Pillar & Scroll Top Clock manufactured by Eli Terry & Sons, ca. 1822. Author's collection.

fallen to about $10 due to Terry's new technology. By the time it was going out of style the hang-up clock sold for $5. Terry's Pillar and Scroll Top shelf clock initially sold for about $12 wholesale and $15 to $18 retail in 1816.[42] It declined in price as more makers moved into the market in the years following. At the end of the wooden movement era, Terry-type clocks sold for $0.375 wholesale.[43]

The market implications of Terry's inventions and designs were far-reaching. Terry made clocks economically available to a much wider market not simply by making them in vast quantities but by making them more cheaply. The technology of mass production lowered unit costs and thus relaxed the price constraint on the purchase of clocks. Clocks were not a normal part of an American household in the early nineteenth century, but by the 1830s and 1840s they were, due entirely to the new technology of mass production employed by Terry and his competitors. At a time when the vast majority of people lived closer to the poverty line than to the luxury line, the mass production of a luxury item was indeed a new and untried concept.[44] The lesson learned by Terry and the Porters in 1807, and subsequently by the rest of the clock industry and many others in the American antebellum economy, was that virtually any well designed and manufactured article could be sold if a demand existed and the cost could be reduced to meet that demand. The lesson was universally known by the 1850s when Henry Terry wrote that his father knew that "the secret of money-making at that time, as well as the present day, was not in manufacturing so expensive clocks as this kind must necessarily have been. The greater demand was, and still is, for a less costly article."[45]

Terry continued to improve the design of the thirty-hour, wooden movement shelf clock. Between 1818 and 1822 he developed and produced a variety of movement designs, finally culminating in the five-arbor, thirty-hour, wooden movement clock (figure 2.8).[46] This design surpassed his earlier four-arbor, strap movement design. Its new features included single piece plates and an adjustable verge, which cheapened the manufacturing process and eased (and cheapened) the assembly processes.

The new shelf clock found a ready market and sold well through the established peddler system. Within ten years it displaced the thirty-hour, "pull-up," tall clock, just as that clock had earlier driven the Cheney-type clock from the market. In the 1820s Terry established his sons in the clockmaking business, licensed the clock design to other manufacturers, and continued to experiment in clock design.[47] In 1826 he gathered all

FIGURE 2.8. The final configuration of Eli Terry's shelf clock movement. This is the five-arbor, thirty-hour movement with the adjustable verge plug located directly below the escapement wheel. Numerous variations of this movement have been catalogued by antiquarians and it was manufactured by at least twelve different makers. This movement is cased in the classic "Pillar and Scroll Top" case shown in figure 2.7. Note the adjustable escapement. Author's collection.

his previous shelf clock inventions in the thirty-hour, wooden movement shelf clock into a single patent. After 1833 he ceased to be active in the clock manufacturing business and spent the remaining years of his life making individual clocks and experimenting in the brass clock field with his son Silas. Eli Terry died in 1852, the "last of the Craftsmen and the first of the Industrialists," who began the revolution in the mass production of consumer durables.[48]

THE CLOCK INDUSTRY'S GROWTH AND STRUCTURE

Terry's new clock designs and the new technology of mass production spread quickly throughout the clockmaking areas of Connecticut and profoundly changed the nature of clockmaking. Almost as soon as Terry perfected the five-arbor, thirty-hour, wooden movement shelf clock, his patents were challenged in court and found inadequate to protect his inventions. Other makers quickly entered the market with thirty-hour, wooden movement shelf clocks. Many were virtually identical to Terry's, while others embodied the essentials of Terry ideas in clocks which were significantly different in outward appearance.[49] Some of the more "adventurous designers were Norris North, Joseph Ives, Chauncey Boardman, Silas Hoadley, and Mark Leavenworth."[50] By the mid 1820s some twenty-two makers were producing wooden movement clocks (see table 2.2).[51]

These clockmaking firms manufactured throughout the 1820s and 1830s until overcome by two forces. First, the superior thirty-hour, brass shelf clock appeared in the late 1830s and began to compete in price with the wooden movement clocks. Second, the Panic of 1837 drove all but the most competitive makers into bankruptcy. By the end of the 1840s only a few makers still manufactured the thirty-hour, wooden movement shelf clock in small quantities. The others had either gone out of business or switched to brass clock manufacture.[52] By the mid-1850s Chauncey Jerome alone was making nearly 280,000 brass shelf clocks per year and the wooden movement era was fast becoming a memory.[53]

Perhaps the most surprising aspect of the new industry was the rapid spread of the new technology, both in clock design and machinery design. Clock design was impossible to conceal. A competitor had only to purchase a new clock and disassemble it to learn its design. Thus, clockmakers had no protection from competitors who sought their new designs. However, there were relatively few designs. Once Terry's thirty-hour, five-arbor design appeared, it became and remained the industry standard for twenty-five years. Antiquarians have noticed this similarity of design and have catalogued the variations. Indeed, so similar are the clocks of some makers that they have been identified only by the style of turnings in certain parts. Samuel Terry, a major maker and successor

TABLE 2.2
Thirty-Hour, Wooden Movement Shelf Clock Manufacturers

Chauncey Boardman and his firms
Ephraim and Anson Downes and their firms
Silas Hoadley
Elisha Hotchkiss and his firms
Chauncey Ives and his firms
Chauncy Jerome and his firms
Herman Northrup and his firms
Eli Terry
Samuel Terry and his firms
Eli Terry, Jr., and his firms
Henry Terry and his firms
Riley Whiting
Eldridge G. Atkins
Olcott Cheney
Luman Watson
Mark Lane
Wm. B. Loomis
Sylvester Root
William Sherwin
Levi Smith and his firms
A. Welton & Co.
John L. Wheeler

to his father Eli Terry, noted the "common width" of clock movements, a feature noticed by antiquarians as well.

Clock manufacturing technology was similarly difficult to conceal. In many cases, clockmakers made no attempt to hide their knowledge and often sold it outright. Workmen learned the business easily and, since the technology was relatively simple, could then begin manufacturing for themselves. The machinery was not only relatively simple but also relatively inexpensive. Thus, there were only the lowest entry barriers to clock manufacturing and a modest investment was sufficient to begin production.[54] One characteristic of the business was its highly competitive nature; firms entered and left fairly easily. This too is indicative of a technology easily acquired.

The social structure of early antebellum manufacturing contributed to the spread of this technology.[55] The clock industry resembled the firearms industry at that time in that technical information was readily available to everyone and was often shared. For example, about 1840

> Seth Thomas sent his foreman over to Bristol . . . to get patterns of movements and cases . . . I allowed my foreman to spend more than two days with his, . . . knowing what his object was. A friend asked me why I was doing this, . . . I told him I knew that . . . if Mr. Thomas set out to get into the business, he certainly would find out, and that the course I was taking was wisest and more friendly.[56]

The sharing of technology found in the wooden movement clock industry is strikingly similar to the pattern of technical information exchange Felicia Deyrup noted in the Connecticut River Valley arms industry at the same time. Innovations made in one shop were quickly transferred to other shops through the visits of foremen.[57] These identical patterns of technological diffusion strongly imply that it was common practice in the early antebellum economy.

There is also evidence that some makers sold the technology itself to other makers. Such was the case of Ephraim Downs and Luman Watson. Luman Watson was born in 1790 in Harwinton, Connecticut, and spent his "boyhood in the center of Connecticut's clockmaking country." By the age of nineteen, he had migrated to Cincinnati and established a clockmaking partnership with Abner and Ezra Read, known as Read and Watson, as early as 1809. In the spring of 1815 the partnership dissolved and Watson continued in business on his own account.

Ephraim Downs was a highly successful Connecticut clockmaker. He sold clocks over his own label and sold movements to other makers. On April 26, 1815, Downs noted in his journal that he "commenced work for Read and Watson." On October 8, 1815, he recorded "Luman Watson, Dr., to 10 1/2 days work at the engine." This implies that Watson was making clock plates, but not cutting his own wheels, hence his contract with Downs.

In 1816 Downs made the first of two trips to Cincinnati from Bristol, evidently to help Watson organize his clockmaking factory. Downs probably brought clockmaking machinery with him or built it upon his arrival in Cincinnati. Anson Downs, Ephraim's brother, also traveled to Cincinnati to help Watson establish his clock business. Anson had served an apprenticeship with Seth Thomas and he remained with Watson for about one year.

Watson prospered with the help of the Downs brothers, and by 1819 he had moved his factory to larger quarters, employed fourteen people, and ran his machinery with a horse-driven treadmill. He clearly had learned his lessons well from his Connecticut teachers.

About this time Hiram Powers (1805–1873), America's first internationally renowned sculptor, came to work for Watson as a debt collector during the lean years of the panic of 1819–1820. Powers soon came to work in Watson's shop where he assumed the "superintendency of all his machinery." Powers described his contributions to clock manufacturing technology.

> There was a machine for cutting clock wheels in the shop, which, though very valuable, seemed to me capable of being much simplified and improved. The chief hands . . . laughed at my suggestions of improvement in a machine which had come all the way from Connecticut, where 'the foreman guessed they knew something about clocks.' . . . I so simplified and improved the plan, that my new machine would cut twice as many wheels in a day, and cut them twice as well.[58]

Thus, Connecticut clockmakers traveled across the country bringing not only knowledge of the new technology but the machinery itself. The Downs brothers were not merely employees seeking better opportunities, but important makers. In addition, technological change occurred within Watson's factory. If Luman Watson, a clockmaker in the wilderness in 1815, had no trouble getting the new technology from Connecticut, Connecticut makers must have found acquiring that knowledge even easier. The ease of procuring the new technology and the low cost of entry led to an industry composed of "independent disaggregated firms."[59]

The industry quickly assumed the shape of a traditional putting-out system with various individuals making particular parts for certain makers. The accounts of George B. Seymour, who made parts for Silas Hoadley (one of Terry's original workmen and a major maker in his own right), Asaph Hall (who made parts and complete clocks), and Elisha Manross all illustrate the relationship between the major makers (assemblers) and their contractors. In what is perhaps the first American industrial time study, George Seymour went so far in his accounts as to compute the time necessary to make various parts. For example, it took him seven days for "turning fitting pinning & tureing (1,000) crown (escape) wheels," but only one day for "drilling (1,000) lift shafts."[60]

The major technical implication of the putting out system is startlingly significant. Because the major makers assembled parts purchased by the thousands from contractors, these parts had to be interchangeable or at least "interchangeable enough." That implies a gauging system of some sophistication, controlled by the major makers/assemblers, and used by the contractors. As the case of the Cincinnati clockmaker Luman Watson indicates, makers could buy parts to order.

Makers often kept large supplies of parts on hand, as Samuel Terry's letter to Norman Olmstead demonstrates. Two surviving collections of wooden movement clock parts—the Seth Thomas material and the Hopkins & Alfred material—support the hypothesis that parts were made in large quantities for future use, as does the inventory of Asaph Hall and the account book of Elisha Manross.[61]

These two surviving collections offer the opportunity to test the interchangeability hypothesis in some detail. Were wooden movement clock parts interchangeable as the assembly process and manufacturing system implies?

To test this hypothesis, I measured unused factory stock. These parts show no wear whatsoever and have never been assembled into clocks. I selected five parts, three from the Seth Thomas material at the Connecticut Historical Society and two from the Hopkins & Alfred material at the American Clock & Watch Museum. The measurements for these five parts were entered on an IBM mainframe computer and subjected to statistical analysis using SAS software.[62] The parts were tested for various descriptive statistics listed in table 2.3.

The results of this analysis are unequivocally clear. These parts were fully interchangeable, especially when the nature and design of the wooden movement clock is considered. For example, pivot diameters, pinion diameters, and wheel diameters all show minimal variance from the mean. The range of values is also quite small in every case. Table 2.4 summarizes some of these statistics, which are found in detail in technical appendix 6.

The conclusion is inescapable. These wooden movement clock parts made in two separate factories or at least under two putting out systems between 1815 and 1840 were fully interchangeable.[63] Even the most strict definition of interchangeability applies. These parts could certainly have been taken at random and assembled into clocks.

The manufacture of wooden movement clocks embodied a new technology which, although built on the earlier craft technology and existing individual skills, completely changed the structure and the nature of the industry, indeed created a new industry. It was a putting-out structure tied directly to the technology of specialized tools for specialized purposes and controlled by the major makers who let jobs to individual contractors and enforced strict standards of measurement and precision in manufacturing, an embryonic form of industrial discipline. The industry's structure fit the agricultural patterns of life in the early antebellum economy, and its general characteristics pointed toward the industries that followed.

TABLE 2.3
Descriptive Statistics

Minimum value
Maximum value
Range
Mean
Variance
Standard deviation

TABLE 2.4
Summary of Wooden Movement Clock Interchangeability Statistics

Part	*Measurement*	*Mean*	*Range*	*Variance*
H & A Pinion 1	Hub Pivot Diameter	0.06374″	0.0045″	0.00000″
H & A Pinion 1	Pivot Diameter	0.06364″	0.0035″	0.00000″
H & A Pinion 1	Pinion Diameter	0.59655″	0.0105″	0.00001″
H & A Pinion 2	Hub Pivot Diameter	0.06518″	0.0050″	0.00000″
H & A Pinion 2	Pivot Diameter	0.06484″	0.0060″	0.00000″
H & A Pinion 2	Pinion Diameter	0.66632″	0.0190″	0.00002″
S. T. Pinion 1	Hub Pivot Diameter	0.06256″	0.0065″	0.00000″
S. T. Pinion 1	Pivot Diameter	0.05848″	0.0035″	0.00000″
S. T. Pinion 1	Hub Diameter	0.31458″	0.0110″	0.00001″
S. T. Wheel 1	Wheel Diameter With	1.63418″	0.0135″	0.00001″
S. T. Wheel 1	Wheel Diameter Agnst	1.64616″	0.0125″	0.00002″
S. T. Wheel 1	Hole Diameter	0.31105″	0.0070″	0.00000″
S. T. Dog Shaft 1	Pinion Diameter A	0.06241″	0.0050″	0.00000″
S. T. Dog Shaft 1	Pinion Diameter B	0.06259″	0.0030″	0.00000″
S. T. Dog Shaft 1	Wire Diameter A	0.06632″	0.0040″	0.00000″
S. T. Dog Shaft 1	Wire Diameter B	0.06559″	0.0070″	0.00000″
S. T. Dog Shaft 1	Wire Diameter C	0.06523″	0.0060″	0.00000″

The description of the new clockmaking technology thus far has been necessarily general. To understand how the technical changes in clock design and the invention of new production machinery influenced the structure of clockmaking and subsequently the entire American economy, we must now turn to an examination of that technology.

WOODEN MOVEMENT CLOCK MANUFACTURING TECHNOLOGY

I am of your opinion, said Samson; but it is one thing to write like a poet, and another thing to write like an historian. The poet can tell or sing of things, not as they were but as they ought to have been, whereas the historian must describe them, not as they ought to have been but as they were, without exaggeration or suppressing the truth in any particular.[64]

Don Quixote, part ii, ch.3;
SAMUEL ELIOT MORRISON, *Admiral of the Ocean Sea*

There are five sources of data from which to reconstruct the manufacturing technology of an "ideal" wooden movement clock factory:

1. surviving machinery,
2. surviving tools and material,
3. contemporary descriptions of the technology,
4. the clocks themselves, and
5. the modern observations of antiquarians and a modern wooden movement maker.

The subsequent description of the manufacturing technology is based on all these sources (described in greater detail in the technical appendix), and follows selected components of the clock movement. The manufacture of the case, glass making, or face (dial) making remains outside the focus of this section.

Generally, wooden movement clockmakers used pattern or model movements, the first instance of a model-based gauging system employed in manufacturing. They employed a sophisticated series of marking gauges and templates in laying out work and an equally sophisticated set of working and checking gauges during the manufacturing process itself. They engineered a series of jigs and fixtures with very specialized purposes. They checked their fixture work with checking gauges. All these tools were highly specialized, single purpose tools developed by the clockmakers for their specific manufacturing needs.

Like all the industrialists of the American System who followed them, the clockmakers understood that technical success depended on specialized tools used to produce large quantities of parts within the necessary tolerances. Indeed, a characteristic feature of the American System is the development of specialized tools, not simply for production of parts, but for the subsequent modification of those parts, their assembly, and testing.

There are two aspects to clock manufacturing, first, clock movement design, and second—based on that design—the manufacturing process itself.

Clock Design

". . . the first need is to develop a sensitivity to the ways of wood. This sensitivity cannot be fully conveyed in a written paper because the three dimensional world of wood has important nonverbal aspects which can be apprehended only by seeing, or even by touching, the objects involved."

BROOKE HINDLE
America's Wooden Age[65]

The mass production of wooden works clocks does not concern all wooden movement clocks. There were two traditions of clockmaking America in the late eighteenth and early nineteenth centuries—brass clocks and wooden movement clocks. The mass production of wooden movement clocks began with the thirty-hour, pull-up clocks made by Terry between 1807 and 1810, and spread to a number of other makers who produced a similar clock between 1810 and 1820. The mass production of wooden movement shelf clocks as an industry did not occur until the 1820s with Terry's development of the five-arbor, thirty-hour wooden movement shelf clock. Thus, mass production of wooden movement shelf clocks primarily concerns what collectors call the "Standard, Terry-Type, thirty-hour, wood movement shelf clock" and its imitators.[66] These movements were produced in great quantities in Connecticut between 1820 and 1840 by several dozen makers.[67] Over eighty-five variations have been identified by avid collectors.[68]

The particular development of this movement illustrates the importance of design technology on manufacturing technology.[69] Terry patent clock movements "are characterized by the following major unique features: escape wheel outside the plates, verge mounted on a steel pin attached to the front plate below the escape (wheel), and motion work between the plates."[70] The "outside" escapement wheel was the most significant design feature of Terry's clock because it eased assembly and allowed for larger tolerances in the manufacturing operation. It was the most important design improvement over Terry's thirty-hour, pull-up clock manufactured from 1807–1810.

The single most important part of a clock is its escapement, the mechanism which harnesses the power stored in the weights and imparts an impulse to the pendulum while simultaneously allowing the train to move at a regular rate. Each escapement was composed of a crown wheel (also called an escape or escapement wheel) and a verge which had to be properly depthed (mounted in relation to each other so that the verge allowed the crown wheel to advance one tooth with each oscillation of the pendulum and at the same time to receive an impulse).

Prior to Terry's shelf clock with the "outside" escapement, the escapement was "inside" or between the plates, mounted as were the other wheels and arbors. These two parts were set up in a depthing tool which held them in position and transferred that measurement to the clock plates where holes were drilled. Each crown wheel and verge was individually matched, a time-consuming job requiring much skill and patience (figure 2.9).

Terry's outside escapement contained a simple, ingenious mechanism which made depthing the escapement an easy process. The escape wheel was mounted in the conventional fashion, except that its arbor was longer than usual and extended through a hole in the front plate to a special bridge. The effect of this was to bring the crown wheel outside the clock plates. Below the wheel, the verge was mounted on a steel pin set off center in a brass plug. During assembly, the crown wheel was secured in position, then the escapement was depthed by simply turning the brass plug which carried the verge on its off center pin closer to or away from the crown wheel. When the proper distance was reached, the brass plug was secured in position with two wire nails (figure 2.10).

This process not only eliminated the old depthing process, but allowed a wider variance in manufacturing tolerances such as the diameter of the crown wheel, since the adjustable verge pin could compensate for some degree of over- and under-sizing.

A second design element incorporated the natural expansion and contraction of wood due to changes in the relative humidity. As wood absorbs moisture, it expands across the grain. Thus, while a wooden movement clock wheel was turned round and its teeth cut round, when the relative humidity changed and the moisture content of the wheel changed, it became slightly oval in shape. Wooden movement clocks were designed to accommodate these changes in relative humidity in two ways. First, the clock plates were always cut with the grain running from top to bottom and the arbors were always located close to a vertical line running through each train. Thus, expansion and contraction of the plates had little impact on the relative positions of the various wheels and arbors. Second, as the wheels expanded and contracted, each remained meshed with its pinion because the pinions were cut especially deep. As the wheel shrank in dry weather, for example, its teeth continued to drive the pinion on the next arbor although the meshing of their teeth was not nearly as deep. This was a problem faced only by the wooden movement makers, since relative humidity had no effect on brass and steel clocks.

These two design features, as well as others, had substantial price ramifications through manufacturing changes. The adjustable escapement depthing design saved time during assembly and eased tolerances of manufactured parts, hence lowering the price. The built-in tolerances for changes in relative humidity expanded the market for clocks by making them salable in the most humid climates in the south, for example, where wooden

FIGURE 2.9. Although this wooden movement clock is unsigned, it dates from 1810–1820 and is clearly patterned after Terry's Porter Contract clocks. This view illustrates the unadjustable nature of its escapement, burried deep within the clock's two plates. COURTESY: Milwaukee Public Museum. Catalogue no. H 42663/26656, negative no. H-672-K-8.

movement clocks enjoyed a wide sale. The expansion of the market effectively spread unit costs over a larger number of units, hence lowering the price. Thus, the technical design of wooden movement clocks, especially Terry's shelf clock, had a direct and substantial impact on clock prices and through prices on the size of the market.

The Manufacturing Process

A complete description of the wooden movement clock manufacturing process would be far too detailed. This section focuses instead on several clock parts as representative examples of the process. From these, the reader may understand the general manufacturing process. Greater detail is available in the technical appendix.

WHEEL MAKING. Cherry was the most commonly used wood for wheels. It was first cut into boards slightly wider than the diameter of the wheel. These boards were then planed (probably by hand) to the proper thick-

FIGURE 2.10. Detail view of Terry's adjustable escapement. Note the brass plug with the off-center arbor that carries the crutch. To put the clock in beat during the manufacturing process, the worker simply turned the plug with the verge in place and some power on the crown wheel. When it was in position, he nailed it in place at the two "V" groves notched in the circumference of the plug. The extra nails in is particular example are later repairs. Author's collection.

ness, giving both sides a smooth, finished surface, and measured for cutting.

The holes in the center of each wheel were first drilled, then the wheel blanks were cut out using a "sweep," probably a cutting tool mounted on a vertical shaft. That shaft ran in the hole previously drilled in the center of the wheel, "about 3/8 of an inch" in diameter.[71] This insured that the hole was indeed in the center of the wheel. A number of these oversized wheel blanks were then placed on an arbor and secured there, probably with common nuts at each end of the arbor which held the wheel blanks in place friction tight. This stack of blanks was then turned to the proper outside diameter on a lathe, checked with a gauge, and then placed in the wheel cutting engine. The stack of wheel blanks was probably coated with linseed oil to reduce the sawdust created by the very fine toothed saws during the wheel cutting operation. The wheel blanks remained on the same arbor throughout the wheel making operation, from turning to size through tooth cutting.

The wheel blank-filled arbor was attached to an indexing device of some sort, and fed into the saws of a wheel cutting engine, probably similar to the surviving Hopkins & Alfred machine (figures 2.11a and 2.11b). The only surviving example of such a machine took power through leather belts and pulleys and transmitted the power to an arbor carrying the saws. A fixture (now missing) carried the indexed wheel blanks into the saws on three wooden rails. As each cut was made, the fixture was withdrawn, the stack of blanks indexed, and the fixture again run into the machine.

There is some debate about the configuration of the saws themselves. One source describes a machine with three arbors, while another machine had only a single arbor. It was certainly possible to cut such teeth using any one of several configurations. Precisely which was used at what time is unclear. There may well have been variations among factories, indeed perhaps variations within factories.

The wheels were then removed from the arbor and individually chucked on a lathe and smoothed with sandpaper before having their grooves cut into them and then lightly oiled with a linseed oil soaked cloth. These finished wheels were ready to be mounted on their pinion arbors.

PINION MAKING. The arbors on which the wheels were mounted had the pinions cut into them. These parts were made of laurel wood (or ivy as it was also known). This wood was quite dense and hard, had great strength, and was difficult to cut compared with the relatively soft cherry wheels. The pinion blanks were first cut to length with a special saw which carried two blades on a single arbor. This insured that the pinions were of the proper length to begin with and that the faces at the ends of the pinion blanks were parallel. This was critical in the next step. The pinion blanks were then placed on end in a drill press (a "vertical lathe") and

FIGURES 2.11A AND 2.11B. Wheel cutting engine from the Hopkins & Alfred wooden movement clock factory, ca. 1832. COURTESY: Smithsonian Institution, Museum of American History. Catalogue no. 326,469, negative nos. 78-15119 and 78-15120.

a single hole drilled in one end. A wire pivot was driven into the hole and the pinion blank then turned upside down and placed again in the drill press. At this time, the first pivot fit into a hole which was in line with the centerline of the drill press arbor. The second hole at the opposite end of the pinion blank was then drilled, assured by the first pivot and its aligning hole that this second hole was properly placed.

Then the second wire pivot was driven into place, minor straightening of the pivots took place at this time, and the pivot wires were left long to be trimmed later. The insertion of the pivot wire and its straightening was a critical step, because all the remaining operations took place based on the pivots as the reference points. Straightening was necessary not because the wire was crooked, but because the drill entering the hard laurel wood had a tendency to "run out" (not to drill straight into the wood but to "run out" at an angle inside the wood). The pinion blank was then completed.

The pinion blank was then placed in a lathe, marked with its particular marking gauge, and turned by hand until its various diameters were reached. The turner employed a drop gauge to indicate when he had turned the blank to its proper dimensions. The drop gauge was mounted on centers parallel with and directly behind the centers holding the working piece. The drop gauge probably consisted of several pieces of wire driven into a wooden arbor with wire pivots at each end. It probably looked much like the lifting shaft or hammer shaft used in the clock itself. The gauge rested against the workpiece as it turned. As soon as the workpiece was turned to size, the drop gauge "dropped," telling the turner that the proper diameter had been reached. The critical diameters—hub diameters for mounting wheels and pinion leaf diameters—were then gauged with the go/nogo gauge cut into the marking gauge.

There were at least two critical diameters on a single pinion blank. One was the top of the pinion leaf diameter and the other was the diameter of the arbor itself on which the wheel was to be mounted.

The diameter of the remaining part of the arbor need only have been thin enough to allow clearance for the nearby working parts. So far as is known the turning was done with hand tools. The semi-finished blank was probably creased or decorated at this point and lightly sanded while still in the lathe.

The semi-finished pinion blank, now turned to its proper diameters and decorated or creased, was now ready for the pinion leaf cutting operation. It was placed in a machine probably quite similar to the surviving example (described in appendix 2.1) to have its leaves cut (figures 2.12a

FIGURES 2.12A AND 2.12B. Wooden movement clock pinion cutting fixture, provenance unknown. COURTESY: Smithsonian Institution, Museum of American History. Catalogue no. 334,012, negative nos. 73-5575 and 73-5576.

and 2.12b). After this, the pinion was essentially finished. It was probably oiled with linseed oil.

WHEEL AND PINION ASSEMBLY. Now the wheel and pinion were ready to be united. The pinion was clamped in some kind of vice which protected its critical parts (the pinion leaves and the pivots) and a finished wheel placed on the arbor. Two small holes were then drilled through the wheel into the pinion and the wheel was then secured to the pinion with wrought iron cut nails which were hammered through the wheel into the pinion arbor and peened over into the face of the wheel. On later clocks, wire replaced the cut nails and, in some cases, holes were drilled and wooden pegs were used (perhaps thorns for pegs). Some wheels were also glued.[72]

This essentially completed the wheel and pinion making operations. There is strong evidence to indicate that wheels and pinions were made in great quantities by the various manufacturers and then held until needed and simply assembled.[73]

DOG SHAFTS, LIFTING SHAFTS, AND HAMMER SHAFTS,. Hammer shafts, lifting shafts, and dog shafts were turned on a lathe much the same as the pinions. The lifting and hammer shafts were relatively simple, requiring no special hubs or diameters. These blank shafts were then marked in a marking gauge for their particular use. This gauge marked the points where holes were drilled in the shafts to take the wires (figure 2.13). The wrought iron wire was driven into these holes, bent in a small "U" shape, and driven into the shaft. The longer end was then bent in special bending fixtures (figures 2.14 and 2.15) and checked in special checking gauges (figures 2.16a and 2.16b).

PLATE MAKING. The plates of wooden clocks were almost invariably made of red oak. There are some examples of mahogany plates and cherry plates, but these are the exceptions. The oak arrived at the factory either in boards or as rough plates, cut oversized in length, width, and thickness. Surviving examples of these rough plates suggest that circular saws were in use by the mid 1830s, although earlier plates were no doubt cut using conventional pit saws. These plates were first cut to size, then hand planed flat on one side, marked around the edge with a marking gauge (figures 2.17a and 2.17b), and then planed flat on the other side. These plate blanks were then placed in a brass fixture (figure 2.18). The fixture and its captive plate blank were then placed in a drill press. The drill press

FIGURE 2.13. This Seth Thomas marking gauge was used to mark the locations of the various wires attached to the dog shaft, the hammer shaft, and lift shaft. COURTESY: Connecticut Historical Society, Hartford, Lewis B. Winton Collection.

table was fitted with a pin (on the same axis as the spindle) that protruded into the steel bushing. Each hole was then drilled into the plate blank, perfectly located by the fixture and the pin.[74] Both the front plate and the back plate required separate fixtures.

ASSEMBLY. Sub assembly of the crutch wire, saddle, and verge occurred before the final assembly of the clock, as did the assembly of the various dogs and shafts. Assembling the wooden movement clocks first required that the pillars be pinned and glued to the back plates. The escapement was then depthed by placing the front plate on a fixture which held the crown wheel in place. This fixture allowed the positioning and installation of the escapement bridge and the verge. One of most significant aspects of Terry's clock patent was the adjustable verge pivot for depthing the escapement easily. The brass plug which carried the pivot for the verge was placed in its hole and the verge then placed on the pivot. By turning the plug with its off-center pivot, the verge could be properly depthed easily and quickly.

The plug was then secured in place with two nails and the crutch wire then bent to bring the clock into beat. The wheels, lifting, dog, and hammer shafts, and motion work could then be placed between the plates and the plates pinned together. This completed the assembly of the move-

FIGURES 2.14 AND 2.15. These two Seth Thomas wire bending fixtures have oddly shaped mahogany bases fit with brass and iron parts around which the wrought iron wire was bent. These show signs of great wear in many cases. COURTESY: Connecticut Historical Society, Hartford, Lewis B. Winton Collection.

FIGURES 2.16A AND 2.16B. This Seth Thomas checking gauge was used to check the wire bending work done on fixtures such as those illustrated in figures 2.14 and 2.15. Although they look much like their bending fixture counterparts, they do not show signs of wear. COURTESY: Connecticut Historical Society, Hartford, Lewis B. Winton Collection.

FIGURES 2.17A AND 2.17B. This hand-held plate marking tool was used to mark the thickness of oak wooden movement clock plates before they were planed to size. COURTESY: Connecticut Historical Society, Hartford, Lewis B. Winton Collection.

ment itself. Oiling the front pivot of the escape wheel and the pivot carrying the verge was the only lubrication required, as the wire pivots ran dry in the oak plates.

Conclusion

> It is a brief step from cabinetmaking to clockmaking: at least clockmakers were used to having their clock cases made by cabinetmakers. However, the return to wooden wheelwork in clocks was made for the purpose of introducing interchangeable, mass-produced components. Wooden clocks proved the bridge between handmade brass clocks and mass-produced brass clocks — in which America took command of the field.
>
> BROOKE HINDLE, *America's Wooden Age*[75]

This reconstruction of wooden movement clock manufacturing technology offers an exciting insight into early American technology because it reveals a surprising degree of technical sophistication. It also illustrates

FIGURE 2.18. Plate drilling fixture for Seth Thomas eight-day wooden clocks. Each hole is fitted with a steel bushing. COURTESY: Connecticut Historical Society, Hartford, Lewis B. Winton Collection.

the essential relationship between the mechanic and the entrepreneur and specifically how that relationship worked in practice.

Wooden movement clockmakers used pattern or model movements as the basis of a model-based system which included gauges, specialized machinery for such purposes as wheel and pinion cutting, specialized tools for such purposes as wire bending, design innovation, and punch press work. All were important components of the American System generally believed to have developed much later and primarily in the arms industry. Yet clockmakers developed independently to the degree needed in their factories.

The clock industry was characterized by simple but specialized tools. Each of the Seth Thomas gauges and fixtures is designed to perform a single function on a single part of the clock. There was clearly a great deal of hand work involved in wooden movement clock manufacturing, but it differed significantly from the hand work of the earlier wood and brass clockmakers. The wooden movement worker was guided by the marking gauges, his work repetitive on a thousand parts before turning to a different part and performing another operation on another thousand parts. This was a logical step in the evolution in manufacturing technology, which we might term embryonic manufacturing.

Most of the work was done by hand, very specialized work, but still hand work. Each worker used a set of marking and checking gauges which sharply defined his task and delineated the tolerances within which he must work. Virtually nothing was left for the worker to decide. Nevertheless, he was still quite skilled. The introduction of specific tasks, repeated thousands of times, did not destroy the skill of the craftsman, but rather channeled and focused it sharply. This, too, was an evolutionary step. The new technology of wooden movement clock manufacturing was designed to employ the existing pool of skilled labor, but it employed those traditional skills in many new and different ways.

This is not to imply that clock manufacturing was simply craft work organized on a large scale. It was not. However, the use of skills common in the craft world adapted to the needs of the new technology was an important step toward greater mass production and the development of new skills. There were, in addition, new tools and new machines to which workers had to adapt. However, these tools seem much closer to the craft tradition than the more sophisticated manufacturing tradition to come.

This was a series of developments and characteristics not confined to wooden movement clock manufacturing. In axe manufacturing, the transition from the craft technology to the machine technology was also an evolutionary process with similarities to and differences from the clock industry.

TECHNICAL APPENDIX: MANUFACTURING TECHNOLOGY

Unless you start with the evidence of the object, you are up a tree.

S. J. FREEDBERG, Chief Curator, National Gallery of Art[76]

1. Surviving Machinery

There are three machines surviving from the wooden movement era at the Museum of American History. One is catalogued as an "indexing device, part of machine for cutting wooden clock gears."[77] The second is a "large wooden framework of machine for cutting gears used in wooden clocks."[78] The third is a wire drawing die. A fourth, not really a machine, is a set of "bench brackets or supports" for work benches from a wooden movement clock factory. A significant section of one of the factories also survives at the Smithsonian Institution's Museum of American History. The first two machines are incomplete but both offer important data which is consistent with the written record. The wire drawing die is complete, but the bench supports and the remaining factory parts add little to our knowledge of the business. These are the only known surviving machines from this era and offer the only opportunity to explore the manufacturing technology through the machinery itself.

PINION CUTTING FIXTURE. The "indexing device" is part of a pinion cutting engine. (As with all the specifics of the technology, I have included only such detail as I felt necessary to explain the general characteristics of the machine and its use.) This conclusion is based on two observations. First, the division of the indexing plate allows only for the cutting of pinions with 2, 3, 4, 5, 6, 7, 8, 13, and 14 leaves (see table 2.5).

Second, the method of holding the workpiece easily accommodates a single pinion blank from a wooden movement clock. The space between the arbor's center line and the nearest edge of the engine's wooden frame is exactly the size needed. There is a hole located in the center of the

TABLE 2.5
The Division of the Index Plate

Circle Number	*Number of Holes*	*Remarks*
1	13	an arch, for cutting a rack[1]
2	70	*unevenly divided,* unusable for wheel cutting
3	13	evenly divided, on scribed circle
4	14	evenly divided
5	8	evenly divided
6	7	evenly divided
7	6	evenly divided
8	5	evenly divided

[a] David Todd, the Smithsonian's clockmaker, pointed out the use of this 13-hole segment and reassured me that my impression of the 70-hole circle being misdrilled was correct. Mr. Todd is a superb craftsman and an avid student of horology. I have benefited greatly from his knowledge and experience. George Bruno also noted the use of the 13-hole segment.

arbor carrying the indexing plate at the end opposite the indexing plate. This hole is perfectly designed to fit the pivot of a pinion blank. Raised in the end of the arbor and surrounding the pivot hole are three sharp notches which are designed to bite into the end of a pinion blank. The opposite end of a pinion blank would easily and conveniently be supported in the adjustable wooden tailstock with its attached center.

This fixture is only part of a machine. The cutting tool and the power source are missing as is the frame in which the slotted sides of the tool ran. Thus, we must speculate about its precise configuration.[79] This part of the engine almost certainly fit into the upper frame of a larger machine. A drum probably carried power through leather belts to a rotary fly cutter. This indexing part held the pinion blank and, sliding in the two slots cut into each side of its frame, was moved forward, carrying the workpiece into the cutter and making the first cut between two of the future pinion leaves. Then the device was pulled back, the workpiece indexed, and then pushed into the cutter for the second cut. In 5 or 6 or 7 or 8 cuts, depending on the number of leaves in the pinion, the job was finished.

THE HOPKINS & ALFRED WHEEL CUTTING ENGINE. This machine is, in many ways, quite typical of antebellum machinery in general and is exactly what we would expect to find if we assume that the wooden movement clock manufacturers followed the existing machine-building technology. At the rear of the machine is a wooden drum with two diameters. The smaller diameter (equipped with an idler pulley) took power by leather belt from a line shaft while the larger diameter transmitted that power by leather belt to an arbor mounted on the two connected vertical arms that extend down from a wooden frame hinged to the top of the two posts comprising the rear of the machine. Here, between the two vertical arms, a tool-carrying arbor almost certainly was held. One dead center remains on the left side vertical arm, and an adjustable center is missing from the right side vertical arm. The hinged frame, and hence the missing arbor, is vertically adjustable through two long, iron, adjusting screws which are anchored in and extend through the upper horizontal frame section of the machine. This allows for the adjustment of the cutter or cutters (circular saws in the case of wheel cutting) which were mounted on the missing arbor.[80]

The upper horizontal stationary frame of the machine contains three small, wooden rails which run at 90° angles to the (now missing) tool carrying arbor.[81] These rails probably guided a fixture of some sort which held the wheel blanks for cutting. We have no clear idea of what this fixture may have looked like, but it certainly held an indexing plate of some sort and had some method of holding a number of wheel blanks on a single arbor. Note the important similarity between the two machines in their methods of moving the workpieces into the revolving tool. Both operate on the principal of holding the workpieces in a moving fixture or frame and then sliding the work into the tool on wooden tracks.[82]

THE HOPKINS & ALFRED WIRE DRAWING DIE. The third piece of surviving machinery is a wire drawing die from the Hopkins & Alfred clock factory in Harwinton, Connecticut. This factory was built about 1830.[83] This machine consists of a large wooden block in the center of which is a tapered hole in the smaller end of which is set a steel die. The die has eight tapered holes of various diameters. The holes in the die fit the various diameters of wooden works clock wire quite well in several cases, strongly suggesting that the machine was used to draw the pivot wires and other iron wires used in the manufacture of wooden movement clocks (figures 2.19a and 2.19b).

THE HOPKINS & ALFRED CLOCK FACTORY BUILDING. One of the more remarkable "surviving" artifacts from the wooden movement clock era is the Hopkins & Alfred Clock Factory itself. It was built about 1830 and served as a clock factory at least as early as 1831. In 1841 or early 1842 clock manufacturing ceased. The building measured 30′ x 28′ and contained three stories plus attic. It was powered by water (figure 2.20).[84]

The donor of the factory mentions moving a pile of bricks located near the factory which he believed was a drying kiln.[85]

2. *Surviving Tools and Material*

THE HOPKINS & ALFRED HAND TOOLS. A number of hand tools survive from the Hopkins & Alfred Clock Factory. They are stamped "A. Alfred Harwinton, Ct." and include a pair of dividers, outside calipers, a carpenter's marking gauge, and a small set of thread cutting dies. While they certainly date prior to January 1864, when August Alfred died, they cannot be tied directly to the clock manufacturing activities of Alfred who was a practicing machinist between 1841 and 1864. Such hand tools as these were no doubt found in most wooden movement clock factories.[86]

An indexing wheel also survives in the Hopkins & Alfred material. It is similar in size to the indexing wheel found on the pinion cutting fixture

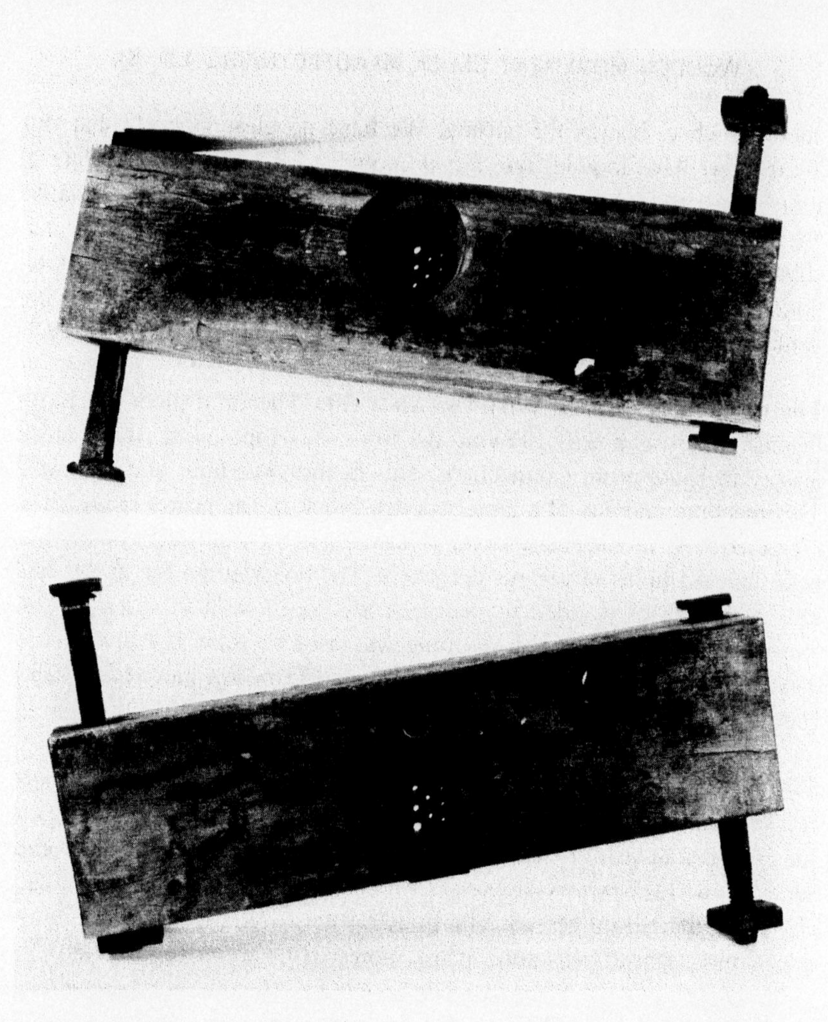

FIGURES 2.19A AND 2.19B. Hopkins & Alfred wire drawing die. COURTESY: Smithsonian Institution, Museum of American History. Catalogue no. 315,806, negative nos. 83-4905 and 83-4096.

at the Smithsonian and can divide circles into 2, 3, 4, 5, 6, 10, 12, and 20 divisions. It has the following hole pattern (counting from the outermost circle):

Circle Number	*Number of Holes*
1	20
2	12
3	10

THE SETH THOMAS GAUGES, FIXTURES, AND TEMPLATES AT THE CONNECTICUT HISTORICAL SOCIETY.[87] In 1978 Lewis B. Winton donated a most re-

FIGURE 2.20. The Hopkins & Alfred Clock Factory, Harwinton, Conn., ca. 1890. The factory was disassembled in 1959 and now survives as parts of two exhibits at the Museum of American History, Smithsonian Institution. COURTESY: Smithsonian Institution, Museum of American History. Negative on file in Division of Mechanisms.

markable collection of clockmaking tools and unused parts to the Connecticut Historical Society.[88] The tools consist of bending fixtures, marking gauges, go/nogo gauges, adjusting gauges, a clock plate template, punch (no matching die), and stacking arbors. The unused material consists of wheels, pinions, lifting and striking arbors, cams, escape wheels, and racks. The tools and material date from the period 1816–1840 and their provenance is impeccable.[89]

This collection of tools and material is not a complete set of tools or clock parts. However, they are the earliest such tools and material known to have survived and they warrant a detailed study.

The largest tool in the collection is a plate drilling fixture. The tool consists of a large (8 1/8″ × 8 5/8″ × 3/16″ thick) cast brass plate with twenty-eight iron or steel plugs set in its surface (seven of which are plugged).

On two opposite sides, wooden edges with brass surfaces rise 5/16″ above the surface of the plate. These hold the wooden clock plate in position for drilling, as does a single permanently attached brass stop at the top of the tool. A large iron leaf spring on the left-hand side of the tool forces the blank clock plate against the right-hand surface.

With the wooden clock plate held securely in place, the fixture could be used during the drilling operation by placing each iron-bushed hole on a pin protruding from the surface of a drill press table directly below the spindle of the drill. This process would quickly and accurately locate the position to be drilled. Such an interpretation suggests that the iron plugs were used to reduce the wear that would have been rapid had the holes been brass.[90]

The pinion gauges are perhaps the most simple tools. Each of the nine different gauges performs the same two functions of marking the pinion blank for turning and then measuring the outside diameter of the pinion blank at the area where the pinion leaves were to be cut in a later operation. These gauges are cut from wrought sheet iron. All the gauges are characterized by two notable features. First, each has a series of sharp points which are located a certain distance from a projecting finger. The projecting finger was placed against one end of the pinion blank while it was turning in the lathe and a circular mark or marks were thus made on the round pinion blank. These marks then served to guide the turner in producing the various widths needed on the blank. (The diameters were gauged by drop gauges described by Hiram Camp.) When the proper diameters and widths were turned, the gauge then served its second purpose, which was to check the critical diameter of one or two areas on the pinion blank. Here, a go/nogo arrangement was used. This was cut into the gauge in some other area and was simply the critical diameter between two horizontal lines with a roughly semicircular area of iron removed to allow clearance. Some gauges have two go/nogo gauges. Each has a small hole drilled in it, perhaps for hanging on a nail near the worker (figure 2.21).

A second set of marking gauges for attaching clicks and click wires also survives. There are five (perhaps six) tools, each of which consists of a wooden handle set in a hemispherical piece of wood with the flat surface located at a 90° angle to the axis of the handle. On the flat surface of the hemisphere, several iron pins protrude. These pins mark the points for locating the click pin, the end of the click spring, and the two ends of the click spring securing wire. These parts are secured to the face of the wheel which is attached to the winding drum.

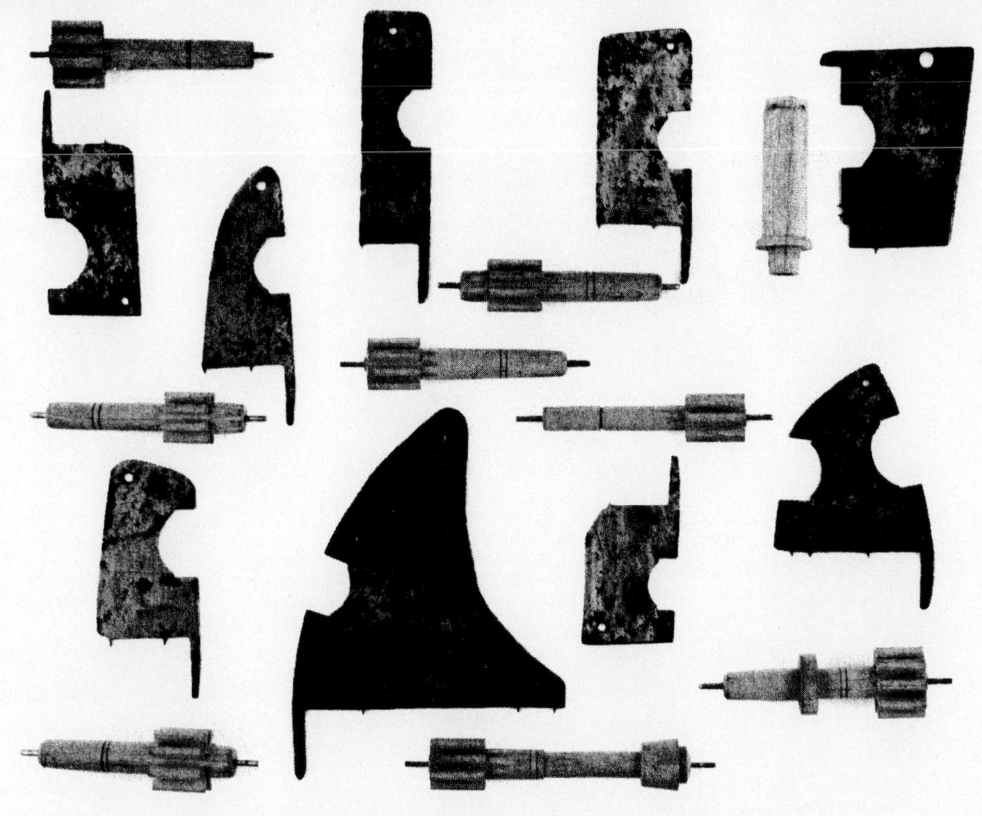

FIGURE 2.21. These nine Seth Thomas pinion gauges, aligned with the parts each was used to produce, are made of sheet iron. They are both marking and checking gauges, having been used to mark each pinion blank before turning, then used to check with a go/nogo arrangement the accuracy of the work. COURTESY: Connecticut Historical Society, Hartford, Lewis B. Winton Collection.

Their location with respect to the center of the wheel, and hence the ratchet wheel, is critical. Therefore the gauge is centered in the completed wheel with an arbor which is part of the handle and protrudes through the hemisphere. The gauge is placed into the wheel's center hole, tapped lightly with a mallet on the end of the handle, and the wheel is then ready for drilling and assembly (figure 2.22).

A gauge similar in principal but different in design and execution is used to mark the locations of the various wires attached to the dog shaft, the hammer shaft, and lift shaft. This marking gauge consists of a block of wood to which is attached a small brass plate. In the working surface of the gauge, semicircular grooves have been cut. Iron pins have been

FIGURE 2.22. Seth Thomas marking gauges for attaching clicks and click wires to main wheels. COURTESY: Connecticut Historical Society, Hartford, Lewis B. Winton Collection.

set into these grooves and the unmarked shaft then placed into the groove and lightly tapped to mark the critical points. The location of the shaft was assured through slots cut in the small brass plate which received the pivot wires already set into the shaft. The shafts were then ready for drilling and assembly.

A plate thickness marking gauge also survives. This marking gauge consists of a simple wooden handle into which is set an iron tool, identical to the gauge described by Hiram Camp (see technical appendix 4).

Six wire bending fixtures survive. These consist of oddly shaped mahogany bases fit with brass and iron parts around which the wrought iron wire was bent. These show signs of great wear in several cases.

Once the wire bending was accomplished, the completed wire parts were checked on a series of gauges. There are three of these gauges in the collection. They look much like their bending fixture counterparts, but do not show the excessive signs of wear that the bending fixtures show. They are designed simply to check the critical dimensions.

There are also tools to cut cams for the strike train, marking gauges to locate these cams on the faces of wheels, and marking gauges to locate clicks and click springs on the faces of wheels (figure 2.23).

FIGURE 2.23. Seth Thomas cam cutting arbor and cam locating gauges. On the surface of a certain strike train wheel there are two cams against which run wire parts. These must be located specifically with respect to the leaves of the pinion. This was accomplished with a cam locating gauge, two of which are illustrated in the lower left and one in position on a pinion at the right. To cut these cams in large quantities, clockmakers first turned small wooden disks, then placed a stack of these disks on an arbor whose ends were divided into three equal parts. The disk-filled arbor was then placed in a tool carrying a circular saw blade and after three cuts, the cams were completed, save for rounding one edge. An uncut disk, three cams, and the arbor (which is actually a fixture as well) are illustrated in the center. COURTESY: Connecticut Historical Society, Hartford, Lewis B. Winton Collection.

PATTERN MOVEMENTS. The surviving pattern movement is the earliest version of Terry's four-arbor, thirty-hour, strap type movement.[91] Thus, such pattern movements existed and were followed in the manufacturing process and were almost certainly in use at the birth of the industry.

3. Raw Materials

The five raw materials used in the manufacture of wooden movement clocks were wood, cast iron, lead, cast and rolled brass, and wrought iron in the form of wire and sheets.

WOOD. Wood was the most important material used in the clock. Wood was purchased by the board from anyone with stock, including merchants who may have taken wood in payment for merchandise or farmers who spent time in the winter in preparing wood for market. This was an age of wood, when virtually everything in daily use was made of wood. Wooden movement clockmakers probably had little trouble finding adequate supplies of good quality, properly dried wood. It was an extremely common commodity in the early nineteenth century.

If the wood did require drying before use, it could be done in two ways: either by air drying, which implied keeping a large supply of wood for several years before it had fully dried (the general rule being one year for each inch of thickness of the wood), or kiln drying. There is evidence to suggest that some manufacturers did some kiln drying, notably Hopkins & Alfred, where the remains of a kiln were found. The inventory of Asaph Hall mentions a kiln. But kiln drying was generally considered inferior to air drying by contemporaries.

Once the wood was properly dried, it was brought into the factory and separated, depending on the kind of wood and hence its purpose in the clock.

WROUGHT IRON WIRE. Wrought iron wire was extensively used in wooden movement clocks. Wire was used to trip the strike train, pivot the wheels and pinions, secure the dial to the movement, hold the count wheel in place, feel the count wheel to stop the strike train at the appropriate place, trip the hammer, drive the hammer through a spring, act as a spring to hold the click in place, to hold other wires in place, and to construct the crutch to deliver the impulse to the pendulum rod which itself was a piece of iron wire. Thus, there was extensive use of wrought iron wire in the clock. Some clockmakers drew their own wire as the surviving wire drawing die from the Hopkins & Alfred factory indicates.

WROUGHT SHEET IRON. There was a small amount of sheet iron used in wooden movement clocks, notably for the hands and the pulley straps. These parts were stamped out using a punch and die of some configuration. Surviving Seth Thomas pulley straps at the Connecticut Historical

Society show unmistakable marks of having been punched, as do a large number of surviving hands from the Hopkins & Alfred factory at the American Clock & Watch Museum in Bristol, Conn. (figure 2.24).

CAST AND ROLLED BRASS. The third critical material in a wooden movement clock is brass. There is relatively little of it, being confined to the escape wheel, the escape wheel bridge (in some cases), the verge saddle, the verge saddle rivet, and the adjustable verge pin. There is a single surviving steel punch in the Seth Thomas material at the Connecticut Historical Society which was used to punch the saddle for early Seth Thomas clocks. Similar punches and dies were used for punching parts as the escape wheel blank from sheet brass. Four surviving escape wheels from the earliest clocks also survive in the Seth Thomas material. Careful examination of the spokes of these wheels reveals identical tool marks, indicating without doubt the early use of punches and dies. All four of these wheels were punched out of sheet brass by the same punch and die.

These parts were finished with wheel cutting engines. The adjustable verge pin was turned from cast brass. From the existing specimens, we can conclude that this brass was primarily narrow rolled brass in most cases and hand-hammered, cast brass in other cases.

CAST IRON. Cast iron was used for the bell and the weights. Cast iron parts, weights and bells, were purchased from local iron foundries which contracted with various makers for specific shapes and sizes.[92]

LEAD. Lead was sometimes used for weights and for the bell hammer head. In the earlier thirty-hour, pull-up clocks, a small lead weight was sometimes used on the pulling end of the winding cord.

4. *Contemporary Descriptions of the Technology*

> In making the wood movements great care was necessary in the selection of the wood. The plates were made of oak split out and then planed up. In the first place the piece was planed level on one side, and then a gauge run around to mark the thickness on the edge, after which it was brought down to mark with a plane which was a slow process.[93]
>
> The wheels were made of cherry which was sawed out in strips of a width and thickness suitable for the wheels, then planed up nicely, then drilled off at sufficient distances to sweep out the wheels, the center of the sweep running in the hole which was about three-eights of an inch, after which the teeth were cut, then the wheels were taken singly and put on the spin-

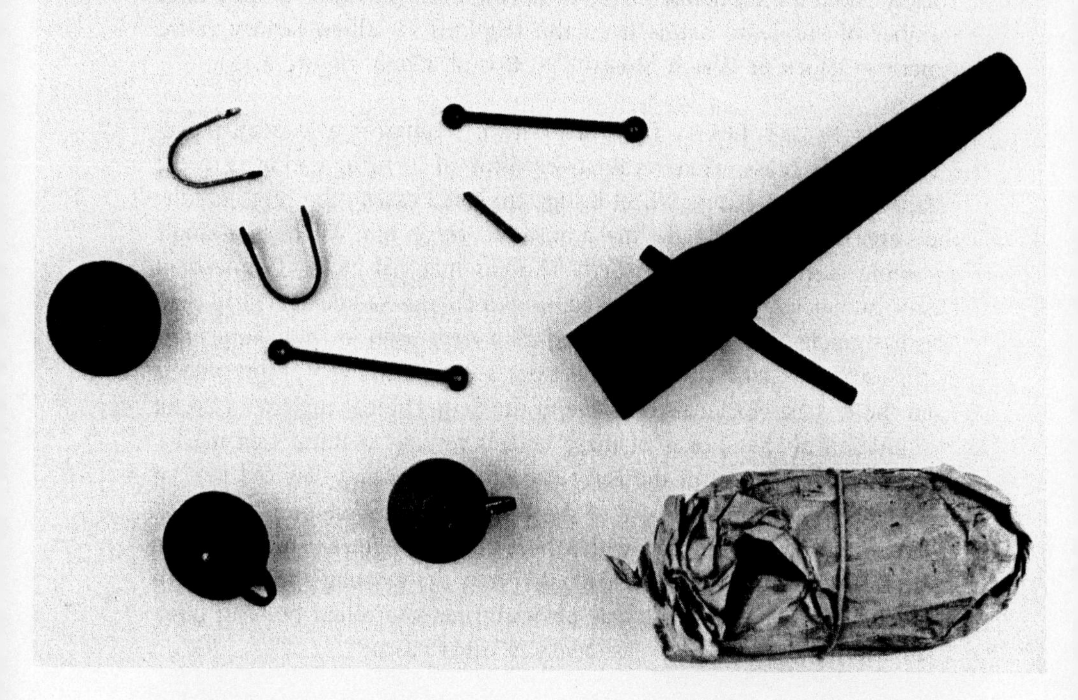

FIGURE 2.24. Seth Thomas pulley bending fixture with various pulley components, including an unopened package of stamped iron pulley parts. COURTESY: Connecticut Historical Society, Hartford, Lewis B. Winton Collection.

dle of a lathe, and a fine piece of sand paper held against the wheel, after which it was creased and ornamented then a rag with a little linseed oil held against it which made a finish.[94]

The count wheels were turned out so as to have a projection on the side in which the spaces were cut for the count.

The pinions were made of laurel or what we call ivy. This is a fine grained wood that is of small growth. It is gathered from the woods (and) thrown under cover to dry. It is a crooked bush, and was sawed into the right length for the pinions with two saws on one mandrell, which made the faces of the ends agree, then these pieces were set under an upright lathe and drilled about an half inch deep of a size right to admit the pivot, and a small piece of wire which was inserted; then this piece was reversed, and this pivot is set in a hole under the lathe, and it then drilled for the other pivot, and when driven is ready for turning. The gauges are hung on some small pivots, and rest on the piece in process of turning, which when the work is brought down to the right size drops by. . . .

The American clock is the cheapest in the world . . . One of my early recollections is going to the Boardman & Wells' clock factory on the Hartford turnpike, where I took my first lesson in clockmaking. I watched the piles of thin boards of cherry, touched by swift saws, falling as clock wheels into boxes below. Then I found that the clocks were not made by one man, but as many sets of men and women as there were pieces; and then they were assembled.[95]

. . . when I saw your son, I was in expectation of sending you a pattern, but have been unable to get one finished.[96]

Plymouth—April 29, 1829
Sir, You wrote me that you wanted the clocks that have been talked about and that you wanted the plate seven inches wide . . . If the plates will answer to be 6-1/2 inches wide, the same as I make, they can be made sooner and for less price. *First, I shall have to make a drill box on purpose and that is a bad job to get it accurate so that the wheels I have now got made will answer*. The pillars I can alter, if you will tell me how in the plate of the room width I am making. . . . but if they (the plates) must be wider, it will be 5 or six weeks and the price will be $2.80, and if they will answer the *common width* (i.e., 6 1/2" plates), I should be very glad.
Respectfully yours,
Samuel Terry[97] (Italics added).

You enquire what I will sell the Machine for making wood mov'ts for, specifying articles, prices, & c. There are such *a multitude of articles fixtures & c* that it would be a task of some difficulty to do so by Letter in a manner to satisfy you or myself. The machinery in Gross ought to bring $700 to $800 & even more, but it would depend something upon how much you wanted to do, as there are in some parts several articles of the same kind. As to whether any improvement can be made in the machinery I am not aware that there can be any very great improvement. I designed while in the business to keep up with the improvements of things in that business. As to where you could procure the Plates I do not know where a good article could be got at present.

There is a lot of them some 30 miles North of here that I selected from, some of the last I used, but they are not a good article, they were probably the cause of a good share of the trouble with my Mov'ts—the last of them—I used to get my plates a good share of them, of A. B. C & the rest of the Alphabet in the Town about to the North of here. *Farmers & others used a leisure day in winter in getting out a few plates*. I have also had a good number & very good ones from Ohio. I usually paid $9 to $13 per M for them.[98] (Italics added).

1826 Dec 12 debit CHAUNCEY JEROME
to a press punch & dies $100.00[99]

This invention or improvement was for the use and introduction of three arbors or mandrels, by means of which one row of teeth on a number of

> wheels were finished by one operation; a machine still in use, *although superseded at the time, by the construction of an engine by Mr. Terry, with only one mandrel, which was used for many years afterwards, and has not been abandoned to this day.*[100]

> It is worthy of mention that all of the several kinds of clocks were made to gauges, so that the parts were interchangeable.[101]

> Mr. Terry . . . being a great mechanic had made many improvements in the way of making the cases. Under his directions I worked a long time at putting up machinery and benches. We had a circular saw, the first one in the town, and which was considered a great curiosity.

> [Eli Terry] was a great man, a natural philosopher, and almost an Eli Whitney in mechanical ingenuity. If he had turned his mind towards a military profession, he could have made another General Scott or towards politics, another Jefferson; or, if he had not happened to have gone to the town of Plymouth, I do not believe there would ever have been a clock made there. *He was the greatest originator of wood clockmaking machinery in Connecticut. I like to see every man have his due* . . . Seth Thomas was in many respects a first-rate man. *He never made any improvements in manufacturing;* his great success was in money making.[102] (Italics added).

Eli Terry is credited with having invented two machines prior to 1806 when he established his first clock factory, "one for cutting the teeth in wooden clock gears and one for cutting the leaves of the pinions."[103]

5. *Wooden Movement Clock Manufacturing Patents*

On August 22, 1814, six United States patents for wooden movement clock machinery and clock design were issued. They are listed in table 2.6[104] There is little evidence to indicate precisely what these patents cov-

TABLE 2.6
Wooden Movement Clock Patents

Part or Machine	*Patentee*	*Address*
Time part of wooden clocks	James Harrison	Boston, Mass.
Pinions	Pharris Bronson and Joel Curtis	Waterbury, Ct. Cairo, N.Y.
Mode of boring plates for clocks	Joel Curtis	Cairo, N.Y.
Wheels, teeth, and pinions	Joel Curtis and Dimon Bradley	Cairo, N.Y. Cairo, N.Y.
Wheels for modern clocks	Asa Hopkins	Litchfield, Conn.
Pointing wire for clocks	Anson Sperry	Waterbury, Conn.

ered and exactly what they described. Descriptions of the machines, patent drawings, and patent specifications have not survived. One of the few pieces of data concerns Asa Hopkins,

> a man of considerable mechanical skill, and a successful manufacturer of clocks. He obtained a patent about the year 1813 or 1814, on a machine for cutting the cogs or teeth of the wheels. This invention or improvement was for the use and introduction of three arbors or mandrels, by means of which one row of teeth on a number of wheels were finished by one operation; a machine still in use, although superseded at the time, by the construction of an engine by Mr. Terry, with only one mandrel, which was used for many years afterwards, and has not been abandoned to this day.[105]

Hopkins' machine was also described by Penrose Hoopes in 1930, "and in 1814, Asa Hopkins patented the well-known three spindle machine for this work, a machine which, with but slight changes, is still used for cutting certain classes of sheet brass wheels."[106] Unfortunately, what was "well-known" to Hoopes' in 1930 is not well known today.

Hopkins "had few superiors as to mechanical skill, however, and really did more in the way of improvements in machinery than others whose names have become a trademark for the prosecution and continuance of the business."[107]

This is all the available data on the six patents issued on August 22, 1814. Neither the patent numbers nor the patent specifications or detailed descriptions of the machines have survived. Murphy notes "that all of these patentees could have had a direct connection with the 1807–1810 activities of Terry. It seems unlikely that the issuance of these six patents on the same day was a historical coincidence. It is more likely that all these inventions were designed for and utilized in Terry's factory."[108]

6. *Wooden Movement Clock Interchangeability Data*

There are two collections of unused wooden movement clock parts, the Seth Thomas material at the Connecticut Historical Society in Hartford and the Hopkins & Alfred material at the American Clock & Watch Museum in Bristol, Connecticut. I examined these parts in great detail in the spring of 1983 and took measurements on a number of parts. I used a dial caliper which reads .001″ on the dial (meaning one reads the last digit as either .0000″ or .0005″). This tool, catalogue number 120, was manufactured by the L. S. Starrett Company, Athol, Mass. I purchased it new in April 1983.

To analyze the data, I used Milwaukee County's IBM mainframe computer, accessed through terminals at the Milwaukee Public Museum. I ran SAS software, a powerful software designed for manipulating large data bases and generating both descriptive and analytical statistics. SAS software is a product of Statistical Analysis System, SAS Institute, Inc., Box 8000, Cary, North Carolina.

Specifically, I programmed the computer to generate several statistics from the raw data, Minimum Value, Maximum Value, Range, Variance, and Standard Deviation.[109]

Data from five parts were analyzed, two Hopkins & Alfred parts and three Seth Thomas parts. There are colored slides of these parts in the research files of the Milwaukee Public Museum. The measurement and the computer's variable name for each measurement are listed below with some additional information on each part.

Part 1. *Hopkins & Alfred Eight Leaf Pinion*

Measurement	=	*Computer Variable*
Overall length	=	TOTLENG
Wood length	=	WOODLENG
Pinion Length	=	PINLENG
Pinion Diameter	=	PINDIA
Hub Diameter	=	HUBDIA
Pivot Diameter	=	PIVDIA
Pivot Diameter, Hub End	=	PIVDIAHB
Hub Diameter, Hole End	=	HOLHUBDI

There are 58 parts, of which I measured 29.

Part 2. *Hopkins & Alfred Nine Leaf Pinion*

Measurement	=	*Computer Variable*
Overall length	=	TOTLENG
Hub to Pinion Distance	=	INNER
Pinion Length	=	PINLENG
Pinion Diameter	=	PINDIA
Hub Diameter	=	HUBDIA
Pivot Diameter	=	PIVDIA
Pivot Diameter, Hub End	=	PIVDIAHB
Hub Diameter, Hole End	=	HUBTHICK
Shoulder Diameter	=	SHOULDIA

There are 58 parts, of which I measured 29. This is a time train part which was to receive a wheel at a later stage of assembly.

Part 3. *Seth Thomas Eight Leaf Pinion*

Measurement	=	*Computer Variable*
Overall Length	=	TOTLENG
Diameter below Pinion	=	SHAFT
Pivot Length, Wheel End	=	PIVLENWH
Pivot Length, Pinion End	=	PIVLENPH
Wheel Thickness	=	WHELTHIK
Wheel Width, Against Grain	=	WHEELAG
Wheel Width, With Grain	=	WHEELWI
Difference in Wheel Widths	=	WHEELDIF
Pinion Length	=	PINLENG
Pinion Diameter	=	PINDIA
Hub Diameter	=	HUBDIA
Pivot Diameter	=	PIVDIA
Pivot Diameter, Hub End	=	PIVDIAHB

There are 31 parts, of which I measured all 31. One part was clearly from a different batch or supplier. This is a time train assembly consisting of an eight leaf pinion and a thirty tooth wheel, the fourth arbor in the standard Seth Thomas thirty-hour wooden movement. It is part number 103 in Winton's inventory.

Part 4. *Seth Thomas Upper Strike Control Arbor Assembly*

Measurement	=	*Computer Variable*
Overall length	=	LENGTH
Pivot Diameter A	=	PIVDIAA
Pivot Diameter B	=	PIVDIAB
Wire Diameter A	=	AWIREDIA
Wire Diameter B	=	BWIREDIA
Wire Diameter C	=	CWIREDIA
Arbor Length	=	ARBORLEN
Arbor Diameter	=	ARBORDIA

There are 40 parts, two are damaged, of which I measured 25. This is a strike train part with several bent wires protruding at various angles. This is Winton's inventory number 139.

Part 5. *Seth Thomas 5th Wheel-Time Train*

Measurement	=	*Computer Variable*
Wheel Diameter, With Grain	=	WELDIAWI
Wheel Diameter, Against Grain	=	WELDIAAG
Difference in Diameters	=	WHEELDIF
Wheel Thickness	=	WHELTHIK
Hole Diameter	=	HOLEDIA

There are 178 parts, of which I measured 90. This is a time train thirty tooth, wheel from an 8-day (E. Terry and Sons Type). It is Winton's inventory number 71.

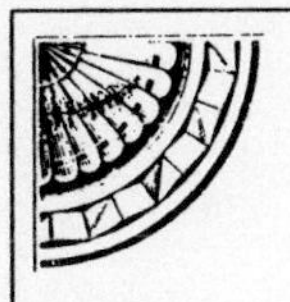

CHAPTER

Elisha K. Root and Axe Manufacturing at the Collins Company, 1830–1849

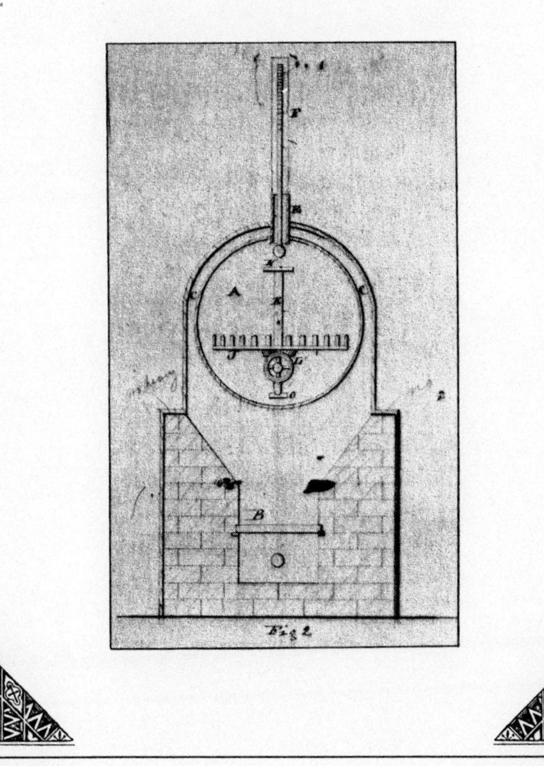

His [Samuel W. Collins] mind was what we term, emphatically, *well-balanced*. He was observant of other men, sought their views, noted the results of their plans, and thus obtained information to direct in matters under consideration. Whatever originated with himself was compared with facts as developed by others, and thus by deliberate, candid examination he decided what course to adopt. He was not hasty in forming his opinions, or reluctant to seek light in matters difficult; nor did he claim to *originate* so much as to *use wisely* what he could gather. Hence it was ever to be seen that he appreciated talent, genius; and aimed to secure the aid and cooperation of those, who were endowed with mental power or mechanical skill; and was ever ready to award them all deserved credit.

Especially is this true of one [Elisha K. Root] who was for several years his chief adviser, and whose genius and skill are manifest in most every work of the establishment. Mr. Collins took pleasure in acknowledging the worth of others, in the possession of those faculties which made them distinguished, and the confidence he reposed in them, as he attracted and bound to himself men of skill and probity. Though acute and discriminating in his estimate of men, he was also generous, and commended the worthy; affording often substantial aid which, in many instances, was gratefully acknowledged.

H. N. Brinsmade
An Address at the Funeral of Samuel W. Collins[1]

UNLIKE the relative anonymity of the wooden movement clockmakers, the manufacturers of axes at the Collins Company are well known and documented. Mechanization at Collins began in 1831 with the work of several mechanics, the most able of whom was David Hinman. Their work was brought to fruition by Elisha K. Root who came to Collins in August 1832 and remained for seventeen years before accepting the offer of Samuel Colt to become superintendent of Colt's Hartford Armory. The Hartford Armory under Root became the showplace of the American System in the 1850s.

The extensive documentation of the Collins Company allows us to explore many of the forces which influenced Root and his employer, Samuel Collins.[2] As was the case with the wooden movement clock manufactures, there is no evidence of government subsidy in any form, either through large purchases of goods or the supplying of technical know-how. There is no evidence of technology transfer from the either the federal or private armories to Collins. Indeed, the transfer—such as it was—was in the opposite direction, from Collins' axe factory to Colt's armory.

The manufacture of axes further points out the independent development of the American System in the private sector, with private capital (not always sufficient—Collins went broke in the early 1830s due in part to the expense of technological innovation) by private entrepreneurs, who were not connected with arms, armory practice, or the federal government.

Unlike wooden movement clock manufacturers, where an industry grew up around the putting-out system and a series of assemblers and manufacturers, axe makers built a complete, power intensive industrial operation at a single location. Of greater interest is the data illuminating important questions about the nature of technological innovation in the private sector. Specifically, how did an individual mechanic function within the firm? What motivated the mechanics in the early nineteenth century, particularly those in the private sector? Why did they act as they did? What forces influenced their thinking and how did they react to these forces? In short, why did they do as they did? At the Collins Company we can explore in some depth the roots of the American System and the role of a single mechanic, Elisha K. Root, in its growth.

Elisha King Root's career falls into three distinct chronological periods. The pre-Collinsville period, beginning in 1808 with his birth and ending in 1832 with his employment at Collins, includes his apprenticeship and his first meeting with Samuel Colt. During the second period of his career, the Collinsville period, which lasted from 1832 to March 1849, Root worked solely on the manufacture of axes for Collins at Collinsville. Finally, the Hartford period, which encompassed his work for Colt's armory, ended with Root's death on August 31, 1865.[3]

During the Collinsville period, in an intellectual and economic climate conducive to innovation, Root: a) developed his own concept of a mechanized, mass production system of axe manufacture, b) then methodically developed the machinery to make the concept work, and c), employing his system-building skill and enthusiasm in the proper economic climate, convinced Samuel Collins to finance the building of his system. Through his machines, Root's mechanical maturation shines as he departed conceptually from traditional axe making methods and moved toward a mechanized, integrated production process. Root's ability to conceptualize axe manufacturing as a system was his greatest contribution to the then developing American System of Manufactures.

The Hartford period of Root's career is a subject outside the focus of this work but of great interest. During this phase of his career, Root supervised the manufacture of Colt's revolver, although he evidently left Colt's employ briefly to manufacture pumps.[4]

THE PRE-COLLINSVILLE PERIOD, 1808–1832

Elisha King Root was born on his father's farm at Ludlow, Massachusetts, on May 10, 1808.[5] The evidence of Root's early career is both sketchy and conflicting. He entered a cotton mill as a bobbin boy at the age of ten or twelve and attended a common district school four months of each year, spending the remaining eight months in the mill. At the age of fifteen he began work in a machine shop in Ware, Massachusetts, as a machinist's apprentice.[6] While he was an apprentice at Ware, Root made the acquaintance of Samuel Colt. They met on July 4, 1829, at Ware Pond, where Colt attempted to destroy a raft with an electrically detonated underwater torpedo.[7] The raft slipped its moorings and drifted away from the torpedo before Colt detonated it and the explosion splashed mud and water on a number of onlookers who sought revenge by attempting to throw Colt into the pond. However, Root, "who had come from Chicopee Falls to see the demonstration," stepped in and rescued Colt.[8] Root's natural curiosity about mechanical and scientific subjects undoubtedly attracted him to Colt's torpedo demonstration.

Root apparently served a conventional apprenticeship followed by some journeyman machinist work in several Connecticut River Valley mill towns including Stafford, Connecticut, Chicopee Falls and Ware, Massachusetts, and Cabotsville.[9] The precise nature of his work in these mills is unclear and additional information on Root's early years is unavailable. In 1832 journeyman machinist Elisha King Root arrived at Canton, Connecticut, and began work at Collins & Co., an axe manufactory.[10]

THE COLLINSVILLE PERIOD, 1832–March 1849

Root arrived at Collins & Co. sometime before August 1832, at the age of twenty-four. He reportedly brought no letter of introduction or recommendation and simply described himself as a machinist. He started work on a "turning-lathe" in the repair shop, but quickly proved himself more than a lathe hand.[11] Although little is known of his nine-year apprenticeship, it is clear that Root possessed considerable skill upon his arrival at Collins & Co., since Samuel W. Collins, president of Collins, appointed him overseer of the repair shop, probably on August 18, 1832, the day Root signed a two-year contract with the axe maker.

Root's contract stipulated that his "business [was to] be building machinery and repairing geering [sic] and machinery, keeping polishing wheels in order Ec, Ec."[12] S. W. Collins, who immediately recognized Root's skill with machinery, hired him to employ that skill, evidence of a progressive attitude toward technical innovation.[13] Root received $546 per year of 312 days, making his daily wage $1.75, compared with the $1.00 per day paid to most skilled workmen at Collins in 1834.[14]

In 1832 Collins employed skilled craftsmen practicing traditional axe manufacturing methods. A careful review of these preindustrial skills helps reveal fully the nature of the mechanical revolution Root wrought in Canton. An understanding of these methods emphasizes the significance of Root's first machine and the nature of his thinking as his mechanical intellect matured.

Axe manufacture in 1832 consisted of welding together two pieces of ferrous metal, the steel "bit" and the wrought iron "poll," to form a wedge-shaped tool with a hole, the "eye," into which the handle was to be placed. The axe was then hardened and tempered, ground, and ultimately polished or painted.

The Collins records leave no doubt that this technique was employed, for they reveal not only the workmen's names but also their occupations. Such jobs as "forging axes," "striking on axes," "wheeling coal," "plating patterns," "grinding on axes," "polishing axes," "tempering axes," and "setting emery wheels" occur repeatedly. The job descriptions reaffirm these findings.[15] The documents prove that by 1834 Collins had a well-developed division of labor in which each laborer specialized in one particular operation at the mill.[16]

Root was not the first inventive mechanic employed by Collins & Co. in the early 1830s who attempted to mechanize the traditional methods. A careful study of Collins' records reveals the presence of at least four (possibly five) other innovative technicians: David Hinman, Erastus Shaw, Benjamin Smith, and Isaac Kellog.

Their presence and the innovative intellectual climate at Collins undoubtedly influenced Root's thinking and his attitude toward mechanization.[17] The following account of these men's work is a necessary prelude to an understanding of Root's intellectual development and his mechanical achievements.

Both Hinman and Shaw received patents for die-forging machinery. Die-forging is a method of giving form to metal by pressing, hammering, or rolling yellow-hot, and therefore plastic (nearly molten), metal into plain or formed dies. Most "die-forgings are produced in formed dies that are shaped more or less closely to the outline of the forging."[18] Two dies are used, each having cut into its working surface half of the shape of the forging so that when the dies are forcefully brought together they create the completed form. Hammering is a violent action and literally pounds the hot metal into the dies, while pressing and rolling exert great force without a violent action.

In 1832 Collins & Co. "contracted with David Hinman to build machines for shaping and welding axe polls."[19] Hinman's die-forging machinery was operating by August 1834, a week after the company suspended business due to financial difficulty. Hinman failed to perfect his die-forging machine as originally scheduled, causing a six-month delay. This delay, combined with the machine's great cost and a financial panic, spelled temporary financial difficulty for Collins.[20] Although the Collins embarrassment was in large part brought on by the expense of constructing the Hinman machines, Samuel W. Collins was certain that the machinery effectively reduced the cost of axe manufacture.[21] By December 1834 Collins & Co. resumed production and Hinman's machines operated successfully.

Hinman's machinery plated out the patterns for axes and welded the polls much faster than a foreman and striker. Collins paid each workman $0.35 per 100 patterns "for plating & turning patterns for axe heads, in machines," and the same price "for welding axe heads in machine." Men could also earn $1.00 per day "for day work" on Hinman's machines, implying that a man could turn out about 300 patterns per day or weld 300 patterns per day using Hinman's die-forging machine.[22] This machine rate of production (300 axe heads per day) was twenty-five times greater than the twelve heads per day that a foreman and his striker could produce using the traditional methods.[23]

Hinman secured patents on two die-forging machines at Collinsville. In the first application, granted on November 2, 1832, Hinman claimed to have invented his machine as early as December 10, 1831, long before Root arrived at Collinsville.[24] Hinman combined water power with the motion of the trip hammer and dies to produce a squeezing effect on the iron instead of a blow, an effect Hinman believed was superior.[25]

Hinman's second patent, June 29, 1833, is far more interesting and important. There he describes what was evidently at least part of the machinery he built for Collins (figure 3.1).[26]

Hinman's second die-forging machine was considerably more advanced mechanically and incorporated more sophisticated features than his first machine. It shows a conceptual leap from simply mimicking the motion of the trip hammer to a creative radical solution to the technical problem of forcing hot metal into dies. In this second machine, Hinman claimed only the application of power to the traditional methods of swedging the pattern in a trip hammer.[27]

The introduction of Hinman's successful die-forging machine violently disturbed the handicraft manufacturing methods by vastly increasing the production of axe polls (heads). This single machine created a tremendous imbalance between the forging and poll welding steps in the process and the bit welding and grinding steps. Essentially, all the later steps suddenly became production bottlenecks, as ever increasing quantities of polls required finishing.

David Hinman was not the only mechanic experimenting with new die-forging techniques at Collins. Erastus Shaw also patented a die-forging machine on June 29, 1833. He described a common, two high, iron rolling mill with the two rollers extended through the heavy iron frame.

In recesses cut in these rollers, Shaw placed dies and geared the rollers to insure their coming together. Shaw claimed only "the application of rollers placed on the outside of the frame to the forging of axes . . . so that they may be more convenient to come at."[28] As in the case of Hinman's first machine, there is no evidence that Shaw's machine was ever built or used at Collinsville.

Benjamin Smith's improvement "in the manufacture of axes," dated November 2, 1832, describes both horizontal and vertical spindle milling machines, two different cutters, and methods of making those cutters.[29] Smith also described a carriage which held the poll and moved it in four directions with respect to the stationary cutter. Since Middletown, Connecticut, was once the center of milling technology and knowledge of John Hall's machinery had reached the Connecticut Valley as early as 1827, it is not surprising to find milling technology creeping into the private sector.[30] It is significant that Smith claimed only to apply a known technology to the manufacture of a different product, milling to axe manufacture. His patent offers nothing new, except perhaps the method of holding and moving the work piece toward the cutter.

Smith's machines are significant in and of themselves, although there

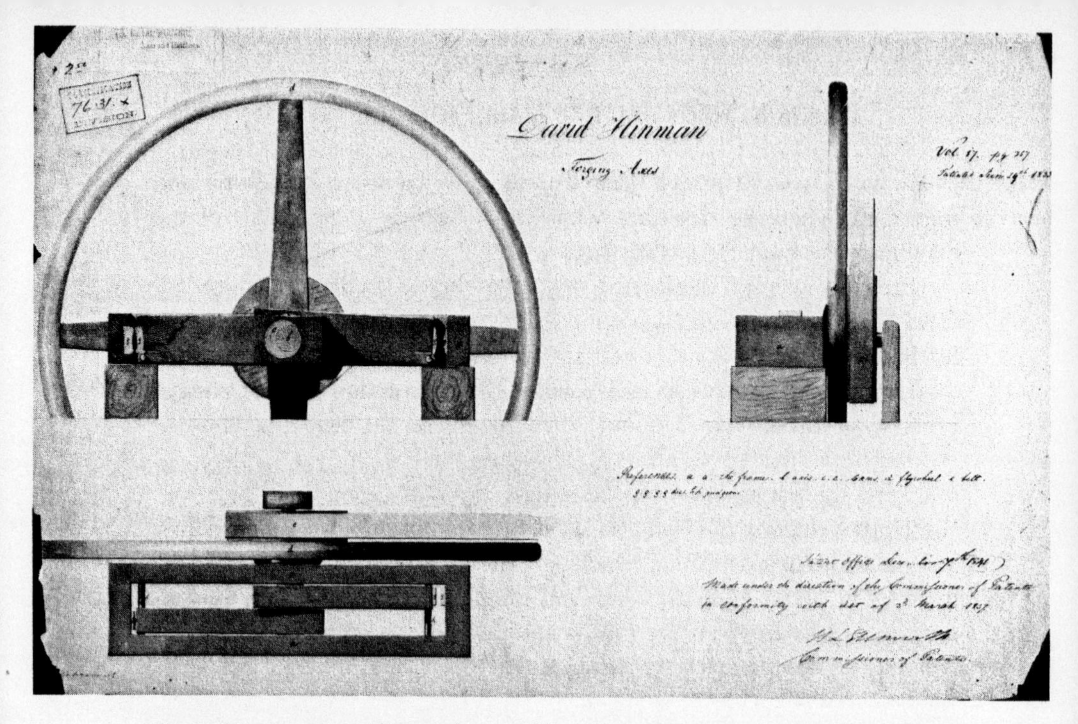

FIGURE 3.1. Reconstructed patent drawing of David Hinman's patent no. 7631 X, for "Forging Axes," June 29, 1833. This second of Hinman's patents shows the machine as it probably appeared in its constructed form at Collins & Co. SOURCE: Records of the Patent Office, Record Group 241, National Archives and Records Administration, Cartographic and Architectural Branch.

is no evidence that Smith built a milling machine or that axes were milled at Collinsville. However, these patents demonstrate clearly that early in his career, Root was exposed to technicians with intimate knowledge of the latest metal working techniques.

Isaac Kellog was the fourth individual who had intimate contact with the innovative mechanics at Collinsville. Although there is no evidence that he applied for or received any patents, he and David Hinman were witnesses to Erastus Shaw's patent, and both Shaw and Kellog witnessed Hinman's second patent. Neither Shaw, Hinman, nor Kellog attested to Benjamin Smith's patent.[31] A fifth person was also associated with this group of innovative mechanics. Known only as "Burke," he was reported to have been connected with Shaw at Canton, Connecticut, in 1832.[32] His contributions, if any, are still under investigation.

What was present at Collinsville between 1832 and 1833, and what Root found when he arrived in 1832, was an active community of closely connected technicians utilizing the very latest metal working techniques. All worked toward mechanizing the traditional methods of axe manufacture with the encouragement of the Collins & Co. management. As a creative and thoughtful individual, Root found an environment for invention at Collins that would stimulate and influence his future work.

In summary, in the late summer of 1832 Root began work at Collinsville, quickly proved his skill and ability, and soon received a promotion to overseer of the repair shop. He signed a two-year contract to build and maintain machinery with Collins & Co. and entered a community of inventive mechanics anxious to mechanize axe manufacturing.[33] They not only had access to the latest technology but also received the blessing and encouragement of Samuel W. Collins. Three of the members of the tiny technical community were actively working on die-forging techniques and a fourth possessed detailed knowledge of milling technology. Most important, Hinman's second die-forging machine had dramatically demonstrated the economic returns to technological innovation. Root's work agreement with Collins & Co. clearly indicates that he joined these innovative people seeking to increase axe production. The following description and analysis stress the influence of these mechanics and the intellectual climate on Root's contributions.

ROOT'S WORK AT COLLINS & CO.

Root's first known innovation in axe manufacture was a pattern rolling machine designed as early as February 6, 1836, and patented on March 30, 1836. Root's machine plated out the pattern, doubled it over and welded it, an improvement on Hinman's machine, which only plated out the pattern and welded it (figure 3.2). Root's machine embodied several elements of the die-forging machinery patented by Shaw and Hinman as well as several new features. Root adopted the rollers with attached dies as described by Shaw, but claimed only the use of reciprocating motion, not the rollers and dies."[34] The idea of using jaws to finish doubling over the pattern prior to welding could well be derived from Hinman's first patent.[35] The method of doubling over the pattern with a punch between two small rollers was novel, as was the combination of the several processes in a single machine and the method of connecting the crank and the roller.

However ingenious this machine was, it neither added nor changed anything in the manufacture of axes. It simply mechanized the hand methods by substituting machines which mimicked the motions of men for men and adding water power. Root built the skill into the machine but did not change the skill itself. This represents an important step in

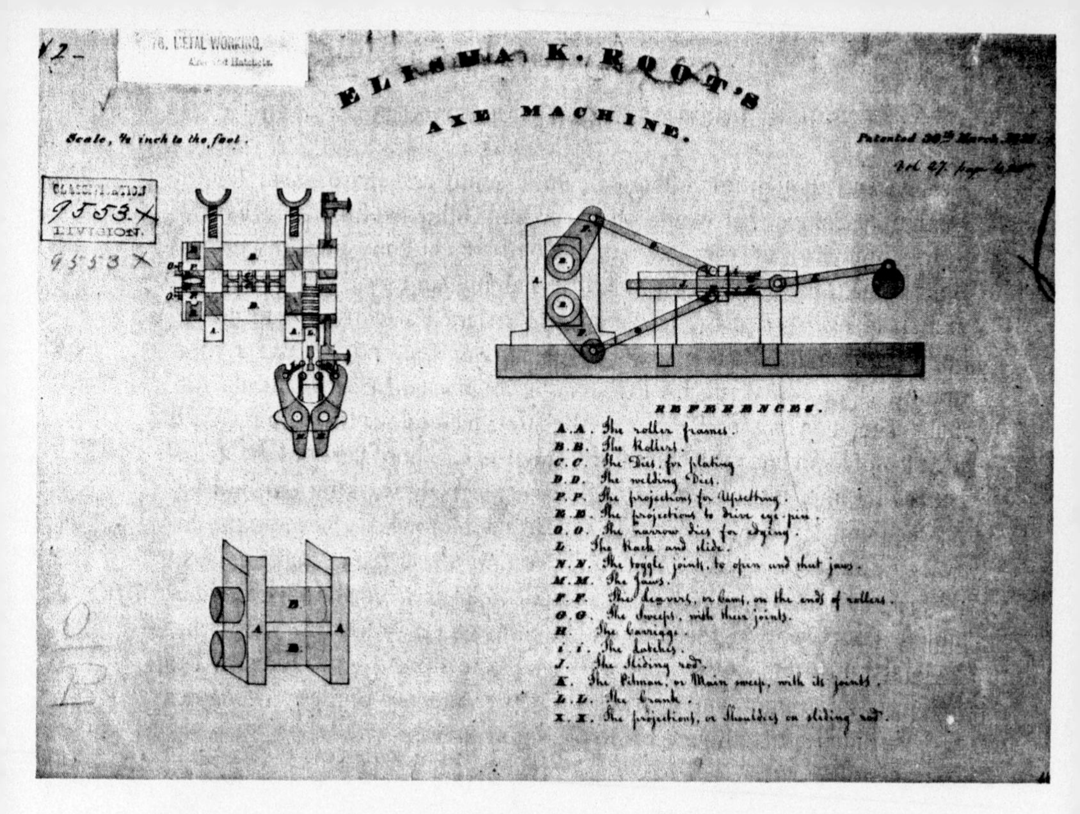

FIGURE 3.2. Reconstructed patent drawing of Root's first "Axe Machine," patented March 30, 1836, patent no. 9553 X. Note the use of dies outside the frame of the machine, "D," a feature patented earlier by Erastus Shaw. SOURCE: Records of the Patent Office, Record Group 241, National Archives and Records Administration, Cartographic and Architectural Branch.

the thinking of a mechanic and in the development of machinery in general. The work of both Hinman and Root displays a common characteristic. Both men began by building machines that embodied existing craft techniques and then moved on to radical departures from these traditional but mechanized techniques. Root moved intellectually toward a machine that departed even more radically from the traditional techniques—an eye-punching machine.

In 1837 Root built a second forging machine which punched the eyes of axes from a solid bar of iron. This machine is the most important of any that Root designed or built at Collinsville, for it was the cornerstone of the axe manufacturing system he was developing. On December 10, 1838, Root received a patent for this important machine which (as Root himself pointed out) embodied several features found in his first machine.[36] Power was supplied through identical methods to an identical frame holding identical reciprocating rollers. However, Root applied cams to the rollers to work steel punches which punched out the eye of the axe instead of plating out the pattern, doubling it over, and welding as on his first machine. Once the eye had been punched from the solid piece of stock, an eye pin was inserted in the eye to preserve its shape throughout the die-forging operations that followed on the same machine.

Of particular significance is the fact that Root's machine punched the eye from a solid piece of wrought iron. This new method, a radical departure from the traditional methods employed at Collins' mill and in Root's first machine, eliminated many plating and welding steps. Root further claimed that his method prevented distortion of the metal and formed a more accurate eye than other forging methods of a violent nature, such as drop-forging or forging under the trip hammer. Root's machine substituted one step for four, effecting a saving in time, labor, fuel, and materials.

Root's eye-punching machine demonstrates not only that he had taken a giant conceptual leap in his thinking about axe manufacture, but that his mechanical thinking had matured. Root's eye-punching machine is the first concrete evidence of his creative ability and his mechanical genius. His choice of techniques shows his knowledge of forging methods, his ability to embody his knowledge in particular machines, and the influence of Shaw and Hinman on his thinking and his machine design (figures 3.3 and 3.4).

Sometime between October 1, 1838, and January 1839, Root considered patenting "a combination of punching and plating machines."[37] He simply claimed the combination of his two previously patented machines as a single machine run by a common power source. All of its features, except combination, had patent protection, and the Collins & Co. management chose not to apply for the patent.[38] Root's "Combination Machine" eliminated unnecessary aspects of his earlier plating machine, displaying a fundamental characteristic of a truly creative mechanic, "the rare gift of improvement by simplifications."[39]

In 1839, shortly after Root had devised his combination machine, the town of Canton rebuilt a bridge across the river running through town on a plan drawn by Root.[40] It is unclear if Root produced an original design for the bridge or simply delineated one of many well-known and commonly used designs. Nevertheless, he carried his engineering interests and abilities outside the machine shop, which is evidence of his technical versatility and his rising status as a mechanic in the community.

In 1842, when business at Collins was slow, company president S. W. Collins sent Root on a grand tour of England to investigate steel making and working techniques.[41] Many American mechanics and engineers took

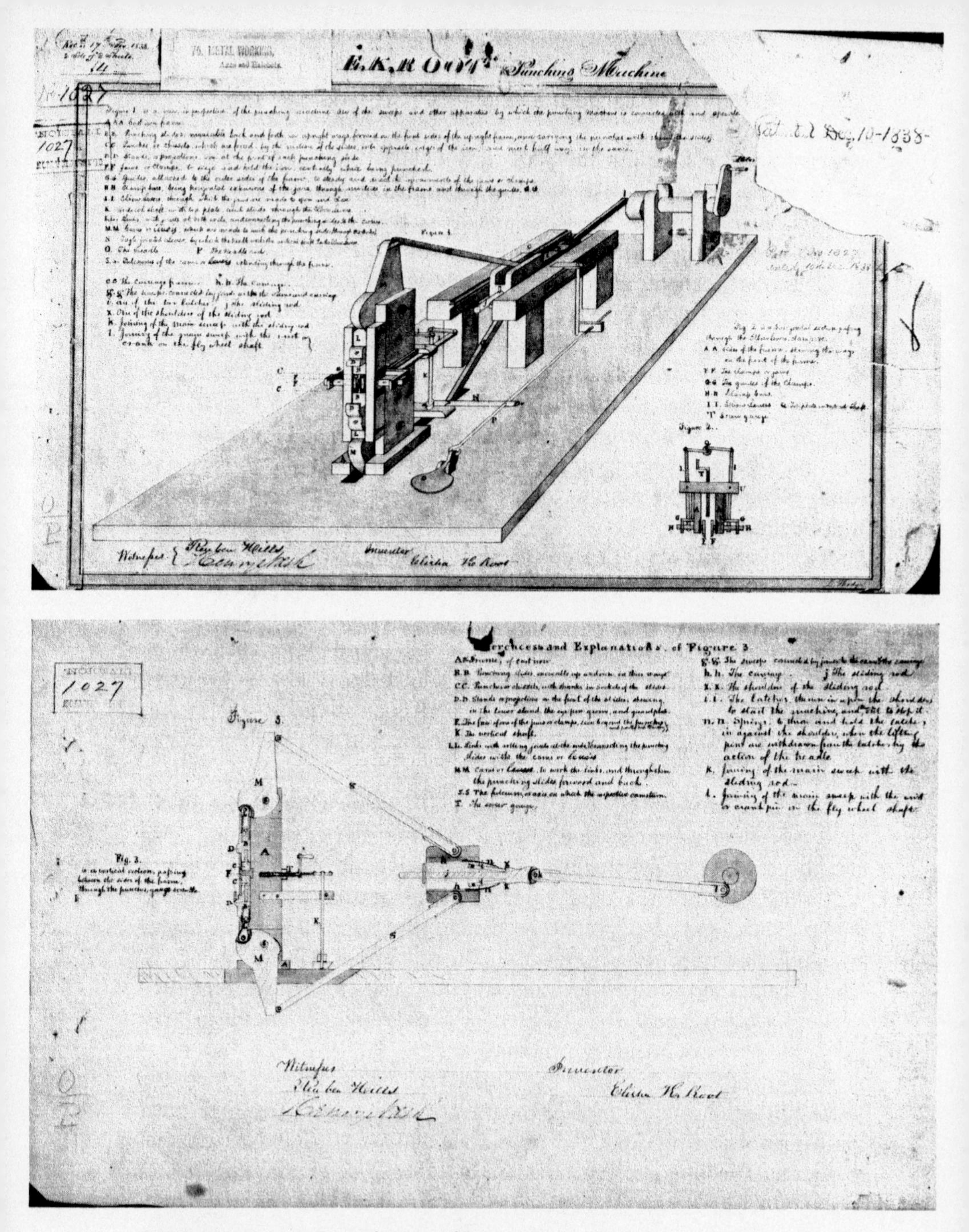

FIGURES 3.3 AND 3.4. Patent drawings of Root's axe punching machine, patent no. 1027, December 10, 1838. SOURCE: Records of the Patent Office, Record Group 241, National Archives and Records Administration, Cartographic and Architectural Branch.

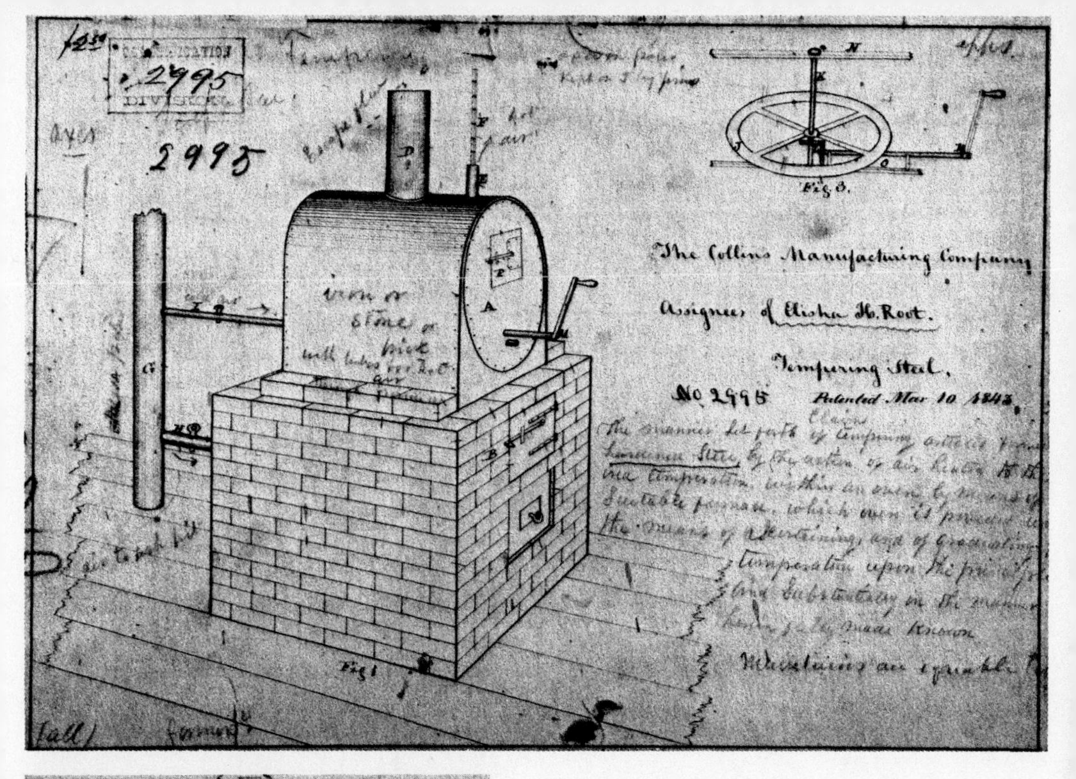

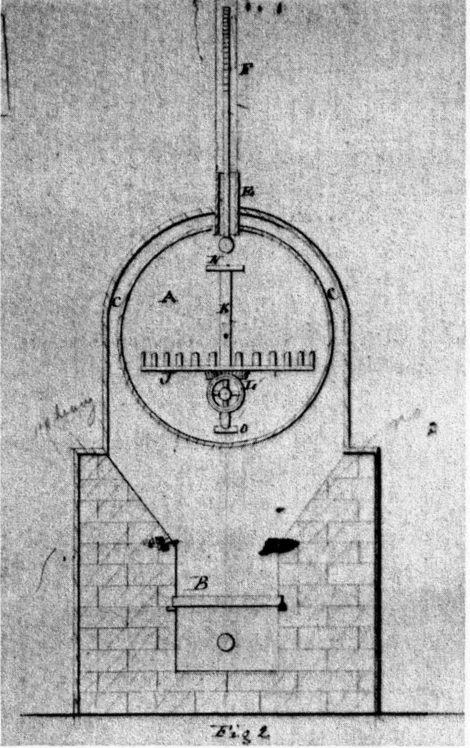

FIGURES 3.5 AND 3.6. Patent drawings of Root's tempering oven, patented March 10, 1843, patent no. 2995. This patent was assigned to Collins & Co. SOURCE: Records of the Patent Office, Record Group 241, National Archives and Records Administration, Cartographic and Architectural Branch (also in Connecticut Historical Society, Hartford, MS. 71890).

FIGURE 3.7. Root's tempering ovens as they appeared in 1871 SOURCE: Horace Greeley et al., *The Great Industries of the United States*, p. 132. COURTESY: Courtesy Milwaukee Public Museum. Negative no. H-638-D.

such trips and Root followed the pattern.[42] S. W. Collins' financing of Root's trip typifies the innovative atmosphere at Collins in the 1830s and 1840s.

Collins found concrete rewards for his investment in Root's patent of March 19, 1843 (#2995), an improvement in tempering steel (figures 3.5, 3.6, and 3.7)[43]. Root's idea was simple but ingenious. He built a large stone furnace which led directly into an iron oven above it. The oven was equipped with a thermometer and an air duct with a valve for admitting cold air to regulate the temperature.[44] A revolving rack in the center of the oven held a large number of axes. Root's tempering oven increased production by allowing many axes to be tempered simultaneously. He built the skill of tempering into the machine and eliminated the need to watch the changing colors of the steel on each axe. The operative simply watched the temperature. While he was in England, Root also learned steel casting techniques, but a revival in business at Collins in 1843 forced him to postpone this development and concentrate on filling current orders.[45]

In 1845 Root "invented a process for shaving axes as a substitute for grinding."[46] Root's axe shaving machine, or an earlier version of it, was evidently in operation as early as 1844 when S. W. Collins complained of "bad shaving."[47] Here Root and Collins responded directly to a particular labor shortage problem at Collinsville due to the health hazards of grinding. As Collins explained, even though the company had the capacity to manufacture 1,000 axes per day in 1845, they averaged only 673 per day because experienced grinders "were not to be had. There had been so many deaths among the grinders that no Yankee would grind, and the Irish were so awkward and stupid that we [the Collins Co.] did not get the quantity needed even by having extra men working at night."[48]

There were two major health hazards to grinders at Collinsville, bursting stones and siliceous particles in the air. Collins used large natural grindstones which contained flaws or inherent weaknesses and could burst during use, killing the grinder (figure 3.8). Grinders also suffered from silicosis, a disease of the lungs caused by inhaling siliceous particles.

> Many of the men ride on "horses" while grinding, thus enabling them to bring their whole bodily avoirdupois to aid the process of abrasion; while the fine dust flies in clouds from the stones in every direction, notwithstanding the stones are all the time completely deluged with water [which also prevented spoiling the temper by cooling the axe poll]. The men in this section are, from their particularly hazardous work, ruled out of all life-insurance companies; since the constant inhalation of the grit and bits of steel thrown off in the process induces "grinders' consumption," as it is rightly termed, from which a premature death is rarely averted. It is said that Americans will not work in these rooms, which are filled by French Canadians, who stop a few years, and then go home to linger a while and die.
>
> But sometimes, the peril of life is of another kind altogether, arising from the rapidity with which the stones must be made to revolve. A flaw in the stone, or possibly a loosening in the clamp holding it upon the shaft, sends the flying fragments hither and thither,—perhaps through the grind-

er's body, or throws him through the roof. It is but justice to add, however, that such casualties happen only at rare intervals.[49]

These conditions drove the Yankees from the grinding room at Collins & Co. and severely limited production.[50] Root's machine accomplished the same task of shaping the axe but removed the hazards of the grinding job and the labor shortage brought on by those hazards, although some grinding was still required. This innovation had at least one other positive effect. The reduction in grinding prolonged the lives of the other machines, particularly the water wheels. The lower dust levels eased the oil hole clogging problem and reduced the number of burned bearings, a problem reportedly caused by the "green Irishmen."[51] Root's shaving machine was eventually patented on August 22, 1848, probably after numerous incremental improvements since 1844 (figure 3.9).[52]

In its patented version, Root's axe shaving machine consisted of a stationary bed of cast iron on which a sliding bed moved back and forth in guide pieces. The axe was securely fastened to this sliding bed and passed under an iron frame which helped guide the shaving knife. The tool worked the entire surface of the axe which had not yet been hardened. This machine was essentially a specialized planer, as its action consisted of moving the work piece past a stationary tool.

Besides developing his shaving machine in 1845, Root took the job of superintendent of the entire axe manufacturing operation.[53] Root refused an offer to become "Master Armorer" at the United States Armory at Springfield. Two large (but unnamed) manufacturing firms in Massachusetts also bid unsuccessfully for his services.[54] S. W. Collins increased Root's pay to retain him, although the evidence fails to indicate if Collins matched the offers of the unnamed Massachusetts firms or the U. S. Armory.[55]

Why did Root reject higher pay and greatly increased status and prestige to remain at Collins & Co., which controlled a large percentage of the axe market and was growing in importance? Part of the answer lies in his long-term project at Collins, still unfinished in 1845. This project was the development and implementation of a mechanized system of axe manufacture. However, until 1845 Root had no opportunity to assemble his system as a system. In August 1843 business revived at Collins and continued strong for several years, well into 1845. Demand for axes had completely outstripped Collins' ability to produce them. In 1846 the Collins company

> decided to build a canal to give room for more shops and *we built in connection with it a three-story stone building and two large breast wheels, that we might put into operation Mr. Root's machinery for punching heads of axes,*

FIGURE 3.8. In 1871 the grinding shop at Collins & Co. still used large natural grindstones, despite the simultaneous use of Root's axe shaving machines. SOURCE: Horace Greeley et al. The Great Industries of the United States, p. 140. COURTESY: Milwaukee Public Museum. Negative no. H-638-C.

> *shaving them and tempering them in ovens* without disturbing or interfering with the present manufacturing business. (Italics added.)[56]

Root probably remained at Collinsville to build his new system of axe manufacture. After spending thirteen years with the firm, he simply could not walk away from the opportunity to build his system under a single roof (figure 3.10). Root could easily see that demand outstripped supply at Collins between 1843 and 1845. He also knew that his system was ready to assemble in a single building, that it would lower costs appreciably, increase output greatly, and solve the production imbalance first created by Hinman's machine. Given this combination of a profitable economic climate and the technical maturity of his system, Root undoubtedly perceived the chance to implement it as a system.

Perhaps S. W. Collins explicitly offered Root this opportunity in ad-

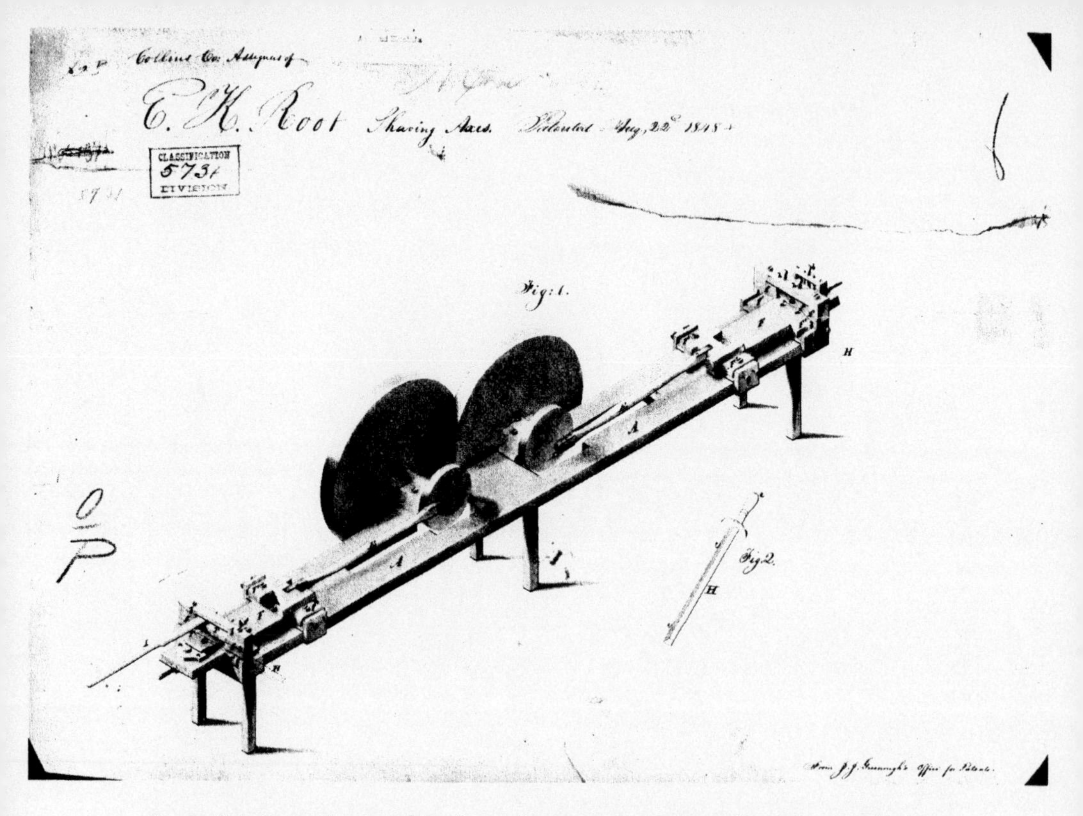

FIGURE 3.9. Patent drawing of Root's axe shaving machine, patented August 22, 1848, patent no. 5731. This patent was also assigned to Collins & Co. The original patent is in the Collins file at the Connecticut Historical Society, Hartford, MS. 71890. SOURCE: Records of the Patent Office, Record Group 241, National Archives and Records Administration, Cartographic and Architectural Branch.

dition to his salary increase in 1845. Certainly Collins was contemplating the major expansion of 1846 in 1845. As superintendent Root would not only have known about it, but was certainly intimately involved to the extent that he designed the three-story stone building himself.

Other factors may have influenced Root's decision to remain at Collinsville in 1845 and build his system. Life as a mechanic at Collinsville had not been easy for Root, who evidently fought a running battle with B. T. Wingate throughout his stay with Collins. Wingate was in charge of quality control at Collins, and, by all accounts, he was a conservative, traditional craftsman. Trouble between Wingate and Root began as early as 1834.[57] In 1845, when Root was appointed overseer of all the shops, S. W. Collins reported that his appointment "ended the quarrel between him [Root] and B. T. Wingate which had lasted more than a year, much to our detriment."[58] In 1847 Root and Wingate argued again, this time about the cause of cracked axes, "attributed by Mr. Root to bad heating of the steel, and by Mr. Wingate to *bad uneven* shaving."

FIGURE 3.10. The Collins & Co. works at Collinsville, Conn., as they appeared in 1871. Root's building is directly in front of the center smoke stack at the left of the illustration. SOURCE: Horace Greeley et al., *The Great Industries of the United States*, p. 132. COURTESY: Milwaukee Public Museum, negative no. H-638-A.

This conflict might best be viewed as an intellectual conflict between the craftsman, Wingate, and the innovative mechanic, Root. Wingate opposed the unproven machinery that appeared to reduce the craftsman to a mere machine tender and produced cracked axes. Root, who was interested in increasing output through invention and innovation and who had been hired for that reason, attempted to mechanize everything. The friction between Wingate and Root stemmed not only from their attitudes about how to produce but from their goals, what to produce and how much.[59] When a problem appeared, Wingate naturally blamed the new technology.

It is entirely plausible that Root refused all offers and remained in Collinsville in 1845 to prove that his ideas were correct, to demonstrate that his machines were well designed and built, and to iron out all the minor, unforeseen problems that plague the builder of new machines, as well as to see the system functioning as a system under a single roof. Root remained in Collinsville at least in part to prove his mechanical virility, which was constantly questioned by Wingate.[60] Only after the system was built and functioning could Root look for other intellectual challenges and mechanical worlds to conquer.

Root's machines were finally installed in the new stone building and were ready for production in late 1847. Now in a single building for the first time, Root had his combination machine for punching the eye of the axe and die-forging it, his shaving machine, and his tempering ovens. The *Scientific American,* which had a reputation for describing the latest and most novel manufacturing methods, portrayed Root's mechanized axe manufacturing system as it stood in 1859, little changed since 1847.[61] The following is their assessment of the "the peculiar machinery and processes belonging to that establishment, the Collins Company." "The main building in which the axes are manufactured is a large stone structure, fifty-five feet deep, by one hundred and thirty in length, and three stories high."

The *Scientific American* reader then learned about machines which compare very closely with Root's combination machine. Here is a description of the stone building described by S. W. Collins.

> On the ground-floor are sixteen very curious and ingenious machines, for forging axes and hatchets. . . . These were invented by Mr. Root. One of these machines somewhat resembles a semi-rotative octagon box, . . . It cuts off a blank for an axe from a bar of iron, punches out its eye, forges it on the face, end and sides, into the proper shape and completes it ready for being trimmed, to receive its steel edge, and all this in a few minutes.
>
> Adjacent to each machine is a furnace, in which the bar of wrought-iron is heated, and from thence it is taken and laid on a proper bearing at one side; the attendant places his foot in a stirrup and makes the machine clutch with its driver, when it makes a semi-rotation, a punch comes down and cuts off the heated skelp designed for an axe. This is now set on edge when another touch of the operative's foot brings the machine into action again, a punch descends from above and another from below, approaching and pushing through the solid metal. . . . Their motion, however, is arrested before they meet, and the hole which they have made forms the eye of the ax with a small piece of metal left in the center. An iron handle . . . is now inserted in this eye, and the machine comes right down upon it, forcing out the small piece of iron left in the hold hole, thus forming an eye. On the machine itself and the anvil on which it strikes, are several dies of the exact size and form for the ax to be made, and by a very few blows, and two heats, it is forged into the required shape. Each machine is under the perfect control of the attendant . . . one man being capable of turning out three hundred and fifty axes per day.

These forged axe polls were then sent to another building where blacksmiths welded in the steel bits before they were returned to Root's shaving machines.

> After the steel edges are welded on, each ax is then hammered off and formed as accurately as possibly can be done under the hammer, when it is fit for the planing or shaving operation. This is executed by an ingenious power shaving-knife, which is under the perfect control of the attendant and is especially adapted for reducing rolling surfaces—not tools with flat sides. There are twenty-six of these shaving machines in the second story of the stone building, and any one of these iron barbers can shave down a rough ax to a pretty smooth face . . . These machines are also the product of Mr. Root's busy brain, and are peculiar to this establishment. Prior to their application the axes were all ground, by which tedious, unhealthy and disagreeable operation, the surplus metal was all washed away with the grit, but the shavings are now all saved and sold for scrap-iron.

The next step in Root's mechanized axe manufacturing system as described by the *Scientific American* was the tempering (hardening) of the polls.

> After planing the axes, the next process . . . is that of hardening. For this purpose, they are placed in a suitable furnace, in which there is a rotating iron wheel to receive them; and here they are heated uniformly, and raised to the proper degree of temperature required. Convenient to each furnace, for this purpose, there is a large bath, containing salt brine, to which a continuous fresh supply is furnished, and in it there is a circular revolving rack, with catches placed around it close to the surface of the pickle. The heated axes are now taken out of the furnace, and set, one by one, on the catches, with their edges left trailing in the hardening liquid. The next process which they pass through is that of tempering. This consists of placing them in an oven (on a rotating wheel, also), the temperature of which is regulated by a thermometer, and here they are kept for several hours, or until the metal is toned to that degree of elasticity and hardness which experience has decided to be the best for all practical purposes. These peculiar manipulations, which are exclusive to this establishment, ensure uniformly accurate results.

There is no doubt that these tempering and drawing ovens were on the pattern patented by Root in 1843, and the quenching apparatus was probably of Root's design and construction since it was used as a part of his system in combination with his ovens.

The evidence suggests that Root planned to expand his system even further. First, in 1853 he patented "a method of chipping edges of axes to take the place of the slow process of grinding."[62] Second, patent specifications for Thomas Blanchard's lathe for turning irregular shapes (such as axe handles) were sent from the Patent Office to Collinsville.[63] Conceivably, Root was thinking of producing the entire axe at Collinsville and not just the head. Perhaps there were still some labor shortages or skill problems at Collinsville that led Root to consider alternatives other than his shaving machine as substitutes for grinding. There were problems with the shaving machine and it evidently required some skill to operate.[64]

In 1849 Samuel Colt offered Root the superintendency of his Hartford Armory at twice the pay of Collins, "the then-unheard-of salary $5,000 a year."[65] On the advice of his friend S. W. Collins, Root accepted Colt's offer, thus ending the Collinsville period of his career and beginning the Hartford period, the period that made Root's reputation as the best mechanic in New England.

THE FACTORS INFLUENCING ROOT

Now that a more complete record of Root's activities and accomplishments at Collinsville has been established, the facts lead to several conclusions and some speculation. There is no argument that Root was an important figure in the rise of the American System. The machines he built for Collins and for Colt influenced not only the financial prosperity of those firms but also the activity of other mechanics. J. W. Roe has noticed the number of important machine tool builders who worked for Root in Hartford.[66] Clearly, Root advanced the mechanic arts of his time.

Of greater significance was Root's ability to envision production as an integrated process, to see the system as a system, to channel his inventive activity to fill the gaps in his system, and to solve secondary effects (technical problems caused by his earlier machines). Root probably was not the first to conceive of the system as a system, or to engage in system thinking, but he was probably one of the earliest to do so and one of the most influential. Although this essential intellectual component in the rise of the American System is difficult to prove, Root could not have built his system without the ability to envision the manufacturing system in total.

The central question of motivation remains. Why did Root do what he did and how was he able to do it? What forces influenced him at Collins? Root was very fortunate, and there was an element of serendipity. The various factors merged in the right way at the right time. In addition, Root was intellectually prepared; he was able to judge the situation well and make the best of his opportunities in the complex atmosphere at Collins.

SUPPY AND DEMAND. Since Collins & Co. introduced the idea of making and selling a finished axe head in 1826, it was unable to meet the demand for its product. The large national and international market Collins attracted kept growing and so did the number of Collins' competitors. Despite the competition, Collins' reputation and the immense size of the market tended to reduce the impact of competing axe makers such as those in Douglas, Massachusetts, Napanoch, New York, and Baltimore, Maryland. Collins simply could not make axes fast enough to fill its orders.[67]

THE LABOR PROBLEM. Collins & Co. experienced a labor problem, but its nature changed between 1830 and 1848. At first, when Collins depended on skilled blacksmiths, its competitors would hire away skilled workers. S. W. Collins complained specifically about the establishment at Napanoch, New York, on the Hudson River only seventy miles west of Collinsville.[68] There was also the seasonal nature of agricultural work for Collins to contend with, specifically harvest and planting seasons. Time and work in the antebellum economy, as Merritt Roe Smith points out, were still tied to agricultural cycles to a great extent.[69] Root's machinery, like David Hinman's, allowed a given number of less skilled men to produce a substantially larger number of higher quality axes per day. Die-forging, as developed by Root, was a simpler, faster method. So effective was the introduction of the forging machinery at eliminating the need for skilled strikers and foremen that a strike in 1838 was unsuccessful. S. W. Collins told those on strike that common farm laborers could work the machinery and that there would be no wage increases.[70]

Later, the nature of the "labor shortage" problem changed.[71] No longer did Collins face a general shortage of skilled workers (there was never a general labor shortage, only a skilled labor shortage), but a shortage of grinders due to the dangerous nature of the work.[72] Grinding became the last bottleneck because it was dangerous and because technological innovation in other areas of axe manufacturing occurred first.

COST SAVING. The technical changes Root introduced saved the company money in materials. With two of the three welds in the axe poll eliminated, workers needed less borax, which suited S. W. Collins well, as he complained about its cost.[73] Furthermore, the machine forging required fewer heats and hence used less coal (Collins started using Lehigh coal in 1831). Since fewer axes were spoiled in working, there were savings in the increased number of finished axes per ton of iron bought. The shaving machine also allowed Collins to recycle the chips. The earlier technique of grinding produced chips too small to collect and sell. So Collins saved on borax, coal, iron, and wages, and gained by the sale of scrap and increased output due to less spoilage. These are individually and collectively substantial economic returns to technological change.

COMPETITION. Collins was affected to some degree by its competition, although the size of the market surely eased the pressure. Competition took two forms. First, Collins had to compete in the labor market for skilled workers. S. W. Collins mentions the unnamed axe manufactory at Napanoch, New York, as a particular source of annoyance, since it waited until Collins trained men before hiring them away at increased wages. S. W. Collins tried several methods to retain his work force, including long-term contracts and the retention of wages. Although he described his methods, S. W. Collins failed to indicate the degree of success he achieved. Second, there was competition to sell products in the market place. Since Collins had a large overseas market (South America), it was not as dependent on the domestic market as were other makers. Statistics as to total sales and exports are not currently available.

WAGES. While Root was certainly concerned about the financial success of Collins & Co., he had no financial interest in the firm, as far as the documents indicate. His only compensation was his wage, which was clearly higher than that of the other employees. There is no evidence to indicate that Root received a bonus or a percentage of any sort for his efforts. However, the evidence does strongly suggest that Root was not really interested in amassing a fortune. To begin with, he turned down numerous offers for higher pay, including the extremely prestigious job as the Master Armorer of the Springfield Armory.

While some of the offers he refused may well have been insecure, this was certainly not true for the armory job. There is evidence in Root's own hand that he preferred the security and friendship of Collins to the risks of a new venture. Indeed, according to S. W. Collins, Root had to be persuaded to take the job with Colt, despite Colt's unusual offer: "I will pay you *again* as much as you are now receiving, and if that is not satisfactory, you may yourself fix such compensation as you think fair and reasonable. The important thing is that you come to me at the earliest possible moment."[74] Despite this offer, S. W. Collins noted that Root "would probably have remained (at the Collins Co.), if I had not advised him to accept this offer as a matter of duty to himself and his family."[75] Root was a man not wholly motivated by economic forces. The technology of axe manufacture itself influenced Root's decisions and his thinking.

A BETTER AXE. Since Root's eye-punching machine produced an axe with two fewer welds than the traditional axe, it was, at least in theory and most likely in practice, a better axe. With fewer welds, there were fewer places for the axe to fail in use. Root also claimed that his new method eliminated distortion in the metal, formed the cheeks more evenly, and the eye more accurately. The concept of improving a product or a manufacturing technique is a noneconomic aspiration of every mechanic. There are financial rewards for the inventor who improves an everyday article, but the mechanic is primarily concerned with the technical development. He derives great noneconomic satisfaction from the problem's solution itself. The more "elegant" the solution, the happier he is.

SECONDARY EFFECTS. Root's shaving machine in particular forced the introduction of several minor innovations. The "shaved" axe, as compared with the "ground" axe, had a surface texture much more susceptible to rust. The small quantity of salt that remained in the eye of the axe after quenching could spread to the rest of the axe if moisture conditions were right. In 1848 this happened to 2,000 dozen axes, which cost $6,000 to rework. Root's solution was the addition of a fresh water bath to remove the salt and an oil swabbing of the eye to inhibit the rust. Both became essential, although minor, parts of the system.[76] Root's machines also required a higher grade of iron and stricter attention to temperature in the tempering operation, hence the control devices on his tempering ovens. The shaving machines caused still another problem. Their use required some skill, and "many axes [were] shaved too thin and flat on steel and failed in consequence. These troubles greatly injured the reputation and sale of our [Collins & Co.] axes and gave our competitors a great advantage," complained S. W. Collins.[77] Root's machinery created new problems as they solved old problems. These "secondary effects" probably caused Root more trouble than the original problems. They were certainly costly to the Collins reputation.

THE INTRODUCTION OF COAL. In 1831 Collins & Co. began to use Lehigh coal in place of charcoal.[78] Alfred Chandler has argued the qualitative importance of coal in expanding the size of manufacturing establishments. The use of coal at Collins probably influenced the pattern of mechanization. The higher, more efficient heat provided by coal would allow more axe polls to be heated more quickly, hence overburdening the hand forgers and their strikers. This improvement in the heating process may have made it worthwhile for Collins to consider a machine that could forge axe polls more quickly. The specific chain of events between coal and mechanization is unimportant—it matters little if Collins was pushed toward machines because coal created an imbalance or eagerly pursued machines because coal suddenly lifted a constraint. The importance lies in the close connection of Collins' first use of coal in 1831 and the first efforts of David Hinman that same year. The coal connection is more than coincidental.[79]

PRODUCTION IMBALANCE. Perhaps the most important technical influence on Root was the tremendous imbalance in the production process caused by David Hinman's second die-forging machine. The firm was suddenly faced with the ability to forge far more axes than could be finished by traditional methods. Aside from a sizable increase in the work force to use those methods and an expansion of the plant to house them, mechanization was the only solution. Rarely have firms abandoned a new technology and reverted to an earlier technology because of the problems caused by the new technology.

This phenomenon was true at Collins & Co. Once Samuel W. Collins and Elisha K. Root perceived the production possibilities suggested by Hinman's machine and seemingly offered by full mechanization, they were unable and unwilling to return to traditional techniques. Once Root's eye-punching machine (and Hinman's die-forging machine) created a bottleneck in the grinding department, Root responded with new technology—his shaving machine.

This pattern, which appears at Collinsville and has been documented in other case studies, is simple. It suggests how a mechanic like Root with a properly prepared mind may have arrived at the idea of a manufacturing system as a system.

THE INTELLECTUAL ATMOSPHERE AT COLLINS & CO. It is clear that Collins encouraged innovation and that there was no penalty for experimenting despite failure. The costs of Hinman's second machine forced the company into temporary financial insolvency in 1832, yet Collins forged ahead with Hinman's machine. The secondary effects of Root's shaving machine and quenching operation cost the firm dearly, yet Collins remained committed to the new system.

THE OTHER MECHANICS. The presence of Hinman, Shaw, Burke, Smith, and Kellog at Collins testifies to more than the innovative atmosphere. They also influenced Root's mechanical work.

MECHANICAL EXPERIENCE AND MATURITY. As Root gained more experience at Collins, his machine design, particularly his die-forging machines, became increasingly sophisticated. Experience exerts a powerful, noneconomic influence on a mechanic, because the mechanic does what he knows how to do regardless of cost (within certain limits). He can't easily compare the cost of what he knows with the cost of what he doesn't know. However, his experience prepares him to solve new and increasingly difficult problems.

ROOT'S MECHANICAL VIRILITY. Each nineteenth-century mechanic experienced great self-satisfaction when he found an "elegant solution" to an engineering problem. Such a solution enhanced his reputation as a mechanic, gave him confidence in his engineering ability, and provided a feeling of accomplishment. The development of his reputation certainly motivated Root. What motivated him more was the threat to his reputation. At Collins & Co., B. T. Wingate, the long-time Collins craftsman in charge of quality control, threatened Root's reputation.

In essence, Wingate questioned Root's entire system and attributed certain problems to Root's machines. Root certainly reacted to that criticism, and it may well have compelled him to remain at Collins not only to insure that the system was built properly and to achieve a sense of accomplishment, but to prove that he was right all along. Root had to vindicate himself, since problems with his system had cost the factory not only time and money, but affected its reputation as well. Root's assertion of his "mechanical virility" was no doubt an important, noneconomic influence. This is not to say that Root was uninfluenced by Collins' financial loss and damaged reputation, but to draw an important distinction between those (economic) forces and Root's (noneconomic) reputation.

FRIENDSHIP WITH S. W. COLLINS. An important aspect of the atmosphere at Collins was the relationship between S. W. Collins and Root. S. W. Collins himself described Root as "not only a superior mechanic, with great inventive faculties, but . . . a man of excellent judgement, and great caution and prudence, which cannot be said of inventors generally."[80] Root found in the company president exactly the kind of employer he wanted, one who demanded a certain level of productivity, but who also understood the need to experiment and innovate. E. K. Root and S. W. Collins were made for each other. Root not only had the ability to design and build the system, but also the intelligence to perceive the proper time to build it. His perception of the economic and social situation at Collinsville was an essential element in promoting his system. Root chose the correct time to push his project and sell S. W. Collins the idea. There were many New England mechanics who equalled Root technically, but few who possessed such broad ranging perception and intelligence. This, among other things, distinguished Root from his contemporaries.

EDUCATION. Root was an educated man, perhaps not in the sense of having had a lengthy, formal education, but insofar as he was aware of developments in the mechanical world around him. He toured industrial England and kept abreast of current U.S. patents. He probably considered the use of Blanchard's patent for turning irregular shapes.[81] He did more than consider the patent of D. C. Stone of Napanoch, New York.

Six months *before* Root received his second die-forging patent on November 17, 1838, Stone received a patent for an axe punching machine (figures 3.11 and 3.12).[82] Stone's patent of April 21, 1838, has all of the features of Root's second die-forging machine, and more. Stone's machine also split the poll of the axe, mechanically preparing it, ready to receive the steel bit, a problem with which Root struggled but never solved. Although the evidence is circumstantial, a strong case can be made that Root learned much from Stone's patent. First, Root took the idea of punching the eye of the axe. Although adapted to the basic design of his earlier machine, the concept was identical to Stone's.

Second, note the patent dates, only six months apart! Since Collins & Co. habitually patented new machines, they certainly kept abreast of other patents, particularly those of a well-known competitor. Six months is ample time in which to secure a copy of the patent and drawing, study it, build its features into a similar machine, and secure a second patent. The apparent duplication of patents, and hence the seeming unpatentability of Root's design, is explained by antebellum Patent Office policy which lacked definition.[83] The important point is that Root learned from a variety of sources and continued to educate himself.

FIGURES 3.11 AND 3.12. Patent drawings of D. C. Stone's "Axe Machine," patented April 21, 1838, patent no. 699. The features of this machine, made public before the development of Root's machine in the Collins & Co. mill only seventy miles east, all appear in Root's machine, with the exception of the device to split the axe poll. The general design of this machine is far more sophisticated and much more developed than Root's machines designed for the same purpose. SOURCE: Records of the Patent Office, Record Group 241, National Archives and Records Administration, Cartographic and Architectural Branch.

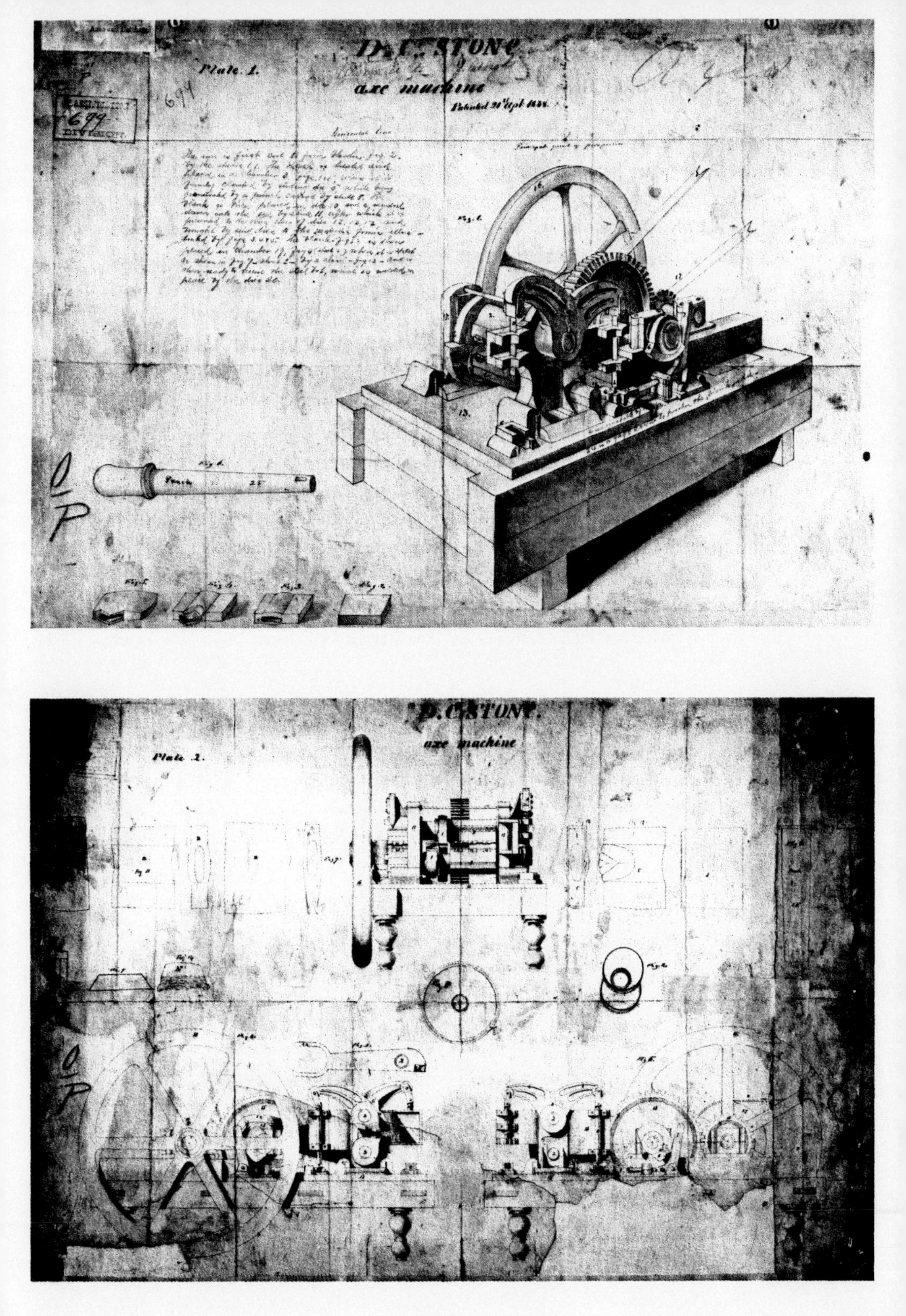
D. C. STONE
Plate. 1.
axe machine
Patented 21st Sept 1838.
D. C. STONE
axe machine
Plate. 2.

SYSTEM THINKING. Finally, Root's most important contribution to the rise of the American System was his ability to think in terms of the entire system, to envision the system as a system, to comprehend "systemness." He could not have done what he did either at Collins & Co. or at Colt's Armory had he not understood that concept. This ability was an essential intellectual element in the evolution of the American System. Root may not have been the first to discover this, but it is clear that he comprehended it sometime before the three-story stone building was built at Collins in 1845. Once he understood the concept of a manufacturing system, Root could perceive the missing aspects of the system and could channel his inventive activity toward filling those needs.

In conclusion, the evolution of axe manufacturing illustrates several important aspects of the rise of the American System in the private sector. The two mechanics, Hinman and Root, reacted to a variety of economic, technological, and intellectual forces throughout the period of industrial evolution during which they invented and built the highly specialized machines of an axe manufacturing system. They understood the economic implications of their work in cutting unit costs, increasing output, and solving labor problems. They also understood the technological imperatives of axe manufacturing and successfully innovated within the economic context of market demand and Samuel W. Collins' outlook on the business. They typify the nineteenth-century mechanics who worked with enlightened entrepreneurs to transform the American economy through technological change.

CHAPTER 4

Typewriter Manufacturing

1853–1924

The early history of the typewriter in America suffers from a lack of investigation which is undoubtedly the result of a paucity of documentation.

TERRY ABRAHAM, "Charles Thurber: Typewriter Inventor"[1]

TYPEWRITER manufacturing presents a particularly interesting American System case study because the typewriter was invented long after the American System was well developed and widely practiced. The typewriter offers the opportunity to study how an industry acts in choosing a production technology.

The history of typewriter manufacturing falls into three phases: First, early designs and failed manufacturing efforts, 1853–1867; second, the typewriter's successful design and manufacture—the Sholes & Glidden Typewriter at Remington, 1867–1881; and third, once Remington's machine had attracted a market, the invention of new designs and the development of typewriter manufacturing by new firms 1881–1924.

The early design and unsuccessful manufacture of typewriters began in 1853, when the Rochester Novelty Works produced some 130 of John Jones' Typographers.[2] In late 1856 or early 1857 the Cooper Hand Printing Machine was produced in limited numbers in Philadelphia.[3] In Europe, about 1867–1870, the Jurgens Mekaniske Establissment began making Reverend Hansen's Writing Ball in the scientific instrument tradition. An Alabamian, John Pratt, had several of his Pterotypes manufactured in England in 1867 and 1868, but failed to attract a market.[4] These four efforts and a host of unmanufactured inventions and patents all failed because they wrote too slowly. Typewriter design was still in its embryonic stage.

The second phase of typewriter manufacturing covers the first successful typewriter design and its manufacture. Between 1867 and 1872 Christopher Latham Sholes and his contemporaries in Milwaukee, Wisconsin, invented a writing machine that embodied the necessary speed of operation.[5] After three failed manufacturing efforts, James Densmore, Sholes' financier, licensed the invention to E. Remington & Sons of Ilion, New York. There, William K. Jenne redesigned the Sholes & Glidden Typewriter and produced it in quantity using American System techniques in 1874. Between 1874 and 1881 the Remingtons had the typewriter market to themselves.

The third phase of typewriter manufacturing encompasses the full flowering of American System manufacturing technology in the private sector. In the early 1880s, as the typewriter became widely accepted, an army of competing machines invaded the field.[6] By the turn of the century most were casualties of competition. The surviving typewriter firms were large manufacturing establishments with highly organized factories brimming with specialized machinery.

The typewriter was the most complex mechanism mass produced by American industry, public or private, in the nineteenth century. Unlike the wooden movement clocks and axes, the typewriter was an extremely complex machine, requiring extensive adjustment and alignment. Realizing this inherent feature of their machines, manufacturers designed their typewriters and associated production technologies accordingly. First, typewriter mechanics designed their machines to be adjusted during assembly, despite the production of fully interchangeable parts. Second, typewriter mechanics employed new materials in typewriters, notably rubber, sheet iron, glass, and steel (for springs, clips, and typebars). Third, while typewriter mechanics used established manufacturing techniques, such as automatic screw machines to produce screws and japanning to decorate their machines, they also developed new techniques including special applications of vulcanizing, grinding rubber, making and soldering type, sheet metal work, and gauging ball bearings. Perhaps the most interesting new technical development was the "exercising" or breaking-in machinery used by many manufacturers to mechanically operate new typewriters. Such machines as the Oliver "Exerciser," the Caligraph typebar "testing" machine, and the L. C. Smith & Bros. "Typebar Exerciser" met a technical need not common to wooden movement clocks, axes, or watches.[7] Fourth, the complexity of the typewriter forced typewriter mechanics to organize highly specialized assembly, adjusting, and aligning departments and to assure the flow of materials to and from those departments.

All of these technological developments were directly related to the nature of the typewriter itself. Typewriter mechanics recognized these problems and, with the support of the entrepreneurs with whom they worked, they designed and built new production machinery and organized their factories to meet the special needs of typewriter manufacturing.

Typewriter manufacturing in America illustrates clearly the close relationship between mechanics and entrepreneurs, particularly in the cases of Sholes and Densmore and later the Remingtons and William K. Jenne.

Like the other American System industries in this study, the typewriter manufacturers received no subsidy from the federal government, whose orders were relatively small (numbered in the hundreds), came late in the period (long after the firms were established), and were scattered among many firms. Hence, what government purchases took place were an insignificant part of the typewriter companies' business. Typewriter manufacturers were independent entrepreneurs who raised their capital in private capital markets.

In sum, typewriter manufacturing was much like other private sector American System manufacturing. Typewriter mechanics worked closely and harmoniously with typewriter entrepreneurs in a fruitful relationship to develop new products designed not only for the end user, but also for ease of manufacture. These manufacturers adopted and adapted existing American System techniques and invented new techniques to produce their typewriters. They were, in this respect, acting just like their counterparts who produced wooden movement clocks, axes, and watches.

TYPEWRITER MANUFACTURING BEFORE REMINGTON

It did good work, but it was too slow.

The Story of the Typewriter, 1873–1923[8]

The Jones Mechanical Typographer

In October 1851 John M. Jones, an inventor in Clyde, New York, read an article in the *Scientific American* entitled "More Improvements Wanted," in which the author stated, "We want a machine which could print as easily as we can now write."[9] The article inspired Jones to return to work on a writing machine which he had begun in 1848 but had laid aside. By February 1852 Jones had a working model, which he used to write to the *Scientific American* to announce his success.[10] So impressed was the *Scientific American* that it expressed considerable pride in yet another article on March 6, 1852, for having inspired Jones.[11] Jones patented his

Typographer on June 1, 1852, and later patented an improvement on May 20, 1856.[12] In 1853 Jones exhibited his Typographer at the New York Crystal Palace where it received an honorable mention.[13]

Jones formed a company (John Jones & Co.) to handle his Typographer and contracted with the Rochester Novelty Works of Rochester, New York, to produce his Typographer.[14] The Novelty Works produced some 130 Typographers which sold for $30 each, before a portion of the factory burned on April 25, 1856.[15] The building destroyed was a five-story, brick structure and, although "a large quantity of hardware, tools, &c. were taken out and saved," the loss still amounted to between $20,000 to $25,000.[16] While 130 machines can hardly be described as mass production, they do represent a serious attempt to produce in quantity, and it is worthwhile to examine Jones' machine in some detail to learn what techniques were used (figure 4.1).

Two of Jones' Typographers survive: one, the New York Crystal Palace machine in the Carl P. Dietz Typewriter Collection at the Milwaukee Public Museum, the second, a conventional production machine in a private collection.[17] The surviving Typographers feature cast iron construction in their frame members and indicator wheels. The Crystal Palace machine has a wooden base while the other machine has a cast iron base. Both machines have a variety of brass parts (springs, levers, type wheels, platen tracks), which show the work of individual bends and filing marks. From these two machines, it is clear that manufacturing in Rochester in 1853 consisted of much hand work and the use of cast iron and brass. The Rochester Novelty Works was probably little more than an oversized jobbing shop.

The Cooper Hand Printing Machine

Considerably less is known about John H. Cooper of Philadelphia and his efforts to produce his "Hand Printing Machine."[18] Cooper received U.S. Patent No. 14,907 for his "Hand Printing Machine" on May 20, 1856, the same day that another writing machine by Ely Beach was patented and John M. Jones improved his "Mechanical Typographer."[19]

Cooper submitted his machine to the Franklin Institute for evaluation by the Institute's Committee on Science and the Arts on April 8, 1857.[20] Incredibly, Cooper wrote to the committee in longhand instead of sending a "hand printed" letter as a product of his machine as Jones had done in corresponding with the *Scientific American*.[21]

The committee reviewed Cooper's invention on June 11 and filed its

FIGURE 4.1. John Jones' Mechanical Typographer, ca. 1851. This machine was shown at the New York Crystal Palace in 1853 where it received an "Honorable Mention." COURTESY: Milwaukee Public Museum, Carl P. Dietz Typewriter Collection. Catalogue no. E 47377/13890, negative no. 423427.

report on July 9, 1857. The committee indicated that while the machine was clever, it had several technical problems yet to be solved. However, the committee believed that an "ingenious inventor" could solve the problems.[22] The committee's report was followed by an illustration of the machine and a "description by the Inventor."[23]

The two surviving specimens illuminate Cooper's work.[24] Cooper's patent model, which is preserved at the Smithsonian, illustrates the mechanism, while a surviving production machine in the Carl P. Dietz Typewriter Collection at the Milwaukee Public Museum is altogether different. The later machine is identical to the illustration of Cooper's machine in both the *Journal of the Franklin Institute* and the *Scientific American.*[25] It does not appear to be the machine exhibited to the committee which was "understood to be only the first experimental model."[26] One of the improvements suggested by the committee was the incorporation of both uppercase and lowercase letters, a feature not part of the model submitted by Cooper. However, both the Milwaukee Public Museum machine and the illustration following the committee's report exhibit an increase in the size of the type wheel and the incorporation of both uppercase and lowercase letters. Clearly, by the time the committee's report and Cooper's description of the machine appeared in September 1857, Cooper had

FIGURE 4.2. John Cooper's Hand Printing Machine, 1856. The intricate cast iron of this machine suggests that it may have been produced in some quantity, although this is the only known surviving example. COURTESY: Milwaukee Public Museum, Carl P. Dietz Typewriter Collection, catalogue no. E 47380/13890, negative no. H-627-5L-2.

constructed a new machine, probably based on the suggestions of the committee.

Further note must be taken of the casting of the Cooper production machine at the Milwaukee Public Museum. This casting is made of several parts and is fairly complicated. One might surmise that Cooper was producing machines in September 1857 or was about to do so. It seems unlikely that Cooper would spend so much time, money, and effort to produce such sophisticated castings merely for the next experimental version. Perhaps Cooper was in the midst of tooling up for production when the Panic of 1857 arrived and put him out of business (figure 4.2).

The Malling Hansen Writing Ball

Americans did not have the typewriter invention field to themselves. The Reverend Hans Johan Rasmus Malling Hansen, whose position as head of the Royal Deaf and Dumb Institute in Copenhagen, Denmark, led him to experiment with a machine that would allow the deaf to write as fast as they could communicate with sign language, began work on his machine in 1865.[27] By 1867 he had perfected an electro-mechanical "skrivekugle" or writing machine on which all his later machines were based. Hansen took his model to Jurgens Mekaniske Establissement in 1869 and, in October 1870, Jurgens began to manufacture it after some additional development. Hansen's machine underwent several modifications over the years but its general configuration remained the same.[28]

The Hansen Writing Ball was a well constructed machine and was used as late as 1909 in some offices. No total sales figures are known. Hansen exhibited the machine in various industrial expositions, including Copenhagen in 1872, Vienna in 1873, and Paris in 1878, where it won a gold medal. The Hansen Writing Ball may safely be said to have enjoyed a worldwide distribution based on sales of forty machines at the Paris Exposition which were shipped all over the world. Its manufacture was licensed to Albert von Szabel in Austria for some time, and it was also manufactured there.[29]

This machine was the only European machine manufactured prior to the Sholes & Glidden Type Writer.[30] It was made in the tradition of scientific instruments, such as microscopes. It featured brass and steel construction and had a lacquered finish to prevent the brass from tarnishing.[31] It was magnificently constructed, each machine individually fit together and hand finished. Although made in large numbers, there is no evidence to suggest that it was made with any but the traditional scientific instrument makers' techniques. There is no data to suggest that it was made with any kind of automatic or semi-automatic machinery. In comparison with its American successors, the Hansen Writing Ball is a vastly superior instrument in finish, fit, construction, and appearance.

The Hansen Writing Ball is technically significant because it incorporates the concept of individual keys for each letter. It was an important idea which occurred independently and somewhat earlier to an American, John Pratt. Pratt developed this idea in the late 1850s or early 1860s and was publicizing the idea in England in the mid-1860s. Perhaps Hansen learned of the general idea from Pratt's work.

John Pratt's Pterotype

John Jonathan Pratt, who spent most of his life in Centre, Alabama, invented two writing machines in the 1860s, one called a "writing machine" and another called the Pterotype. He was born in Union, South Carolina, on April 14, 1831, and died at the age of seventy-four on June 24, 1905, in Chattanooga, Tennessee.[32] Pratt was an educated man. He attended Cokesbury College and read law with Judge B. F. Porter of Greenville, Alabama, whose daughter he married.[33] Pratt was a local lawyer and schoolteacher. Like Sholes who followed him, he edited a newspaper, *The National Democrat,* which he started with J. W. Ramsey and Thomas B. Cooper. His *National Democrat* favored the policies of Stephen A. Douglass, and Pratt evidently foresaw the inevitable approach of the Civil War. Pratt became county Register in Chancery in 1857 and is reported to have conceived the idea for a writing machine after suffering from writer's cramp, brought on by his multiple jobs as editor, lawyer, and Register.[34]

Accounts of Pratt's earliest inventive activity vary, but sometime after 1857 he began work on a writing machine.[35] As the editor of a political newspaper, Pratt realized that following the outbreak of the Civil War, a Southern inventor had no chance of securing a patent from the Northern federal government. So, in 1863, he took "his Negroes to Selma and sold them for gold, and set sail from Mobile to Glasgow [where] he was marooned for the duration of the war."[36] There, in political safety and in an entirely different economic and technological environment, Pratt continued to work on his writing machine. He protected the concepts for his first machine with a provisional patent in England ("British Prov. Specn. No. 1,983") in 1864.[37] This machine evidently contained the general concept of a radial plunger type machine, as Pratt is reported to have used large numbers of knitting needles and type set in small wooden blocks during his early work in Alabama. Together with an unknown pianoforte manufacturer and an unknown scientific instrument maker in Glasgow, he constructed the first machine in 1865.[38]

On December 1, 1866, Pratt patented a second, far more sophisticated machine, his Pterotype, in Britain.[39] He returned to the United States in 1867 and patented this machine on August 11, 1868.[40] Some number of these Pterotypes were produced in London in 1867 and 1868 by E. B. Burge, a mechanical engineer.[41] Nothing is known of Burge's techniques of manufacture. The *Scientific American* opined that "The case of the instrument is small and compact, the parts are mostly of wood, and it could be manufactured and sold on a large scale for about $15 with a handsome profit."[42]

Pratt failed to get his machine produced in the United States. He had first attempted to secure financial backing in 1860 in Alabama, but without success.[43] In the mid-1870s, after his return to America, he drew the attention of James Densmore who inquired about the mechanical specifications of his machine.[44] He evidently reached some understanding with the Remingtons and/or Densmore, but this agreement was flagrantly broken by the Remingtons according to Pratt's sister.[45] Pratt continued unsuccessfully to seek capital in Alabama in the late 1870s before selling the rights to his invention to James B. Hammond on December 6, 1879, for $700 and "other consideration" for his patents including a royalty of $.50 per machine.[46]

Pratt edited the *Gadsden Times* in 1873 and again from 1883 to 1885.[47] In that year Pratt and his wife moved to Brooklyn, New York, where he went to work for the Hammond Typewriter Company at its factory. There his full-length portrait hung and it "was always recognized and pointed out to visitors as the likeness of the first inventor of the practical typewriter."[48] In return for future typewriter improvements and inventions, Pratt received an annuity of $2,500 from Hammond. He patented at least three inventions between 1882 and 1892.[49]

It is unclear exactly how well Pratt perceived the need for manufacturing expertise in the initial stages of his invention. His trip to England in 1863 seems to represent an attempt to develop the machine further and find such production knowledge which was unavailable during the Civil War. The fact that he remained in Alabama seeking capital and negotiated with James Densmore and the Remingtons in the 1870s suggests that he may not have understood the need for such expertise or thought he could develop it independently. As Sholes and Densmore discovered in 1872, after three unsuccessful attempts to manufacture their "Type Writer" in the Midwest, they needed manufacturing expertise. Both Pratt and Sholes and Densmore eventually came to similar solutions to their similar dilemmas—licensing agreements with manufacturers.[50]

With the exception of Hansen's Writing Ball, these early manufacturing attempts failed. Both the Jones Typographer and Cooper Hand Printing Machine were far too slow in practice to succeed as practical writing machines. They belong to that class of machines known as "index" or "indicator" machines. The central technical feature of these machines is a selecting lever of some sort used to position a type wheel in order to bring each letter into proper place for printing or typing. This process is necessarily slow because it requires at least two motions to print each letter, one to position the wheel and a second to print the letter. Every machine which featured it failed because it could not type as fast as a person could write. Until the design of the writing machine advanced beyond the index or indicator stage, no amount of manufacturing know-how could make it successful. Only the Hansen Writing Ball was manufactured with any success. Its technical feature—separate keys for each letter—clearly distinguish it from the Jones and the Cooper.

The Pratt Pterotype had design potential, but like Sholes and Densmore, Pratt faced the monumental problem of finding both financing and manufacturing expertise. Pratt finally succeeded in joining with James Hammond, but only after the Remingtons had begun manufacturing the Sholes & Glidden Type Writer.

Pratt's influence extended beyond the Hammond Typewriter Company. In July 1867, nearly ten years after his initial efforts in Alabama, the *Scientific American* published an article describing Pratt's Pterotype, an article read in the Kleinsteuber Machine Shop in Milwaukee, Wisconsin, by Christopher Latham Sholes.[51] Inspired by this article, Sholes advanced typewriter design with the development of individual typebars, each carrying a single type and each operated by a single finger. With this design (simultaneously invented but never produced by a German named Peter Meitterhofer about the same time),[52] a typist could achieve great speed, notably faster and more accurate than handwriting. Success did not come to Sholes and his contemporaries based on design alone. They failed three times to manufacture the machine before combining proper design with manufacturing experience on their fourth attempt to produce "The Sholes & Glidden Type Writer, Manufactured by E. Remington & Sons, Ilion, New York."[53]

THE SHOLES & GLIDDEN TYPE WRITER AT E. REMINGTON & SONS

In 1867 Christopher Latham Sholes, a former newspaper editor, was collector of the port of Milwaukee (figure 4.3). He was also a part-time inventor, who did much of his experimenting in Charles F. Kleinsteuber's machine shop (figure 4.4).[54] Here, while working on his page-numbering machine, Sholes met Carlos Glidden, a fellow inventor, who first suggested the idea of making a writing machine. Several months later the *Scientific American* of July 6, 1867, carried the article on John Pratt's "Type Writing Machine," which inspired Sholes to invent the typewriter.[55]

Working together with Glidden, another inventor named Samuel W.

FIGURE 4.3. Christopher Latham Sholes, the inventor of the typewriter, was born in Columbia County, Penn., on February 14, 1819. He died on June 7, 1890. COURTESY: Milwaukee Public Museum, Carl P. Dietz Typewriter Collection, Gift of Miss Lillian Sholes. Catalogue no. E 16766/4345, negative no. 72075.

Soule, and the machinist/clockmaker Mathias Schwalbach, Sholes began to invent the typewriter. By the fall of 1867 they had a working model and searched for capital with which to manufacture it. Sholes contacted an old newspaper associate, James Densmore, who, without ever having seen it, bought a quarter interest in the machine in November 1867.

When he did see it, in March 1868, Densmore was keenly disappointed and demanded that the three inventors keep inventing. They did, and by July 1868 Densmore had secured patents on two "type writing machines." Densmore took the first of these machines to Chicago, where in association with Soule and E. Payson Porter (a telegraph school operator) he spent $1,000 manufacturing fifteen machines. Under testing at Porter's school, however, the machine's design proved defective and Densmore gave up the manufacturing effort with that early machine.

Densmore continued to prod Sholes to finish inventing the typewriter. By March 1871 Densmore felt the time had arrived to attempt manufacturing again. This time he used Kleinsteuber's machine shop where some twenty-five machines were produced in the summer of 1871. Densmore was still dissatisfied. He lacked both capital and skilled workmen. His

FIGURE 4.4. Here, in Kleinsteuber's machine shop in Milwaukee, Wis., Christopher L. Sholes, Carlos Glidden, Samuel W. Soule, and Mathias Schwalbach began work in 1867 that would culminate in the Sholes & Glidden Typewriter. COURTESY: Milwaukee Public Museum, Carl P. Dietz Typewriter Collection. Negative no. 413432A.

workers had no manufacturing experience and knew only how to make machines by hand. They could not produce a fine finish nor could they make the machines accurately. The mechanics in Kleinsteuber's shop, and Schwalbach in particular, were simply not up to the task of manufacturing.

While this second manufacturing effort was failing, Sholes continued to invent and perfect his machine. By the spring of 1872 the machine was technically ready for the market, but, despite his efforts, Densmore could find no one to purchase the invention or to invest in its manufacture. In desperation, he returned to Milwaukee and attempted its manufacture for a third time, this time under his own direction. He rented an old mill with some water power between the Milwaukee River and an abandoned canal turned mill race. There, by the end of June 1872, he installed secondhand tools and hired Mathias Schwalbach as superintendent of several laborers to manufacture typewriters.

Each Milwaukee-made typewriter was a handmade machine. Densmore personally inspected each one. He often required that much work be redone before the machine was shipped. In addition to making new machines, Schwalbach's men repaired and rebuilt old machines. Densmore realized the limitations of the Milwaukee operation and envisioned manufacturing on a large scale. "I am anxious to get every thing in such shape that the various parts can be made by machinery, without this everlasting filing and fitting, which makes but a botch after it is done."[56]

Botched or not, the Milwaukee-made typewriters were sold and their users were pleased with their performance. The Dawes brothers, Fox Lake, Wisconsin, attorneys, were pleased. Their machine, manufactured after November 8, 1872, survives, and an analysis of it reveals the early techniques employed by Schwalbach and his laborers in the refitted Milwaukee mill (figure 4.5).[57]

The Dawes machine consists of a cast iron frame supporting a thick walnut skin. On top is another casting on which are mounted the various mechanisms. The type levers are wood with porcelain knobs with hand-painted letters and numerals. Throughout its mechanism, the Dawes machine shows evidence of hand work. It was, as Densmore complained, made only by "filing & fitting."

Despite their technical success, the Milwaukee-made typewriters were financial failures. "The fact is that as we are now making them they are costing more than we ask for them. And *until we cheapen the making, we are losing all the time,*" complained Densmore (italics added).[58] The venture was still losing money in December 1872 when George Washington Newton Yost came to Milwaukee.[59] During his one-day visit, Yost advised his long-time friend and business partner, James Densmore, to take the typewriter elsewhere to be manufactured. Yost recommended the arms makers E. Remington & Sons of Ilion, New York (figure 4.6).

In mid-February 1873 Densmore and Yost arrived in Ilion with a Milwaukee-made typewriter. Over the next two weeks they negotiated with the Remingtons to manufacture the machine, signing a contract on

FIGURE 4.5. This Milwaukee-made Sholes typewriter, ca. late 1872—early 1873, was sold to the Dawes brothers, lawyers in Fox Lake, Wis. Julius H. Dawes later became a member of the board of trustees of the Buffalo and Erie County Historical Society and donated this machine in 1889. B. and E. C. H. S. accession no. 58-240. SOURCE: George Iles, *Leading American Inventors*, opposite p. 328. COURTESY: Smithsonian Institution, Museum of American History, negative no. 83-3724.

March 1, 1873, on terms greatly favoring the Remingtons. So badly had Densmore wanted to get the machine manufactured that he agreed to take most of the risk. Densmore and Yost paid the Remingtons $10,000 in advance and agreed to pay a royalty of $.50 per machine to Jefferson Clough for design services. In return, the Remingtons agreed to redesign the machine and to manufacture 1,000 (plus 24,000 more at their discretion).[60]

Jefferson Clough, superintendent of the Remington armory, worked with William K. Jenne, assistant superintendent of the Remington Sewing

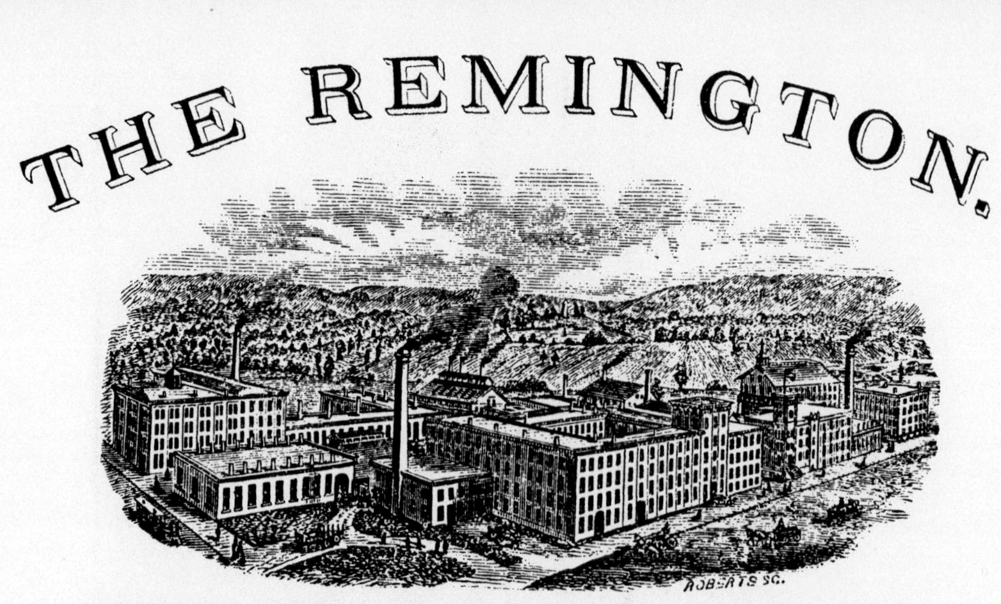

FIGURE 4.6. The Remington manufacturing complex in 1874. SOURCE: *The Remington* (Ilion, N.Y.: Remington Sewing Machine Co., July 1, 1874), an advertising broadside. COURTESY: Smithsonian Institution, Museum of American History, Division of Textiles, Sewing Machine Research File. Negative no. 83-9516.

Machine Company, to redesign the Milwaukee-made machine. This they accomplished by the summer of 1873 when they had "got done the pattern machine, by which the rest [were] to be manufactured."[61] The first "Sholes & Glidden Type Writer" was shipped to Densmore on April 30, 1874, and by July the Remingtons were shipping machines to authorized agents in major cities (figure 4.7).

The story of the manufacture of the typewriter at E. Remington & Sons is clouded not only by the lack of documentary sources but also by the confused state of affairs at Remington itself. The earliest description of the manufacturing process at Remington comes over ten years after the Sholes & Glidden typewriter was first made at Ilion. Both typewriter design and the manufacturing technology had changed dramatically during those ten years.

The decade of the 1870s was reportedly the golden age of E. Remington & Sons.[62] The firm was one of the major private firearms manufacturers in nineteenth-century America, capable of producing 1,500 guns daily in 1872. The firm held numerous government arms contracts during the Civil War, following which they adopted a dual strategy for economic growth; first, it expanded firearms sales to foreign governments and second, it diversified into nonmilitary or consumer products, notably

FIGURE 4.7. The Sholes & Glidden Typewriter manufactured by E. Remington & Sons, Ilion, N.Y., ca. 1875–1877, serial no. 2044. This is the second model. The typewriter table and the foot return have been abandoned in favor of a side arm return. COURTESY: Milwaukee Public Museum, Carl P. Dietz Typewriter Collection. Catalogue no. E 41871/12124, negative no. H-627-13-J-1.

agricultural equipment and sewing machines but also into fire engines, cotton gins, pumps, and bridges as well as the typewriter.[63] In reality, poor management was leading the company into bankruptcy in the 1880s.[64]

The Remington firm fell on hard times in the late 1870s and early 1880s. The Egyptian and Mexican governments defaulted on large orders, despite their delivery.[65] In addition, the Remington Brothers Agricultural Works suffered badly from mid-west competition and by 1885 "they [were] not d[oin]g much except in house pumps & Remington Fire Engines."[66] According to R. G. Dun & Company reports: "This Co. have no existence apart from E. Remington & Sons . . . no new credit can be recommended."[67] The Remington Sewing Machine Company did no better. By the early 1880s the firm was deeply in debt and unable to meet its obligations. The Remingtons struggled on until April 22, 1886, when it was placed in the hands of a receiver who ran it until liquidation two years later. By April 22, 1888 the firm of E. Remington & Sons was "out of Bus.[iness]."[68]

In an attempt to avoid bankruptcy, E. Remington & Sons sold the typewriter business in March 1886 to the firm of Wyckoff, Seamans & Benedict. This company was the exclusive sales agency for the typewriter, having begun business on August 1, 1882, succeeding a long line of unsuccessful typewriter sales agencies.[69] Wyckoff, Seamans & Benedict paid $197,000 for the Remington typewriter business, which included plant, $50,000; stock and material $25,000; and franchise and patents $100,000.[70] In addition, they had $50,000 in cash with which to continue the manufacturing operation on their own. They immediately created the Standard Typewriter Manufacturing Company in Ilion with an authorized capital of $225,000 (figure 4.8).[71]

The Standard Typewriter Manufacturing Company produced the typewriters in Ilion and sold the entire output to Wyckoff, Seamans & Benedict, which distributed the machines to existing agencies in cities across the country.[72]

Prior to its sale to Wyckoff, Seamans & Benedict, the typewriter manufacturing operation was split between the Remington Brothers Agricultural Works, the Remington (Empire) Sewing Machine Company, and almost certainly extended into the armory itself.[73] The Agricultural Works had a foundry where the side panels, the top, and the frame as well as other pieces, were cast.[74] The sewing machine department did the wood work which consisted of the table tops for the first models (soon abandoned due to a major design change), shipping crates, and, on the machine itself, the wooden typebar levers.[75] The japanning and decoration are almost certainly products of the sewing machine operation where similar tasks were performed on the sewing machines themselves. The forged typebars, screws, and other machined parts were almost certainly products of the armory.

The Remingtons never formed a company to manufacture the typewriter as they did for their sewing machine. Indeed, with most of their later products—bridges, pumps, fire engines—the Remingtons seem to have used existing facilities rather than attempt to create separate firms and facilities. Typewriter manufacturing was spread throughout the complex of shops at Ilion.

The Remingtons followed the general concept of "armory practice,"

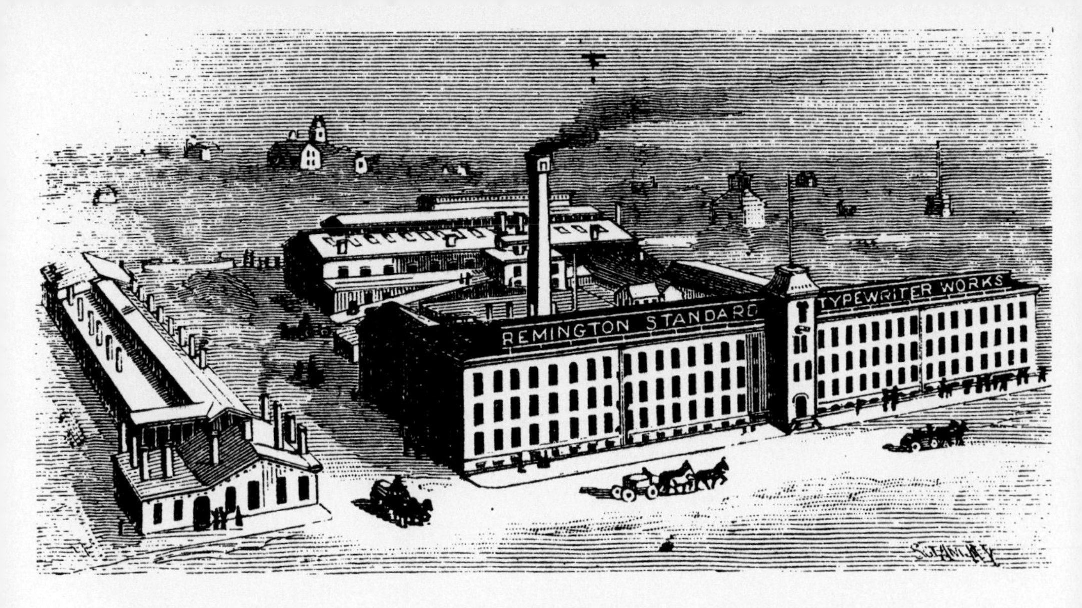

FIGURE 4.8. "The Works" of the Standard Typewriter Manufacturing Company in Ilion, N.Y., in 1888. SOURCE: "The Remington Typewriter," *Scientific American* (December 15, 1888), N.S., 59(24):375. COURTESY: Smithsonian Institution, Museum of American History. Negative on file in Division of Mechanisms.

with manufacturing centered on a model machine on which the production machines were based and against which gauges were checked. This machine was one of three built between March 1 and the summer of 1873, when Densmore reported that the Remingtons "have already got done the pattern machine, by which the rest are to be manufactured . . ."[76] Yost, who by then had become active in the business, hoped that in July 1874 the Remingtons' production expectations of one hundred machines made per month would be met. "They are going to try hard at it," Yost noted.[77] The Remingtons never produced one hundred machines monthly in the 1870s. In December 1884 they were averaging only twenty to twenty-five weekly, despite a very strong demand which they were unable to meet.[78]

Despite beginning their manufactured life in one of the great American armories, the early typewriters were far from perfect. James Densmore recognized the manufacturing problems at Remington and, at his urging, the Remingtons did gradually improve their technology. In 1877 Densmore visited the factory and later complained that "one reason the machines have not been made better heretofore is that so large a part has been 'made by hand.'" However, he also noted that "they have nearly got a tool for making every different part, and the manufacture in consequence is getting more and more perfect all the time."[79]

Notwithstanding this progress and the healthy demand for machines after 1880–1882, the typewriter manufacturing process under the

Remingtons drew little notice. The Dun reports on E. Remington & Sons fail to mention the typewriter until December 14, 1880, six years after its manufacture began. Even then, the references were sporadic.[80]

In view of the structure and organization exhibited by E. Remington & Sons with their other consumer products, it clearly approached the typewriter as just another item. Its manufacture was intertwined with all the other Remington products and processes, which were badly disorganized. Dun reported that at the Remington Brothers Agricultural Works "matters are so complicated that no one outside the parties concerned know their condition."[81] Similarly, Dun recorded that in December 1876 the Remington Sewing Machine Company was actually two Remington Sewing Machine Companies. One was a manufacturing firm so intermixed with E. Remington & Sons that it was "not understandable alone."[82] This was clearly the case with the typewriter, completely intertwined both economically and technologically with E. Remington & Sons. Once Wyckoff, Seamans & Benedict acquired the typewriter operation, they focused all their energies on typewriter manufacturing and properly organized the business and the process.

TYPEWRITER MANUFACTURING COMES OF AGE: 1881–1924[83]

Typewriter Design

> It is well to recall that although produced by mass production methods, the typewriter is essentially a precision instrument. Its thousands of parts must work together with exquisite exactness, yet withstand hard usage. It must do fine work, fast work, yet require a minimum of attention.
>
> WILLIAM P. TOLLY, *Smith Corona Typewriters & H. W. Smith*[84]

Typewriter manufacturers realized that their extremely complex product required adjustment and they designed adjustability into their machines. The design ingenuity in the Remington typewriter is evident in the typebar hanger, which was secured with a screw (figure 4.9). Another clever design aspect of the typebar hanger was its reversibility. The same hanger suspended all the typebars, but, to provide clearance for the typebar arms, every other hanger was installed at 180° from the hanger on either side of it (figure 4.10).

This is not the only typebar adjustment. The "Transverse Section of the Typewriter" shows another adjustment via the steel wire connecting

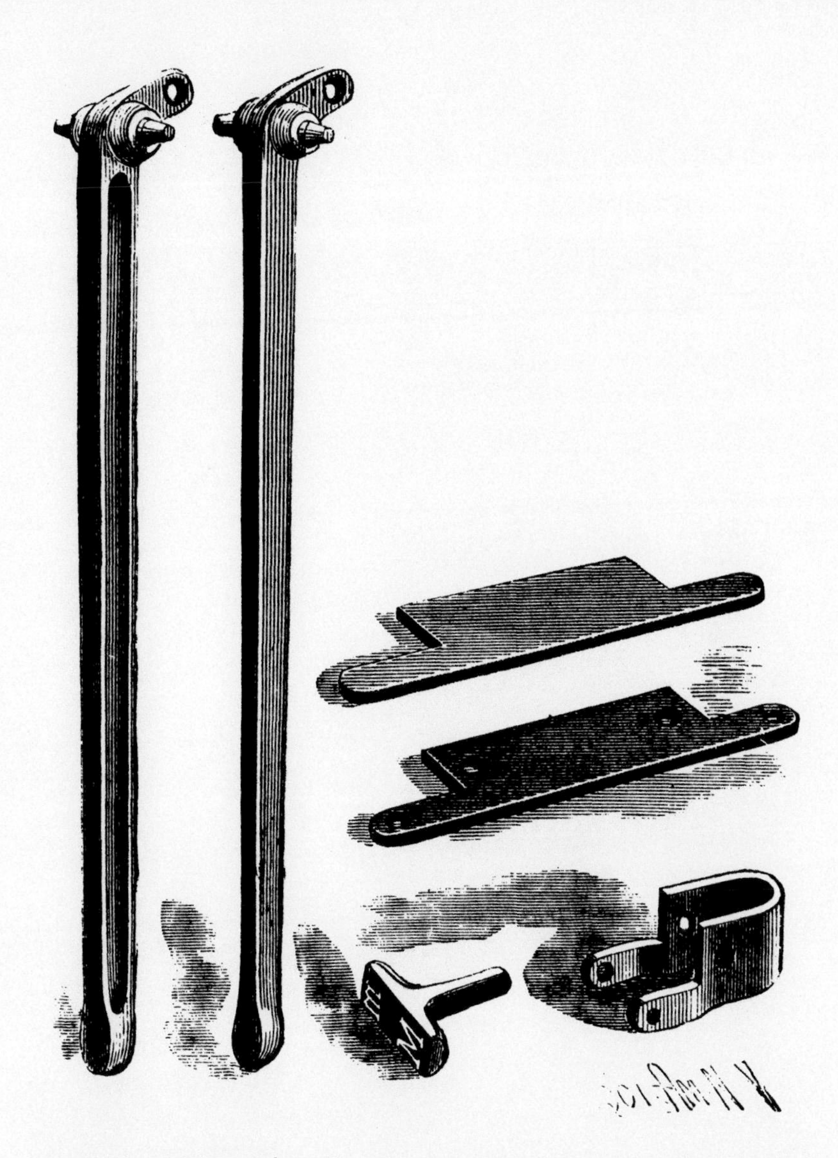

FIGURE 4.9. "Type Arms and Type." SOURCE: "The Remington Typewriter," *Scientific American* (December 15, 1888), N.S., 59(24):374. COURTESY: Smithsonian Institution, Museum of American History. Negative on file in Division of Mechanisms.

the typebar to the key lever (figure 4.11). This adjustment assured that the keys on the keyboard could be adjusted to the same height without affecting the typebars themselves. The Remingtons advertised their earliest machines as being adjustable by the user.[85]

Typebar assembly and adjustment were especially important because

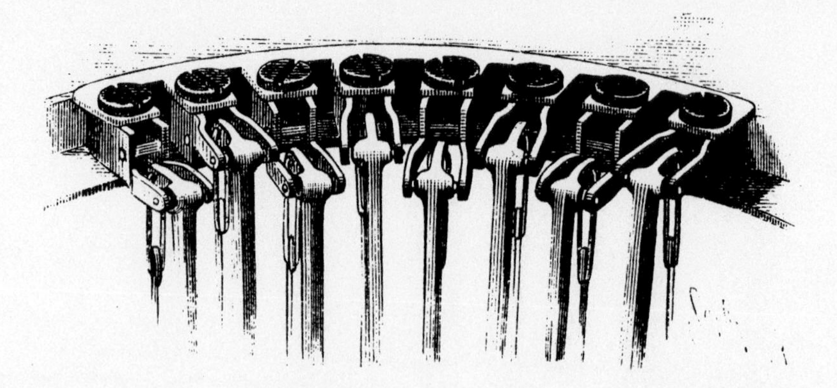

FIGURE 4.10. "Pivots of the Type Arms." SOURCE: "The Remington Typewriter," *Scientific American* (December 15, 1888), N.S., 59(24):374. COURTESY: Smithsonian Institution, Museum of American History. Negative on file in Division of Mechanisms.

improperly aligned type produced an unevenly typed line. To avoid this, the Remingtons developed special tools to insure the proper alignment of the type. "Centering the Type Arms" shows the use of a special typebar adjusting fixture, which insured that when the type was hung in its hanger, each type would strike a common point (figure 4.12). Here, the typebar ring is placed in a frame supported on four legs, similar to the typewriter top plate. The typebars hang as they will when finally assembled. Across the back of the machine is an adjustable iron bar on which is mounted an indicating arm which marks the center point of the typebars. The hands in the illustration are holding a typebar in position against the indicating arm while the typebar screw is tightened after an adjustment. Notice particularly that the two tools used are a screw driver and a hammer.

Despite specialized machinery, a gauging system, and its dedication to the interchangeable system, the American Writing Machine Company also found it necessary to build adjustability into its writing machine. The examination of a Caligraph reveals four separate adjustment points between the steel type face and the wooden type lever. First, there is a turn buckle joining the "Long Connecting Rod" with the "Short Connecting Rod" between the typebar and the wooden type lever, which allows the keys of each bank to be adjusted to a single height.[86] Second, there is the typebar hanger, held in place by the "Hanger Washer" and a machine screw. By loosening the screw and washer, the typebar hanger and assembly could be moved toward, away from, and in an arc with respect to the printing point. Third, the typebar hanger itself was adjustable through the conical bearings on the typebar, the holes in the hanger, and the

FIGURE 4.11. "Transverse Section of the Typewriter." SOURCE: "The Remington Typewriter," *Scientific American* (December 15, 1888), N.S., 59(24):374. COURTESY: Smithsonian Institution, Museum of American History. Negative on file in Division of Mechanisms.

adjustment screw (figure 4.13—see letter "A"). Fourth, the type itself was forced into a tapered hole in the steel block brazed onto the end of the typebar. If needed, it could be loosened, turned in its hole, and then driven back into the steel block. These were not the only adjustments built into the machine. Additional adjustments included the rack adjustment, dog adjustment, carriage tension, paper feed, ribbon feed, and finger-key tension (figure 4.14).[87]

These multiple adjustments were an important advertising point for the American Writing Machine Company and its Caligraph. The firm took special pride in promoting its adjustable features, notably the typebars, both in its advertising and its instruction manuals.

> *DURABILITY*. This is an important consideration, as writing machines are expensive and subject to continuous use and in this respect we claim the Caligraph is far ahead of all competing Machines. Its type-bars are adjustable, and in event of any lost motion in the journals, it can be taken up easily, no other machine possesses this unquestioned advantage. The paper feed bands on the Caligraph are of tempered steel and always adjustable, a great improvement over machines using rubber bands for this purpose [the Sholes & Glidden and the Remington Nos. 1 & 2].[88]

FIGURE 4.12. "Centering the Type Arms." SOURCE: "The Remington Typewriter," *Scientific American* (December 15, 1888), N.S., 59(24):374. COURTESY: Smithsonian Institution, Museum of American History. Negative on file in Division of Mechanisms.

> *ALIGNMENT*. The following cut, which represents the new type-bar hanger, has an adjusting screw and shows how the wear can be taken up, from time to time, by the operator. Remember this is the only machine that can be aligned by users, and shows the best work under hard strain and rapid manipulation. The parts which move, in any kind of machinery, will wear. The faster they move the faster they wear. This is common to all; and the Caligraph alone is adjustable. Look out for durability![89]

Adopting and Adapting Technology

Typewriter manufacturers used a number of existing technologies and adapted others to their particular needs. For example, in 1906 the Remington Typewriter Company used "twenty-six Tabor pneumatic molding machines" which had replaced hand molding. The new machines were said to be "cheaper, cleaner, and more accurate," and were part of Remington's newly built factory. Remington also used hundreds of automatic screw machines which produced a wide variety of small turned parts (figure 4.15).

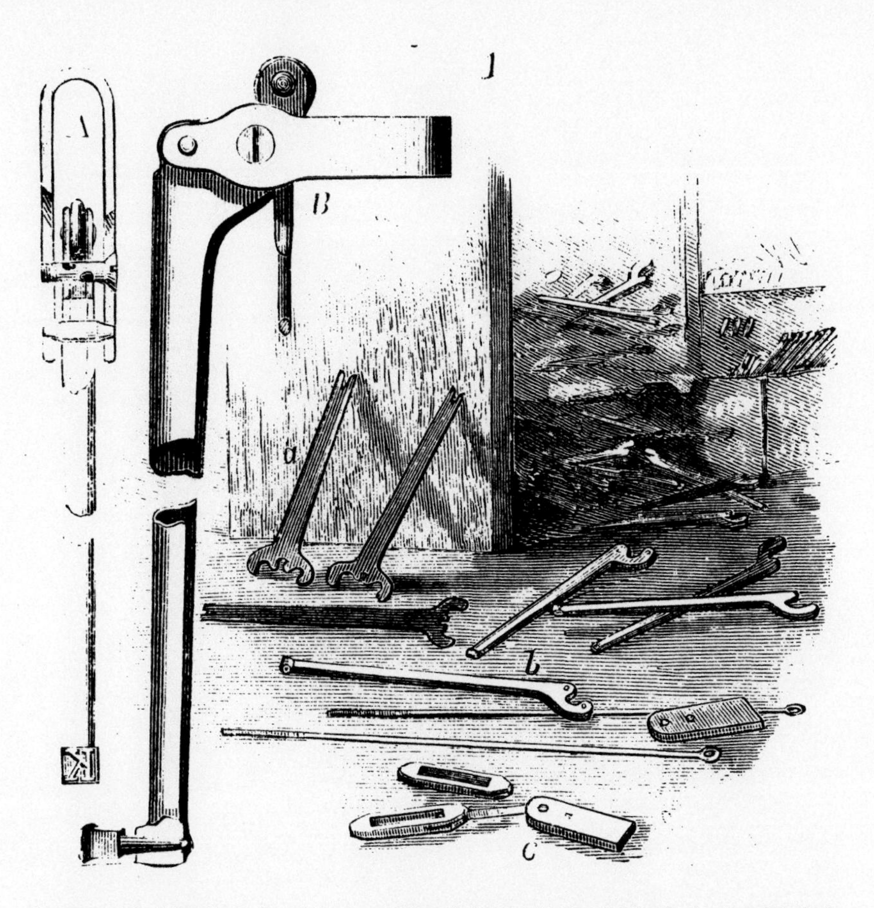

FIGURE 4.13. "The Type Bars." SOURCE: "Caligraph Writing Machine," *Scientific American* (March 6, 1886), N.S., 54(10):150. COURTESY: Smithsonian Institution, Museum of American History. Negative on file in Division of Mechanisms.

The Hall Type Writer Company manufactured an index machine, that sold successfully through the 1880s and into the 1890s. It was a distinctly different machine than the Remington and the Caligraph (figure. 4.16). Hall also developed specialized machinery, such as "a graduating machine for the bell rods and 'clips.'"[90] It was a semi-automatic machine that marked the bell rod, a horizontal positioning bar indicating where the typist was on a particular line of type. The bell rod was secured on a movable carriage which slid on the bed of the machine.

Power was supplied to Hall's "Graduating Machine" by a belt engaged by a hand operated clutch. A visible shaft operated the spring activated marking tools. A second shaft, located behind the machine, ran off the primary shaft and drew the carriage past the stationary tools. A vertical post on the carriage engages a sleeve mounted on a shaft attached to the

FIGURE 4.14. "The Caligraph." SOURCE: "Caligraph Writing Machine," *Scientific American* (March 6, 1886), N.S., 54(10):150. COURTESY: Smithsonian Institution, Museum of American History. Negative on file in Division of Mechanisms.

FIGURE 4.15. "Automatic Screw [Machine] Department—800 Foot View" of the Remington Typewriter Company, December 1906. SOURCE: "Producing a Wonderful Writing Machine," *Office Appliances* (December 1906), 5(1):22. COURTESY: SCM, Milwaukee Public Museum Photo. Negative no. H-627-15K-8-6A.

FIGURE 4.16. The Hall Type Writer. SOURCE: "The Hall Type Writer," *Scientific American* (July 10, 1886), N.S., 55(2):25. COURTESY: Smithsonian Institution, Museum of American History. Negative on file in Division of Mechanisms.

clutch handle and automatically stops the machine when the work is completed (figure 4.17).

"Fig. 7—Vulcanizing Type Plates And Rubber Rolls" illustrates the process used by Hall to manufacture the type plates (figure 4.18). "The type forms are interchangeable," claimed Hall, and were available in fifteen different styles of English and "many other languages."[91]

Hall's use of rubber type—made with conventional vulcanizing technology by forcing liquid rubber into a mold at high temperature and pressure—shows the adaptive use of non-firearms technology by a private sector American System industry. It is particularly noteworthy that Hall designed his typewriter to take advantage of the flexible properties of rubber. All the typewriter manufacturers handled rubber in some form, primarily in platens and rollers, but also, as in the case of Hall, as type itself. The machining of rubber was a technology far removed from armory practice and developed entirely in the private sector. All the manufacturers used a vulcanizing process or purchased parts from firms that did.

The Oliver Typewriter Company also had a battery of automatic screw machines to produce parts for its typewriters including screws and es-

FIGURE 4.17. Hall's semi-automatic "graduating machine for the bell rods and 'clips.'" SOURCE: "The Hall Type Writer," *Scientific American* (July 10, 1886), N.S., 55(2):27. COURTESY: Smithsonian Institution, Museum of American History. Negative on file in Division of Mechanisms.

capement wheels (figures 4.19 and 4.20), as did L. C. Smith & Bros. (figure 4.21). These machines, like those in the Remington factory, were specially adapted to produce the particular parts that Oliver and L. C. Smith & Bros. needed, but were nevertheless a commonly available technology.[92]

New Materials and New Technologies

> The Caligraph is now fitted with a new Patent Steel Type-Bar. . . . this bar is struck up from sheet steel. This bar embodies the principle used in making bicycles and other fine machines where great strength must be combined with lightness. This bar will resist more strain than a solid steel one of like dimensions, and, of course, its extreme light-ness materially adds to the speed of the machine.
>
> *The Caligraph Manufactured by the American Writing Machine Co.*[93]

Typewriter manufacturers employed a number of new materials and necessarily developed new technologies to work those materials. A machine as expensive and as complex as the typewriter needed to be pro-

FIGURE 4.18. "Vulcanizing Type Plates and Rubber Rolls." SOURCE: "The Hall Type Writer," *Scientific American* (July 10, 1886), N.S., 55(2):25. COURTESY: Smithsonian Institution, Museum of American History. Negative on file in Division of Mechanisms.

tected from everyday dust and most typewriters were sold with dust covers.

These early covers were usually made from sheet iron. Later, rubberized cloth was used. Some manufacturers, notably Hammond, made extensive use of wood in their cases, usually a veneer over a base wood. Hammond is particularly interesting as its first machines included much wood, notably ebony keys and exterior veneer panels.

A part of the Remington's iron dust cover making process was illustrated in "Case Making" (figure 4.22). The Remingtons departed from the earlier sewing machine practice of making wooden covers and substituted a simple sheet metal cover which was painted black and decorated with the firm's name applied in gold lettering. This 1886 illustration shows the main section of the case being bent around a pattern by rollers.

In 1906 the Remingtons completed construction of a new factory which covered eighteen acres, employed 1,600 men, and produced a typewriter every working minute. It was equipped with "the most ingenious modern machinery, including many machines devised especially for the work and not to be found elsewhere."[94] Some of the "ingenious modern machinery" included a top plate drilling machine which could drill one hundred holes

FIGURE 4.19. "Screw Machines." SOURCE: Oliver Typewriter Company, *The Oliver Typewriters, The No. 3 Models*, p. 18. COURTESY: Smithsonian Institution, Museum of American History. Negative no. 83-13657.

in a top plate simultaneously in three and a half minutes. One man tended three of these automatic multiple spindle drilling machines (figure 4.23). For other operations, additional specialized machines were designed and employed, for example, the reducing pantograph for type making.

Grinding the platen was extremely important, since imperfections in it would be readily apparent in the typed material. Hence it is not surprising to find Hall and the American Writing Machine Company showing off their platen grinding machines. In 1886 the American Writing Machine Company was more interested in establishing the technical superiority of its polygonal platen (vis-à-vis the round Remington platen) and illustrated a "machine that automatically grinds . . . the rubber cylinder" (figure 4.24).[95] The platen was held between centers and indexed by a plate visible at the left end of the machine. As each face of the platen was ground flat, the machine indexed automatically and the next surface of the platen was brought into position for grinding.

The Hall Type Writer Company also developed "a machine for grinding the rubber rollers," a platen grinding machine (figure 4.25).[96] The platen was held between centers mounted on a carriage which slid in the main frame of the machine. It was carried back and forth by a wire or belt looped around the large gear at the right end of the machine and its worm gear mounted on the end of a belt driven shaft. This shaft had two belts running in opposite directions and an automatic mechanism for changing the direction of the shaft and hence the carriage carrying the

FIGURE 4.20. A sample of automatic screw machine work, the escapement wheel of the Oliver typewriter. SOURCE: Oliver Typewriter Company, *The Oliver Typewriters, The No. 3 Models*, p. 17. COURTESY: Smithsonian Institution, Museum of American History. Negative no. 83-9802.

platen. On the opposite side of the machine a stationary grinding wheel was forced against the platen as it is carried past. This machine was drawn to show its self-acting, automatic features and the pains to which Hall went to have the platen properly ground.

The complexity of the typewriter and its many moving parts forced the typewriter mechanics to develop "exercising machines" to work their mechanisms mechanically before final adjustment and aligning. This idea of making parts work together during the manufacturing process appears not to have been unique to the typewriter industry, but was apparently not universally practiced. Only a sewing machine manufacturer, Wheeler & Wilson, is known to have adopted the practice of "breaking in" its machines with a machine as a step in the manufacturing process.[97]

FIGURE 4.21. "Automatic Screw Machine Department." L. C. Smith and Bros. Typewriter factory, Syracuse, N.Y., ca. 1924. SOURCE: L. C. Smith & Bros. Typewriter Company, *A Visit to the Home of the L. C. Smith Typewriter*, p. 8. COURTESY: SCM, Milwaukee Public Museum Photo. Negative no. H-627-15-G-27.

The Hall Type Writer Company used a special "device for easing the "motions," that they may run smoothly" (figure 4.26).[98] Four separate but identical parts of the machine exercised two Hall parallel movement devices. This parallel movement device carried a rubber type plate and positioned the proper type above an aperture in a lower plate through which it printed.

The American Writing Machine Company exercised its Caligraph typebars in 1886 before final assembly. After the fabrication and assembly steps, the completed typebars were placed in a "working jack" to be broken in. On this machine the typebar and hanger assembly was screwed into place as it would later be in the machine itself (figure 4.27). A reciprocating rack and pinion arrangement rapidly moved the typebars

FIGURE 4.22. "Case Making." SOURCE: "The Remington Typewriter," *Scientific American* (December 15, 1888), N.S., 59(24):367. COURTESY: Smithsonian Institution, Museum of American History. Negative on file in Division of Mechanisms.

through the arc of a circle, forcing the bearings to wear into each other. The result was to "obtain an accurate and easy movement of the type bars when . . . inserted in the machine."[99]

In 1903 the "Oliver Exerciser" worked the mechanism of the completed machine rather than a particular sub-assembly as was the practice with the Caligraph and the L. C. Smith & Bros. typewriter. After it was "exercised," each Oliver typewriter was again subjected to another rigid inspection and alignment (figure 4.28).[100]

In 1924 the L. C. Smith & Bros. "Typebar Exerciser" worked each ball bearing typebar for two hours, the equivalent of 36,000 keystrokes, before the typebar was assembled into the machine (figure 4.29). It had already been through at least two sub-assemblies including type soldering and ball bearing assembly.[101]

FIGURE 4.23. This "Automatic Top Plate Drilling Machine" was a highly specialized multiple spindle drilling machine used in the Remington Typewriter Company's factory, December 1906. SOURCE: "Producing A Wonderful Writing Machine," *Office Appliances* (December 1906), 51(1):21. COURTESY: SCM, Milwaukee Public Museum Photo. Negative no. H-627-15-K-17-25A.

FIGURE 4.24. "Grinding the Printing Cylinder." SOURCE: "Caligraph Writing Machine," *Scientific American* (March 6, 1886), N.S., 54(10):150. COURTESY: Smithsonian Institution, Museum of American History. Negative on file in Division of Mechanisms.

The L. C. Smith & Bros. Typewriter Company took special pride in its ball bearing manufacturing operation. "The process of manufacturing balls for the typebars is not only interesting but brings forcibly to the attention of one going through the plant the great care and skill used in our manufacturing department." The process began with steel rods which were heated, cut to proper length, and formed into rough balls while still hot. They were then placed in the "lappers" after a first inspection and ground with emery and oil for eight minutes in the first grinding. This reduced the balls to nearly the right size and made them round. After the inspection of a sample of balls, they were "hardened to such an extent that they can be driven into a piece of steel without flattening." The hardened balls were returned to the "lappers" again and ground smooth to the proper size. Following this, they were "tumbled" for three weeks in a barrel with a "certain substance" which further smoothed them and made them appear to be nickel plated. Balls which did not assume the nickel plated look had not been properly hardened and were rejected. The balls were then ready for the two ball sorting machines which gauged

FIGURE 4.25. "A machine for grinding the rubber rollers." SOURCE: "The Hall Type Writer," *Scientific American* (July 10, 1886), N.S., 55(2):24. COURTESY: Smithsonian Institution, Museum of American History. Negative on file in Division of Mechanisms.

the ball bearings twice. The first machine contained two steel rollers, which were further apart at the lower end than at the upper end (figure 4.30, left). One of the steel bars turned at a slightly higher speed than the other, thus giving each ball a slight rotation. As each ball was released individually onto the rolling bars, it moved down until it fell past the steel rolls into a canister corresponding to its diameter. In other words, each ball was automatically sorted by diameter into fifteen different diameters. These sorted balls were then run through the second gauging machine four times to insure that they were indeed within specifications. The second gauging machine consisted of two steel knife edges which were .0015 of an inch further apart at the bottom than at the top (figure 4.30, right). Specifications for L. C. Smith & Bros. typebar ball bearings were .0001 inches.[102]

Each L. C. Smith & Bros. typewriter had at least forty-two typebars and each typebar had fifteen ball bearings, making a total of 630 ball bearings in each typewriter.[103] The L. C. Smith & Bros. factory produced some 30,000 typewriters annually and thus over 18,900,000 typebar bearing balls per year.

FIGURE 4.26. A "device for easing the "motions," that they may run smoothly." SOURCE: "The Hall Type Writer," *Scientific American* (July 10, 1886), N.S., 55(2):24. COURTESY: Smithsonian Institution, Museum of American History. Negative on file in Division of Mechanisms.

Factory Organization

The complexity of the typewriter forced typewriter mechanics to organize the assembly, adjustment, and alignment of the machine. Each sub-assembly took place in a separate department. As early as 1886 the Remington factory had at least three different departments, an assembly department for "Putting in Connecting-rods and Levers," an "Aligning Room," and an "Adjusting Room" (figures 4.31 and 4.32).[104]

In 1906 shortly after its "recent enlargement," the Remington Typewriter Company's factory was divided into two sections, production and assembly.[105] Parts production required 70 percent of the factory complex with separate buildings for the brass foundry, iron foundry, inspection, cleaning and grinding of castings, japanning, tin work, wood working, rubber working, forging, and annealing and hardening.

The remaining 30 percent of the plant was used to assemble, adjust,

FIGURE 4.27. "Testing the Type Bars." SOURCE: "Caligraph Writing Machine," *Scientific American* (March 6, 1886), N.S., 54(10):150. COURTESY: Smithsonian Institution, Museum of American History. Negative on file in Division of Mechanisms.

and inspect the individual machines. In the "great machine hall" some 3,000 machines were in process of assembly at any one time by "several hundreds of skilled assembling experts." "After receiving a registered number," the machine "rapidly [grew] . . . to a frame . . . " The various components of the machine (the type basket, carriage, ribbon mechanism, etc.) were added at various stages of the assembly process as the increasingly complete machine progressed through the factory. As it traveled, it was temporarily stored on shelving before being carried independently to each work station. There were several sub-assembly areas in which the various components were assembled, while some minor assemblies were put together in the production wing. Then, after its assembly, the machine was ready for its first adjusting, followed by its second or "touching up" alignment, and then its "ordeal of final inspection

FIGURE 4.28. "The Oliver 'Exerciser.' " SOURCE: "The 'Oliver' Growth in Facts and Figures," *Office Appliances* (January 1908), p. 36. COURTESY: Smithsonian Institution, Museum of American History Warsaw Collection. Negative no. 83-13302.

and adjustment." "Seldom (was a machine) passed without criticism" (figure 4.33).[106]

It is significant that, in 1906, the Remingtons illustrated their promotional article with only three production machine illustrations and six assembly and adjusting illustrations. Clearly, assembly and adjusting and aligning were the highly skilled, labor intensive aspects of typewriter manufacturing at Remington in the early twentieth century. Interestingly, they seem to have approached the concept of an assembly line—the typewriters worked themselves down the length of the building as they gradually grew from frame into completed machine—and perhaps William K. Jenne (who designed the factory and much of its machinery) considered such an idea only to discard it due to the problems of intricate assembly. Perhaps in the interest of quality, assembly was not rushed.

Little is known of the Hartford factory organization of the American Writing Machine Company, except that some automatic machinery was used. In addition, the firm advertised that the

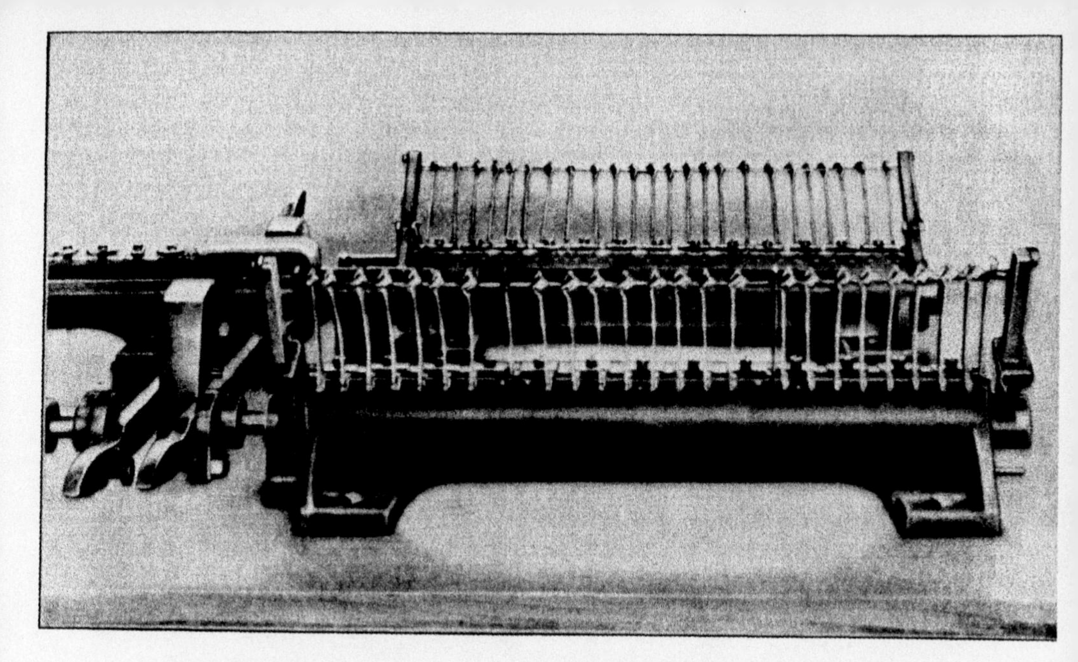

FIGURE 4.29. "Typebar Exerciser" in the L. C. Smith & Bros. typewriter factory, Syracuse, N.Y., ca. 1924. SOURCE: L. C. Smith & Bros. Typewriter Company, *A Visit to the Home of the L. C. Smith Typewriter*, p. 21. COURTESY: SCM, Milwaukee Public Museum Photo. Negative no. H-627-15-G-14.

works of this company are provided with special machinery for making the various parts, and constant care is exercised to insure the smooth and reliable working of each part. The several parts are made by the same machines, and to the same gauges, thereby obtaining that most essential and desirable feature, interchangeability.[107]

The Oliver Typewriter Company also developed an extensive assembly, adjustment, and alignment organization in its factory in Woodstock, Illinois. Its organization consisted of at least six separate departments: Type Bar Department, Carriage Department, Assembly Department (figure 4.34), Tabulators and Adjustment, Inspection Department, and Aligning Room.[108]

L. C. Smith & Bros. employed precision gauging throughout its factory. They, like the Remingtons, divided their factory into production and assembly departments. In their production departments, they made extensive use of punch presses and automatic screw machines, by then standard in the typewriter industry. Like the Remingtons, L. C. Smith & Bros. used highly specialized machines such as multiple spindle drilling machines (for such operations as drilling holes in the top plate and the base) and specially designed milling machines for cutting the slots in the typebar segments. There were departments for electroplating parts and for baking the enamel finish (some eighteen different operations in the

FIGURE 4.30. Ball Bearing Sorting Gauges in the L. C. Smith & Bros. typewriter factory, Syracuse, N.Y., ca. 1924. SOURCE: L. C. Smith & Bros. Typewriter Company, *A Visit to the Home of the L. C. Smith Typewriter*, p. 19. COURTESY: SCM, Milwaukee Public Museum Photo. Negative no. H-627-15-G-18.

enameling process) and special machines for cutting type dies and making type, as well as special jigs for soldering type to the typebars.

In assembling their machines, L. C. Smith & Bros. divided the process according to the parts and mechanisms of the machine itself. Each department employed numerous operatives for each assembly. For example, at least forty men assembled ball bearings into the typebars. In the process of assembly, the machine passed through a series of assembling departments and "in each department certain parts [were] added until finally the machine [was] complete." In the first assembly department, "Sub-Lever and Key-Lever Assembling," the machine was serial numbered for the first time. The machine then passed through: Segment Assembly Department; Universal Bar, Space Key; Shift Lock Assembling Department; Ribbon Mechanism Assembling Department; Spring Drum Department; Carriage Assembling Department; Aligning Department (figure 4.35); Final Adjusting Department (figure 4.36); Tabulator Department (when applicable); and Final Inspection Department.

These ten assembling, aligning, and inspecting departments demonstrate just how complex the typewriter was as a mechanism. The

FIGURE 4.31. "Corner of the Aligning Room." SOURCE: P. G. Hubert, Jr. "The Typewriter, Its Origins and Uses" (*Century Magazine*?) (April 25, 1888), p. 29. COURTESY: Smithsonian Institution, Museum of American History. Article and negative on file in Division of Mechanisms.

complexity of the product itself, the "thousands of parts," forced the typewriter manufacturers to develop highly complex and sophisticated manufacturing operations in order to achieve mass production.[109]

Typewriter manufacturers faced problems unknown to arms manufacturers, particularly in assembling and adjusting. It is interesting to observe how L. C. Smith & Bros. seemed to treat as routine the actual production of most of its parts and focused so heavily on its ball bearing operation and the assembly procedure.

In the early 1920s the typewriter manufacturers had developed the American System just short of the assembly line. They had fully developed automatic machinery for producing their parts, had evolved some form of statistical quality control (at L. C. Smith & Bros. for testing ball bearings), and had vast numbers of assemblers, aligners, and inspectors. The need for adjusting assembled and aligned typewriters continued at least through the early 1950s. In the Royal Typewriter Company's factory, the final adjuster was an especially skilled person who adjusted only three or four typewriters per day.[110]

Like other private sector American System manufacturers, the typewriter industry faced a technological imperative in producing its ma-

FIGURE 4.32. "Corner of the Adjusting Room." SOURCE: P. G. Hubert, Jr. "The Typewriter, Its Origins and Uses" (*Century Magazine*?) (April 25, 1888), p. 30. COURTESY: Smithsonian Institution, Museum of American History. Article and negative on file in Division of Mechanisms.

chines. The typewriter had special assembly problems, problems that were related to the nature of the typewriter itself, not the process of manufacturing its various parts.

CONCLUSION

The typewriter industry provides the opportunity to study the response of the manufacturing sector to a new product, invented after the American System was fully developed and armory practice had been in place

FIGURE 4.33. "Machine Assembly Room—800 Foot View," in the Remington Typewriter Company's factory, December 1906. SOURCE: "Producing a Wonderful Writing Machine," *Office Appliances* (December 1906), 5(1):22. COURTESY: SCM, Milwaukee Public Museum Photo. Negative no. H-627-15-K-8-12A.

FIGURE 4.34. "Section of Assembling Room." SOURCE: "The "Oliver" Growth In Facts and Figures," *Office Appliances* (January 1908), p. 37. COURTESY: Smithsonian Institution, Museum of American History, Warsaw Collection. Negative no. 83-13306.

FIGURE 4.35. The "Aligning Department" in the L. C. Smith & Bros. typewriter factory, Syracuse, N.Y., ca. 1924. SOURCE: L. C. Smith & Bros. Typewriter Company, *A Visit to the Home of the L. C. Smith Typewriter*, p. 31. COURTESY: SCM, Milwaukee Public Museum Photo. Negative no. H-627-15-G-5.

for several decades. How did the typewriter manufacturers respond in deciding how to mass produce their machines?

The typewriter manufacturers took several routes to typewriter production. First, given the great complexity of the typewriter, manufacturers designed a great deal of adjustability into the machine itself. The American Writing Machine Company advertised that its machine could be adjusted and aligned by the owner, while the Remington No. 6 machine had numerous adjustments built into its design. These machines were superb examples of nineteenth-century industrial design because they were designed not only for the ease of the operator, but also for ease of manufacturing and assembly.

Second, typewriter manufacturers adopted existing technologies which suited their purposes, for example, automatic screw machines. Every factory in this study had a screw machine department which produced a

FIGURE 4.36. A "Part of Final Adjusting Department" in the L. C. Smith & Bros. typewriter factory, Syracuse, N.Y., ca. 1924. SOURCE: L. C. Smith & Bros. Typewriter Company, *A Visit to the Home of the L. C. Smith Typewriter*, p. 32. COURTESY: SCM, Milwaukee Public Museum Photo. Negative no. H-627-15-G-3.

wide variety of small turned parts such as screws and escapement wheels.

Third, these firms adopted new materials for the manufacture of their machines, for example, the use of rubber for platens and rollers, glass for key tops, and sheet iron for cases, paper tables, and sides. The introduction of these new materials required the use of new technology such as the Remington's case forming machine and the adaptation of existing technology as in the case of Hall's type vulcanizing process.

Fourth, the typewriter manufacturers developed new technologies, such as "exercising" parts of their typewriters. The Oliver "Exerciser" and the similar machines used by Hall, American Writing Machine Company, and L. C. Smith & Bros. all performed the same function of "breaking-in" the machine or some part of it as an integral part of the manufacturing operation.

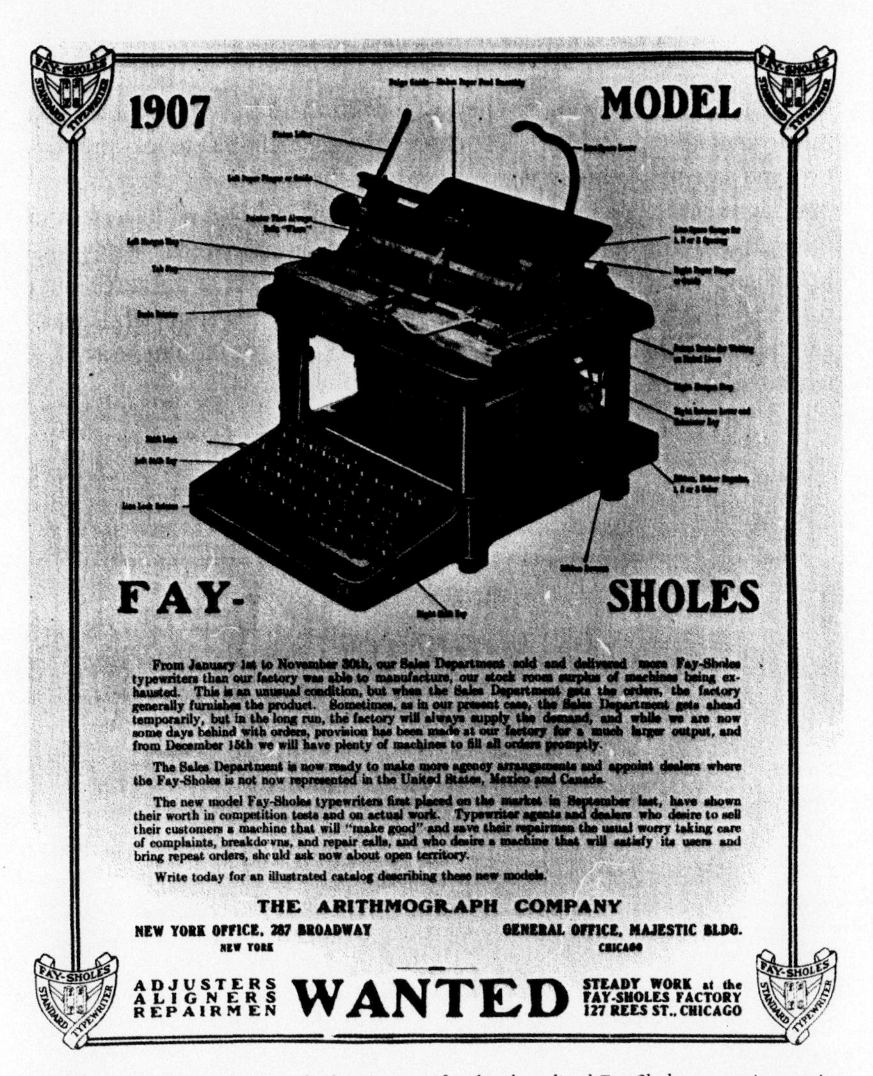

FIGURE 4.37. This trade journal advertisement for the short-lived Fay-Sholes typewriter carries a plea for "adjusters aligners and repairmen" in the lower lefthand corner. SOURCE: Office Appliances, (1907), p. 12. COURTESY: SCM, Milwaukee Public Museum photo. Negative no. H-627-15K-8-6a.

Fifth, all the manufacturers faced the problem of assembling, aligning, and adjusting a very complex mechanism consisting of "thousands of parts." All the manufacturers responded by subdividing the assembly process into the various components of the typewriter and hiring vast numbers of highly skilled people to do the complex assembly work (figure 4.37).

In sum, the typewriter manufacturers responded just like the manufacturers in other private sector American System industries. They adopted and adapted existing technology and developed new technology to produced interchangeable parts. They assembled these interchangeable parts into complete typewriters. They chose this general technique—production and assembly of interchangeable parts—because it was the most economical way to mass produce typewriters. Their techniques differed from other American System manufacturers only to the extent that their product differed from others. In philosophy, they were close to the antebellum wooden movement clock makers. In practice, they equaled the arms makers, but they trailed the watch industry which brought the American System to its height in the nineteenth century.

CHAPTER

The Development of Watch Manufacturing at the Waltham Watch Company, Waltham, Massachusetts,

1849–1910

The American system of manufacturing by interchangeable parts means much more than making a part under the roof of a factory, buying other parts in the market and obtaining other parts by the piece from work people who live in their own cottages, for which they are paid at piece work rates. It means the establishment of working facilities for the entire manufacture. That everything is made on the premises, not according to the plans or ideas or methods of work of individual workmen, but under the direct supervision of a company's foreman, according to gauges the company furnish, under conditions of time, cleanliness and care which the company prescribe.

LEONARD WALDO, "The Mechanical Art of Watchmaking in America."[1]

. . . there is probably no line of manufacturing in the world demanding such a high grade of business and mechanical ability, and such unremitting care and oversight, as well as technical skill and individual dexterity and judgment as is indispensible in systematic watchmaking.

EDWARD A. MARSH, *The Original American Watch Plant; It's [sic] Planting, Growth, Development and Fruit.*[2]

THE history of American watch manufacturing and particularly manufacturing at the Waltham Watch Company is a history of continuous technological change from 1849 to 1910. These changes occurred both in watch movement design and in production machinery design. The dominant firms, the Waltham Watch Company and the Elgin National Watch Company, pioneered most of the technical innovation, and the smaller firms continually scrambled to keep abreast.

This chapter closely examines several aspects of technical change at

Before my dissertation won the Nevins Prize, I arranged to publish an expanded version of the watch manufacturing chapter as the main historical essay in *The Time Museum Catalogue of American Watches,* which I wrote for The Time Museum in Rockford, Illinois. It reappears here with the kind permission of The Time Museum and its director, Patricia Atwood, in somewhat altered and expanded form. The sections on gauging and screw manufacturing have been summarized, while new sections on balance manufacturing, automatic transfer, and pneumatic control have been added. The economic analysis has remained unchanged.

Waltham. These particular changes, notably technical advances in production gauging and screw manufacturing, are typical of the general trend initiated at Waltham and subsequently copied by the rest of the industry, a trend from hand production to fully automatic machine tools. The chapter also scrutinizes the production and assembly of the watch balance, with particular attention to matching the escapement and adjusting the watch as a part of the assembly process.

Three forces propelled Waltham's technical progress and unparalleled success in manufacturing vast quantities of watches with interchangeable parts. Once a minimum technical capability enabled the company to produce watches successfully, about 1857, Waltham mechanics were driven for twenty years by their desire to lower prices and increase output to satisfy demand. In the late 1870s new watch factories began to compete seriously with Waltham for the domestic market, and through the end of the century they forced Waltham to lower its costs through increasingly efficient production technology. Throughout the entire period, the Waltham Watch Company employed a number of exceptional mechanics, who, inspired by successful early automatic machinery, strove to automate every aspect of production. By the turn of the century they were designing and building pneumatically controlled, self-feeding, self-acting, self-gauging, automatic machine tools to produce fully interchangeable watch parts. In addition, they had installed miniature railroad tracks in front of some benches which allowed an operative to slide her wheeled chair past a row of automatic machines, inserting wire into one end of each machine and collecting finished, fully interchangeable parts from the other end (figures 5.1 and 5.2).[3]

Waltham manufactured its watches on the "interchangeable system." Aaron L. Dennison (1812–1895), the "Father of American Watchmaking" who promoted the company and designed some of its earliest machinery, was inspired by a visit to the United States Armory at Springfield, Massachusetts.[4] However, the manufacturing technique known as armory practice, and particularly its gauging techniques, had only a minimal influence on watch production before 1860 and virtually none thereafter. The most important transfer of technology from the armories to the watch factories was the imposition of a rigid system of organization and the elevation of the machine shop to a position of supremacy.

Having adopted these two concepts, Waltham's mechanics quickly diverged from armory practice and developed new techniques to meet the peculiar needs of watch manufacturing. These mechanics adopted the general concept of armory practice—a model-based system, gauges made

ONE OF THE WORKING-ROOMS OF THE WATCH-FACTORY AT WALTHAM—INTEGRATION OF THE WATCH INDUSTRY.

FIGURE 5.1. This 1870 view of the Waltham Watch Company shows individual operatives, primarily women, working at individual machines. SOURCE: American Watch Company, "II. The Watch as a Growth of Industry," *Appleton's Journal of Literature, Science, and Art* (July 9, 1870), 9(67):29. COURTESY: Milwaukee Public Museum. Negative no. H-627-15K-25.

to fit the model, interchangeable parts, manufacturing to fit the gauges—and combined it with English watchmaking techniques and their own creative ideas to produce a new manufacturing system. This system of production, which had already been applied to such consumer goods as wooden movement clocks and axes, was sharply focused and greatly constrained by the technical requirements of the watch itself. For example, each watch required careful adjustment of its escapement and its balance, which necessitated special tools and much skilled hand labor for its assembly and adjustment. It is particularly interesting that Waltham's mechanics, like their counterparts in the typewriter industry, perfected the techniques to mass produce parts, but still relied heavily on hand assembly and adjusting (figures 5.3 and 5.4).

While the conceptual inspiration for this system originated in Dennison's visit to the Springfield Armory and much of the early technique was transferred from England and copied from American clock manufactures,

ONE OF THE "AUTOMATIC ROOMS" SHOWING ROLLER CHAIRS

FIGURE 5.2. In 1916, and perhaps as early as 1890, these banks of automatic machine tools were tended by women who sat in the "roller chairs" and propelled themselves from machine to machine. Their only tasks were to insert material, remove finished products, and check the accuracy of each machine. Compare this factory room with the drawing in figure 5.1, 1870. SOURCE: Edward A. Marsh, *"Workers Together:" A Story of Pleasant Conditions in an Exacting Industrial Establishment*, p.15. COURTESY: Milwaukee Public Museum. Negative no. H-668-A-22. MPM H 45301/26877.

Waltham Watch found it necessary to design and build its own production machinery. The Waltham machine shop became an important source of technological change throughout the nineteenth century. Factory employees originated virtually all the new production machinery and watch movement designs.

Perhaps the most significant aspect of watch design outside of production was the first appearance in American commerce of the full product line. As early as 1867 Waltham offered twenty-four different grades of watch movements. This was relatively easy in watch manufacturing, as many parts (for example, plates, wheels, and screws) were interchangeable within a particular model, and jeweling, balance type, and regulator type usually determined the grade.[5] The higher grades had more jewels, a temperature-compensated balance (sometimes called an expansion or chronometer balance), and a micrometer regulator.

ASSEMBLING ROOM WHERE WATCHES ARE "PUT TOGETHER"

FIGURE 5.3. Despite Waltham's monumental advancements in watch part production, assembling and especially adjusting remained very skilled and highly labor intensive processes. Compare this illustration of the "Assembling Room" with the illustration in figure 5.2 of an automatic production room; the former is crowded with skilled people, the later is not. SOURCE: Edward A. Marsh. *"Workers Together:" A Story of Pleasant Conditions in an Exacting Industrial Establishment,* p. 32. COURTESY: Milwaukee Public Museum. Negative no. H-668-A-9. MPM H 45301/26877.

In the process of reaching the low-priced market, watch factory entrepreneurs and mechanics taught themselves not only to manufacture watches but also to merchandise them. Waltham mechanics and entrepreneurs learned their lessons well, judging by the production of the Waltham Watch Company. Between 1849 and 1857 the company produced fewer than 5,000 watches. In 1864 they produced 38,103 watches, and in 1905 they sold 752,941 watches.[6] Between 1849 and 1910 the Waltham Watch Company manufactured some 18,000,000 watches.[7]

This achievement was primarily the result of the interaction between Waltham's entrepreneurs and mechanics, entrepreneurs who understood production needs and perceived the marketing potential of increased mechanization and lower costs, and mechanics whose ingenuity, enthusiasm, and creativity fueled technological change through the design of new automatic machinery.

FIGURE 5.4. "The Master Assembler" appeared in a 1919 Waltham trade catalogue. While Waltham was proud of its automatic production machinery, it also made much of its skilled assemblers and adjusters. SOURCE: Waltham Watch Company, *Waltham and the European Made Watch*, p. 42. COURTESY: Milwaukee Public Museum. Negative no. H-627-19-L-81. MPM H 45866/27141.

WATCHMAKING BEFORE WALTHAM

In 1849 the watch industry was dominated by the English and the Swiss. Although their watch designs had been refined and "thinned" somewhat by the 1850s, the English manufacturers were conservative and retained such features as the fusee and chain, maintaining power, and intermediate dial plate as late as the 1880s. British makers employed the classic "putting out" system to produce their watches, which Aaron L. Dennison himself described at mid-century.

> the party setting up as a manufacturer of watches bought his Lancashire movements—a conglomeration of rough materials—and gave them out to A, B, C and D to have them finished: and . . . A, B, C and D gave out the different jobs of pivoting certain wheels of the train to E, certain other parts to F, and the fusee cutting to G. Dial making, jeweling, gilding, motioning, etc, to others, down almost the entire length of the alphabet.

Dennison further noted that "promptness of delivery" and

> uniformity of products [were] utterly unattainable under the prevalent methods of what is called the factory system outside of the United States, where the work of operatives is collected from their local habitations to be made up into timepieces at a central putting together establishment.[8]

The English watchmaking industry actually consisted of separate industries, each of which produced particular parts. Ebauches were finished in central towns (e.g., London and Prescot) from parts made in the country. In addition to the putting-out system for watch production and assembly, a substantial horological tool manufacturing business flourished in England. It too was arranged like the watch industry on the putting-out system.[9]

The English industries were important sources of early watchmaking technology for the Americans. Imported English watches, often engraved with American jewelers' names, were the patterns on which the Americans designed their watches. (For example, see figure 1.3 in the Introduction for an English watch engraved "Brown & Sharpe, Providence, R. I.") The English tool and material manufacturers supplied the embryonic American factories, and English workmen migrated to America to work in American factories.

Americans were not the only people developing watch manufacturing technology in the mid-nineteenth century. The French and the Swiss, including such firms and individuals as Japy, Ingold, and (later) Roskopf also did. Ingold's story is particularly interesting, but the debate concerning his influence on American manufacturing has yet to be settled. Pierre Frédéric Ingold, born in Switzerland, established a watchmaking firm in England based on a series of English patents. His firm failed in the early 1840s and Ingold evidently attempted to start a similar firm in America between 1845 and 1846. It too failed, and afterward he apparently settled in France. Ingold's son-in-law reportedly worked for Dennison at Roxbury, presumably before 1854 when Dennison moved to Waltham. Exactly what, if any, influence Ingold had on the American watch industry is unclear. What is certain is that the Americans did mass produce watches and the rest of the world learned American techniques after 1876.[10]

In America, there was one notable effort to manufacture pocket watches prior to 1849.[11] The brothers Henry and James Flagg Pitkin produced some 400–800 watches in Hartford, Connecticut and perhaps in New York City between about 1838 and 1841 (figures 5.5a and 5.5b).[12] The concept of manufacturing watches was conceived by Henry Pitkin, who drew heavily on the ideas developed in the brass clock industry in the Connecticut River Valley. The Pitkin watch dispensed with the English fusee. In its place was a going barrel, which carried a much longer mainspring and drove the train directly.[13] Techniques borrowed from the brass clock makers included plate stamping, an embryonic gauging system of some kind, and an ingenious method for making the escapement

FIGURES 5.5A AND 5.5B. Henry and James F. Pitkin Watch, serial no. 148, ca. 1838, dial and movement views. This 3/4 plate movement features an English style ratchet tooth escapement. Note especially the steel "jewels" that are screwed into the top plate, effectively providing an adjustment for endshake. Note the American flag engraved on the top plate. COURTESY: Smithsonian Institution, Museum of American History. Catalogue no. 324,002, Negative nos. 74-132 and 74-135.

pallets. There were several design flaws in the Pitkin movement, notably the use of lantern pinions (also copied from the brass clock makers) and conical steel jewels (in place of stone jewels) which screwed into the plates.[14] The Pitkins imported English dials, hands, balance jewels, and main and hairsprings to combine with their American-made parts.[15]

The Pitkins probably were capable of solving their technical problems and becoming the first successful American watch manufacturers. The pallet-making machinery certainly illustrates the inventive capacities of the Pitkins and their mechanics, several of whom eventually found work at the Waltham Watch Company, where their experience with the Pitkins proved useful. But the Pitkins failed primarily because they attempted to perform two distinctly different functions simultaneously, entrepreneurial and technical. The Pitkins were successful manufacturing jewelers in Hartford when they attempted to become watch manufacturers. They did not nor could not anticipate the massive investment in time and capital needed to design not only a watch but also the machinery for its manufacture. They failed to comprehend that these immensely time-consuming technical necessities required total dedication, backed by a financier willing to invest until all the technical problems were solved. The Pitkins, although they could have been either silversmiths or watch manufacturers, could not have been both simultaneously.

It was several years later that Aaron L. Dennison, a New England watchmaker, independently conceived the idea of manufacturing watches on the interchangeable system after a visit to the Springfield Armory. Unlike the Pitkins, Dennison sought and found a sympathetic entrepreneur to fund his enthusiasm, Edward Howard, a clockmaker and scale maker.[16]

THE TECHNOLOGY OF WATCH PRODUCTION

> Certainly, so far as the idea of making watches by machinery in America on the interchangeable plan was concerned, Mr. Dennison was its father . . . He went so far as to make a pasteboard model of a factory, having in view that the different departments should be connected by moving bands passing along in front of the workman to transfer the work from one to another.[17]

The production lessons at the Waltham Watch Company fall into three phases: (1) 1849–1857, learning and experimenting; (2) 1858–1870, refinement, gauging, and early automatic machinery; and (3) 1871–1910, complete automation of watch production.

1849–1857: Learning

In 1849 Aaron L. Dennison joined with Edward Howard to start the company that, after a financial failure, eventually became the first to manufacture watches successfully in America. They were joined by several investors, including Samuel Curtis (a mirror manufacturer in Boston) and D. P. Davis (Howard's partner in the clock business). Their earliest efforts took place in a corner of the Howard & Davis clock factory, but the company, needing more space for its machinery, soon moved and erected a new, steam powered building across the street. In 1854 the company moved again, building its new factory on a farm in Waltham, Massachusetts, where it was periodically expanded throughout the nineteenth century.

In 1857 the original firm failed financially (not technically) and was sold at auction to Royal E. Robbins. Robbins supplied the necessary capital and leadership to carry the company through the early years of the

Civil War. Sales generated by the wartime cutoff of English imports and the desire of every Union soldier to carry a watch (preferably American) generated sufficient sales to return a profit (figure 5.6).[18]

In the early 1860s the firm found itself in a position to capitalize on the new demand for watches because it had spent twelve very difficult years learning to manufacture them. Edward Howard commented on the early years, saying;

> When I look back and bring to mind what I went through physically and mentally to start and perfect the watch business, I am astonished at the endurance and perseverance with which I stuck to the task. . . . Could I have seen beforehand the trials and tribulations I never should have made the first movement. Millions would not tempt me to go over the same ground. Mr. Dennison had given the business his whole time and energy, and he was determined . . . to make this, the effort of his life, a success.[19]

The relationship between Dennison and Howard illuminates the crucial joining of technical expertise and financial wherewithal, the mechanic and the entrepreneur. As the projector of the effort, Dennison's enthusiasm first sold Edward Howard on the idea (Howard had originally come to Dennison to discuss steam fire engine construction), then carried it nearly to success before the Panic of 1857. Edward Howard noted his relationship with Dennison. "One could not have accomplished it [the watch company] without the other. Mr. Dennison never could have made watches without me, and I never would have attempted to make watches without Mr. Dennison.[20]

Although he was the originator of the watch manufacturing idea, Dennison was not the only mechanic at the Waltham Watch Company during its learning phase. Nelson P. Stratton and John R. Proud, both of whom had worked for the Pitkins, joined the Waltham company in 1850.[21] Later in the 1850s James T. Shepard, Stratton's brother-in-law, left his job at the Springfield Armory to come to Waltham.[22] James L. Baker came to Waltham as an escapement maker and screw maker, having previously "enjoyed the distinction of having worked on the first sewing machines manufactured in America."[23]

Other important mechanics included Charles S. Moseley, who developed the split collet, Charles Vander Woerd, who later developed the first automatic screw machine at Waltham, and James H. Gerry and Belding D. Bingham, who in 1859 led the first group of mechanics to break away from Waltham to form a new watch company, the Nashua Watch Company.[24] In 1857 Ambrose Webster came to Waltham via the Springfield Tool Company after having served an apprenticeship at the Springfield

FIGURE 5.6. This particular Waltham magazine advertisement appeared in *National Geographic* in 1923. It was one of several which drew nostalgically on Waltham's connection with the Civil War. COURTESY: Milwaukee Public Museum. Negative no. H-627-19-L-1. MPM H 42454/26656.

Armory.[25] Webster's apprentice at the Springfield Tool Company, Edward A. Marsh, followed him to Waltham. Marsh later distinguished himself not only as one of the most inventive mechanics at Waltham but also as Waltham's most important historian (figure 5.7).[26]

Ambrose Webster was an especially important employee for the new watch factory owner, Royal E. Robbins. As Marsh noted,

> aside from Mr. Webster's abilities as a machinist, he possessed the valuable qualifications or ability to realize the imperative need of "system" in creating and maintaining a successful manufacturing enterprise.
>
> Here, he had his first opportunity to urge the adoption of an initial system, in the work belonging exclusively to the machine shop in which he was employed. That led to the standardization of sizes of certain "spindles" and "bushings" which were common to a variety of uses. He also endeavored to emphasize the vital dependence of the entire factory to the Machine Department, and to demonstrate the fact *that that department should not be regarded as a burden which had to be carried, but rather as the means through which the entire factory could be made productive*. (Italics added.)[27]

Webster apparently took charge of the Waltham machine shop in 1859 and through this position made his greatest contribution to the Waltham Watch Company (and to the industry that would follow it).[28] He imposed a system of manufacturing on the group of mechanics at Waltham. Webster was evidently able to structure the work within the factory and to insist on a series of standard measurements to which the individual operatives were required to conform. Webster eventually left the watch factory to join the American Watch Tool Company in 1876 where he teamed with Elihu Whitcomb to design the Webster-Whitcomb lathe, based on tools manufactured and used in the Waltham watch factory.

In addition to American watchmakers and machinists, Dennison also lured a number of Englishmen to Waltham, including a Mr. Brown, an English balance maker, and John Todd, a Scottish dial maker. Mr. Todd's efforts were evidently somewhat less than satisfactory, as his dials reportedly had a tendency to crack at the spots where the dial feet attached.[29] To learn dial making, Dennison and Howard sent John T. Gold to Liverpool for ten months at company expense to acquire the trade secrets. Nelson P. Stratton also traveled to England to learn the art of gilding and Dennison himself went at least twice to buy English material and tools (such as pinion wire, files, broaches, hands, jewels, mainsprings, and hairsprings), learn techniques, hire workmen, and establish suppliers. The Americans remained dependent on European sources for certain parts for several decades. Eventually, they perfected the manufacturing techniques themselves or American subcontractors appeared to

FIGURE 5.7. Edward A. Marsh. SOURCE: Edward A. Marsh, *The Evolution of Automatic Machinery*, p. 131. COURTESY: Smithsonian Institution, Museum of American History. Negative no. 17969.

meet the demand. J. H. Winn, for example, began manufacturing watch hands in Massachusetts in the mid-1860s.[30]

This group of early mechanics made numerous important technical advances during the learning period at Waltham, among them the spring chuck with hollow draw-in spindle, invented by Charles S. Moseley sometime before October 1854 (figure 5.8). The principle of drawing a miniature split (spring) vice into a hollow tapered spindle to tighten it was well known and commonly employed in hand tools. His hollow ta-

FIGURE 5.8. Charles S. Moseley. SOURCE: Marsh, *Automatic Machinery*, p. 51. COURTESY: Smithsonian Institution, Museum of American History. Negative on file in the Division of Mechanisms.

pered spring chuck or collet was closed by drawing it into a tapered spindle with a hollow inner, or draw-in, spindle that ran inside the tapered spindle. This allowed a long work piece, such as a piece of wire, to be fed through the lathe spindle to the chuck that held it during the machining operations (figure 5.9).[31]

Although Moseley's invention was a substantial improvement, it was not without fault. The problem with this configuration was that the spring chuck moved into and out of the tapered spindle as it closed and opened. When the work piece varied in diameter, the collet closed at different positions vis-á-vis the cutting tools. This introduced errors.

FIGURE 5.9. Moseley's general idea, to split the collet into equal thirds and pull its tapered head into a matching tapered spindle through the headstock of a lathe, remains unchanged to this day. This collet is from Charles Stark's 1902 tool catalogue, but could have come from any one of number of catalogues. It illustrates the basic features of Moseley's invention. The collet is partitioned into three sections, the base is threaded to fit into the lathe spindle, and a groove has been milled or ground into the side to fit on a projection inside the spindle. SOURCE: Stark Tool Company, *Stark Tool Company Manufacturers of Precision Bench Lathes and Fine Tools of Every Description*, p. 17. COURTESY: Milwaukee Public Museum. Negative no. H-668-A-38. MPM H 43766/26878.

Either Moseley or John Stark found the solution to this problem in 1865.[32] The idea was to keep the collet stationary and move the spindle onto or off of the collet. Thus, variations in the diameter of the work piece had no effect on its position in the machine. This idea was subsequently modified in the three-bearing, *sliding spindle lathe*.[33] In this arrangement, an outer tapered spindle slid onto and off of a stationary chuck to open and close it. This ingenious invention was quickly applied to a number of operations and soon became standard practice not only at Waltham but also throughout the watch industry and precision manufacturing generally (figure 5.10).[34]

A foot-operated, spring-activated lever improved this lathe and greatly increased its productivity by making the collet self-closing, thus freeing the operative to use both hands in manipulating the work piece and tools. The self-closing feature of the sliding spindle lathe was later adopted and adapted easily to automatic machinery by substituting a cam for the operative's foot.[35] Faced with the problem of producing large quantities of

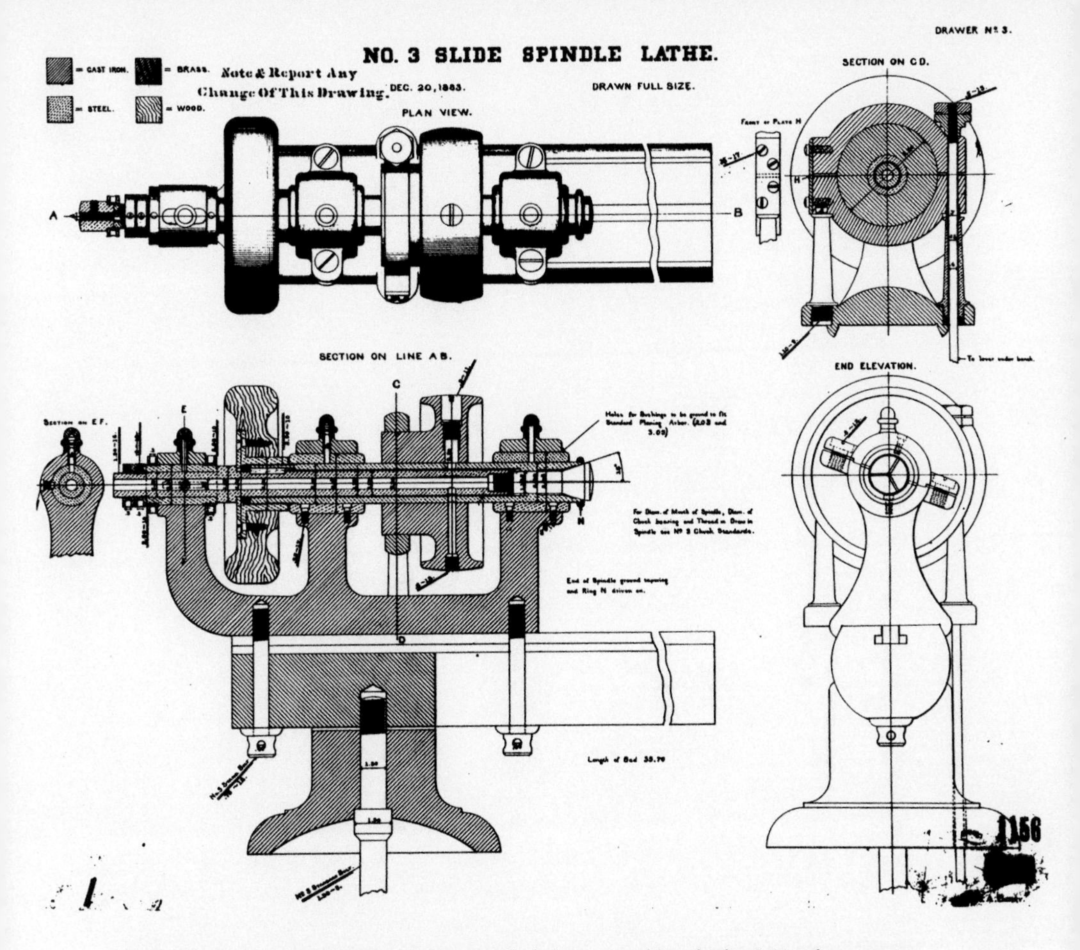

FIGURE 5.10. The three-bearing, sliding spindle lathe invented by Charles S. Moseley sometime before October 1854. This drawing is dated December 20, 1883, and shows some refinements in the thirty years since its invention, but technically it is virtually identical to the machines used in the 1850s. SOURCE: Waltham Watch Company Drawings, Smithsonian Institution Specimen. COURTESY: Smithsonian Institution, Museum of American History. Negative no. 83-12688.

precision collets and spindles, Waltham mechanics developed techniques of precision grinding.[36]

Thus, by the late 1850s and early 1860s the Waltham Watch Company had a cadre of skilled workmen, including traditional watchmakers trained both in England and the United States, as well as a number of experienced machinists with no horological training. The company had some ten years of experience designing watches and building the first generation of watch manufacturing machinery. Most important, they had learned the essential concepts making up the American System of Manufactures; interchangeable parts, produced on specialized machinery arranged in sequential operation, and gauged with specialized gauges from a model watch.

They fully comprehended the role of specialized machinery and had established the Machine Department as the "means through which the entire factory could become productive."[37] The company was still dependent on English suppliers for certain materials and parts, but it was fast becoming independent, and from that position it moved into the second stage of its technical development, refining its technology.

1858–1870: Refining

> The plan of manufacture is highly philosophical, comprehensive, complete and peculiarly American, resembling that which Eli Whitney first applied so successfully to the manufacture of fire-arms, and which has been since most thoroughly tested and demonstrated at the Springfield United States Armory; by Col. Colt, at Hartford; at Enfield in England; and which has been more lately introduced at Bridgeport, Ct., in the manufacture of sewing machines.[38]

The period between 1858 and 1870 marked a new phase in Waltham's growth and development. Financed by wartime sales and inflation, the company added new models and grades of watches, expanded its production facilities, adopted a new measuring system, developed new gauging techniques, and produced its first automatic machine tool. In doing so, through the efforts of such men as Ambrose Webster, Charles Vander Woerd, Edward A. Marsh, and later Duane Church (to mention only the most prominent), the company continued to refine its old manufacturing methods and invent new ones, thus embarking on a path that ultimately led its mechanics to design and build mechanically and pneumatically controlled, self-acting, self-feeding, self-gauging, fully automatic machine tools.

By 1910 they had completely transformed watchmaking (and through the spread of their techniques, all precision manufacturing) from an activity based on craft skills to a business based on tool building and tolerance maintaining skills. The Waltham watchmakers of 1910 were no less skilled than their predecessors of 1860, they simply had different skills.[39]

The demand for American watches created by the Civil War made the Waltham Watch Company profitable beyond its owner's wildest imagination, earning 42 percent of its capital in 1864 and 164 percent in 1865.[40] The strong demand for watches continued after the war, serving the dual function of providing profits to fund expansion of plant and equipment and exerting pressure to expand output of the lower price goods. Waltham's cheaper grades, especially its *Wm. Ellery* grade, had been popular with Union soldiers, and their reputation as well as their affordable price made them desirable and available to less affluent buyers at the lower end of the American market, most of whom had never owned watches.[41] The mechanics at Waltham, responding to the incentive of strong demand and under the able management of Royal E. Robbins, began to refine their system of manufacture and invent new techniques.

In 1862 Waltham acquired the bankrupt Nashua Watch Company and with it the defunct company's material, unfinished movements of a new 3/4 plate design, and some newly designed machinery.[42] The Nashua Watch Company attempted to manufacture the first mass produced *precision* American timekeeper. One very important feature of the Nashua 3/4 plate movement was its straight line lever escapement and club tooth escape wheel, eventually adopted by all the successful American manufacturers.[43]

Three years later, in 1865, Waltham mechanics developed their first automatic machine tool, an automatic pinion cutting engine.[44] This machine is evidence of a general trend not only toward automatic machine tools in general but also away from the traditional English techniques on which the factory had been founded.

In 1868 Waltham mechanics, probably at Ambrose Webster's insistence, abandoned the English measuring system of a 30th of an inch and adopted the centimeter as the standard unit of measure throughout the factory.[45] The Waltham Watch Company was the first American manufacturer to adopt the metric system. In connection with the standardization of measurements, the company necessarily redesigned its gauges, organized the manufacturing operation into various operating departments under the inside contract system, and established several supervisory positions above the department foremen.[46]

Gauging

> We have often heard people who were desirous of conveying a clear idea of nice workmanship say that this or that was not a hair's breadth out of the way, but what figure of speech shall we adopt when we come to the thirtieth part of a hair? Clearly, we must invent some new extravagance to represent such attenuation as this. Yet in this factory [Waltham Watch Company] we find power machines cutting metals to this fineness, and gages [sic] to determine whether the work is well done![47]

The key to interchangeable manufacture was (and is) producing large numbers of parts to specified standards of precision. These methods were first developed and adopted by Eli Terry and the wooden movement clock manufacturers in the Connecticut River Valley between 1807 and 1816 (see chapter 2). They were subsequently used by the federal armories at Harpers Ferry, Virginia and Springfield, Massachusetts, and spread throughout American industry in the nineteenth century.

In practice, the American System consisted of first producing a model clock or musket, then producing a set of master go/nogo gauges to fit the model, and finally making a set of inspector's or working gauges based on the model and the master gauges for use on the shop floor. As each clock or musket part was finished, it was accepted or rejected by testing it against the go/nogo gauge.[48] It was this general process of gauging and the specialized production machinery that Aaron L. Dennison saw on his tour of the Springfield Armory. With the Springfield Armory as a very general model, the Waltham mechanics adapted that plan to watch manufacturing.

The mechanics at Waltham understood the concept of producing a model movement and gauging from it. The first model watch was finished in the summer of 1850 by two brothers, Oliver B. and David S. Marsh (figures 5.11a and 5.11b).[49] The design of this watch proved inadequate, notably its eight-day feature. After the first seventeen were produced, it was abandoned in favor of a conventional thirty-six-hour movement, which was manufactured until about 1856.

Then they designed a new model, the Waltham *Model 1857*. It was this watch that was first manufactured successfully. It is unclear if the watches made before the *Model 1857* were made with a model watch and gauges.[50] However, beginning with the Waltham *Model 1857* and every succeeding production model, Waltham mechanics produced model watches with specified tolerances and developed special gauges to ensure that tolerances were met. In doing so, they diverged sharply from the armory practice that had so excited Dennison in the 1840s.

The inventions in gauging at Waltham were especially significant. While based on the same general armory practice concept of producing a model and gauging from it, the methods used were quite new and radically different. "Here are a different system of gages [sic] from any we have seen before," noted an observer in 1863.[51]

This divergence illustrates in general how private sector American System industries shared broad concepts with the federal armories, yet developed quite independently. The development of gauges at Waltham also illustrates how existing European watch making technology was adopted

FIGURES 5.11A AND 5.11B. This is one of two model watches designed and made by Oliver B. and David S. Marsh for Howard and Dennison between 1849 and 1851. It is an eight-day movement with dual mainsprings. Note the double American eagles engraved into the barrel bridge and the American eagle engraved on the balance cock. Like the Pitkins, Dennison and Howard showed nationalistic sentiments. They apparently considered it a matter of national pride to begin watch manufacturing in America. COURTESY: Smithsonian Institution, Museum of American History. Catalogue no. 334,625, Negative nos. 82-3763 and 74-6146.

and adapted to the peculiar needs of watch manufacturing on the "interchangeable system."

Mid-century armory practice gauging consisted of the making of a highly specialized series of hardened steel gauges, both a master set and a series of inspector's gauges. The master gauges were used to test the inspector's gauges which were actually used to test the various parts. Parts were tested in a go/nogo arrangement in which gauges had to fit into certain orifices (for example, the gun barrel) or parts had to fit into various parts of the gauge itself. Each gauge had a specific use and was used to gauge a particular part of the gun. None of these gauges was designed to produce a measurement, only to indicate if the part was within tolerances. Each gauge would tell the inspector if the part was acceptable, but not what its measurements were (figures 5.12 and 5.13).[52]

Although this process was perfectly adequate for the large, relatively crude parts of a firearm, it was totally unsuited for gauging watch parts.

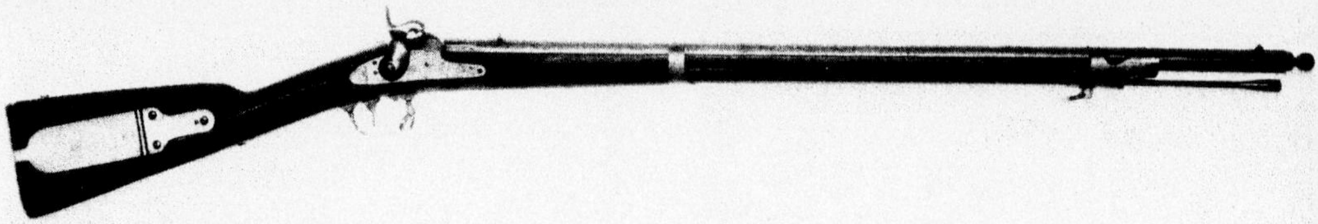

FIGURE 5.12. This U.S. Model 1841 rifle was manufactured under contract with the United States Ordnance Department by Robbins and Lawrence of Windsor, Vt. The government set the standards to which contractors were held in producing these weapons. COURTESY: Milwaukee Public Museum, Nunnemacher Arms Collection. Catalogue no. N3472, Negative no. H-513-1-J.

First, in 1863, "it [was] obvious that these standards which insure accuracy by fitting could not be used in watch-work, the pieces [were] too small to be tested in that way . . ."[53] Second, the process of producing a go/nogo gauge for each of some 150 watch parts was prohibitively expensive, as was reproducing those gauges in large quantities for factory use. The manufacture of watches required very precise measurement, speed of measurement, ease of reading the measurement, and adjustability of the gauge to take up wear.

In the process of this development and adaption, the mechanics at Waltham capitalized on European ideas practiced over the previous century. They combined European measuring tools with the basic concepts of armory practice to produce a measuring and gauging system unrivaled in the nineteenth century. Instead of laboriously constructing hundreds of go/nogo gauges, they developed a series of gauges to measure parts and set tolerances in certain sizes. In other words, they specified precise standards of measurement in a clearly defined unit of measure rather than standards falling within a measure expressed in a physical object, the go/nogo gauge. Their ability to design, make, employ, and enforce such a system represents the most important development in precision measuring since the introduction of go/nogo gauges by the wooden movement clock makers and later by the federal armories.[54]

The watchmakers and mechanics at Waltham adapted the European douzième gauge to their particular needs (figure 5.14). It is unclear exactly who was responsible for this development, but it evidently occurred as early as 1859, when mechanization had developed to the point that each machine had "its peculiar office to perform, doing its special work to a gauge or pattern, with an exactness which handicraft [could not] equal."[55] By 1863 these gauges were used throughout the factory.[56] Drawings of a "Balance Guage [sic]," dated April 10, 1866, and at two versions of a "Fine Gage [sic]," dated June 27, 1868, typify the adaptation of the douzième gauge to watch manufacturing (figures 5.15, 5.16, and 5.17).[57]

All three of these gauges were adjustable. When closed, each "Fine

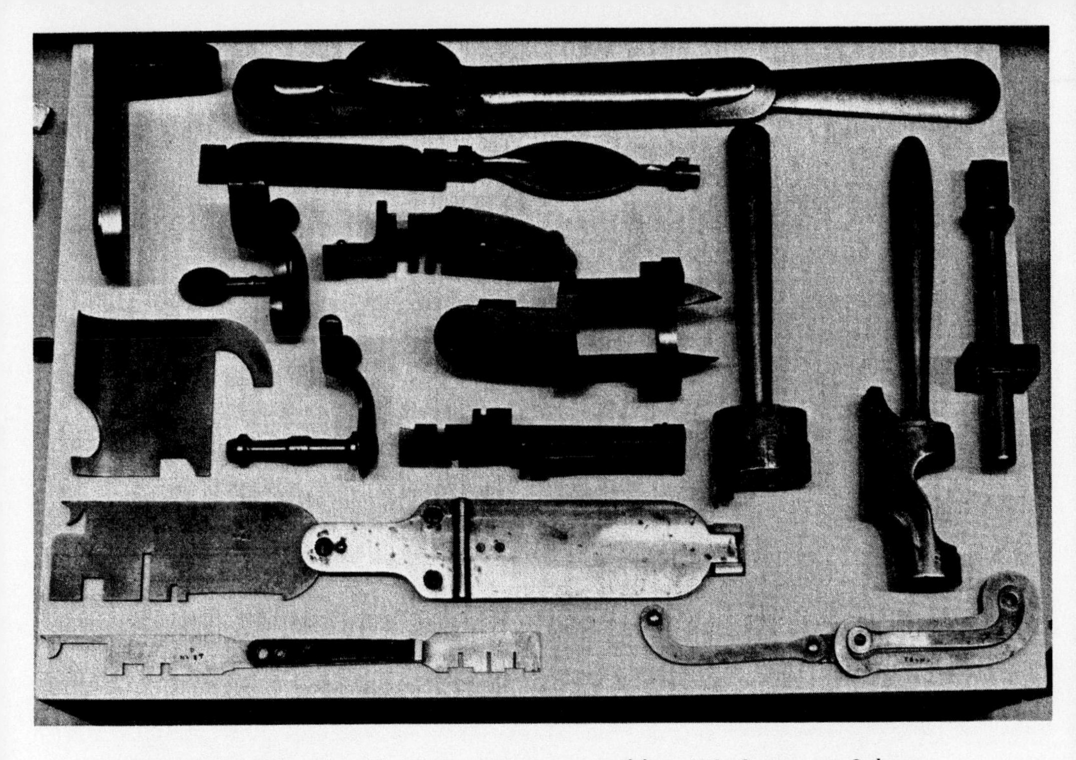

FIGURE 5.13. These cast steel go/nogo gauges were used by a U.S. Government Ordnance Department Inspecting Officer to accept or reject the work of private American arms contractors. This is a partial set of "snap gauges" for the U.S. Model 1841 rifle, such as the example manufactured under contract by Robbins & Lawrence illustrated in figure 5.12. Individual contractors used identical sets of working gauges to check their own work during the manufacturing process. The letter "D" indicates that this is gauge set D (for the 1841 rifle) and the number stamped on each number gauge corresponds to written instructions for its use in the U.S. Ordnance Manual (see note 52). COURTESY: Milwaukee Public Museum, Nunnemacher Arms Collection. Catalogue nos. N 30995-31026/26409, Negative no. H-654-30-H.

Gage [sic]" indicated 'o.' If it were out of adjustment it could be recalibrated by adjusting its anvil with a screw and then securing it with a locking screw. The "Balance Guage" [sic] could also be adjusted so that its indicating needle rested in the center of its scale. Each gauge was designed to function while secured to the surface of a work bench, thus freeing both hands of the operative to use the tool. The operative's left hand moved a knob or lever while his or her right hand placed the part between the jaws. The measurement was read on a finely graduated scale. Thus, these three gauges have all the essential features of manufacturing gauging: adjustability, rapid use, extreme accuracy, and ease of reading the results. As early as 1863 Waltham promoted its new gauging system, noting that the diameter of a "rather delicate" human hair was magnified to 1 1/4" on one particular gauge.[58]

For other gauging purposes, the Waltham watchmakers developed the

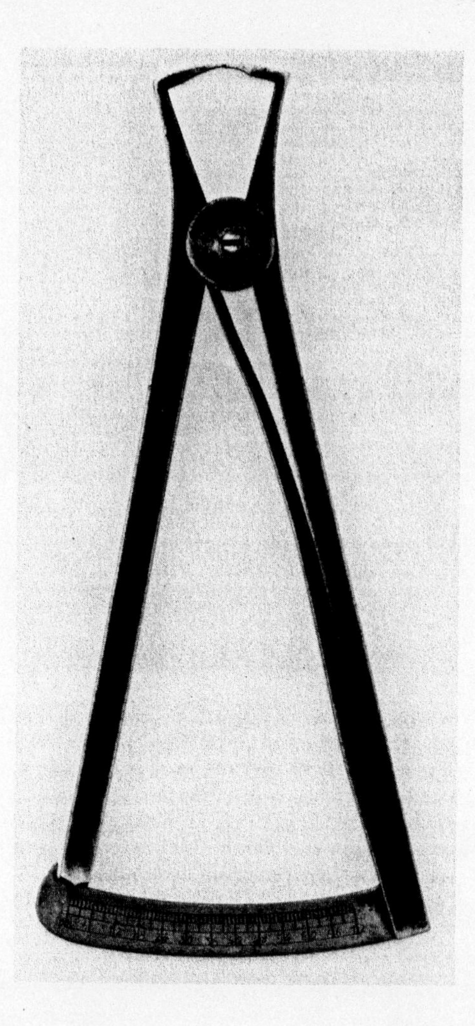

FIGURE 5.14. This is a typical nineteenth-century European-style douzième gauge. COURTESY: Milwaukee Public Museum. Negative no. H-627-19-L-2. MPM H 44405/26877.

upright gauge. This gauge seems to have no European antecedent, although certain of its features are found in European tools, notably the use of a dial for indicating the measurements. The "Upright Gage [sic]" was fully developed and used in the 1860s (figures 5.18 and 5.19).[59]

Needle gauges were used to measure the inside diameters of hole jewels. Each gauge had a long tapered needle, on which the jewel was placed. The needle "was then pushed back into a perforated holder until the jewel touche[d] a collar. . . . At the inner end of the needle is an index, which shows on a scale the distance which the needle has been pushed into the collar, and thus shows the diameter of the needle at the point where the jewel stopped. Nothing can be simpler or more easily legible, and it will indicate to the naked eye even differences in the aperture of 1-250,000ths of an inch" (figures 5.20 and 5.21).[60]

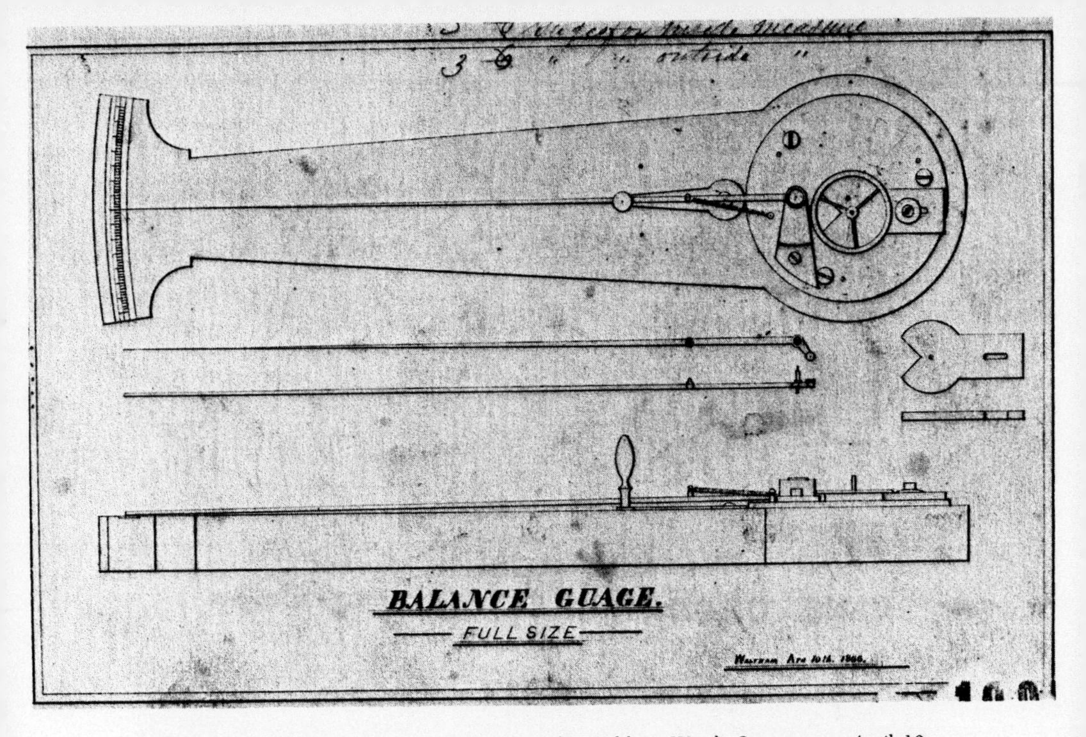

FIGURE 5.15. A "Balance Guage" [sic] drawn at the Waltham Watch Company on April 10, 1866. This gauge was used to measure the outside diameter of low grade balance wheels. This was an important measurement, since the diameter of the balance wheel directly affected the rate of the watch. Notice how the very small movement in the lever touching the balance wheel is translated into a large movement on the scale at the left end of the gauge. There is an adjustment in the part of the gauge which holds the balance wheel. SOURCE: Waltham Watch Company Drawings, Smithsonian Institution Specimen. COURTESY: Smithsonian Institution, Museum of American History. Negative no. 83-12699.

These general types of gauges were produced in three degrees of accuracy, 1/250", 1/1,250", and 1/17,000" and used throughout the Waltham watch factory.[61] They were adapted to every aspect of the watchmaking operation—for example, measuring the angles of the pallet stone slots and the teeth of escapement wheels.[62] These gauges were refined and improved incrementally over the nineteenth century, but in principle remained unchanged.[63]

Their measuring and gauging problems solved, the Waltham mechanics turned their attention to other aspects of the watch manufacturing operation, notably increased parts production. They turned to the machinery itself and began the evolution of automatic machinery.

1871–1910 Automation

No one who examines the operations of this Company can doubt that a revolution impends in the watch-manufacturing of the world; for, by the American machinery, watch movements without cases are already pro-

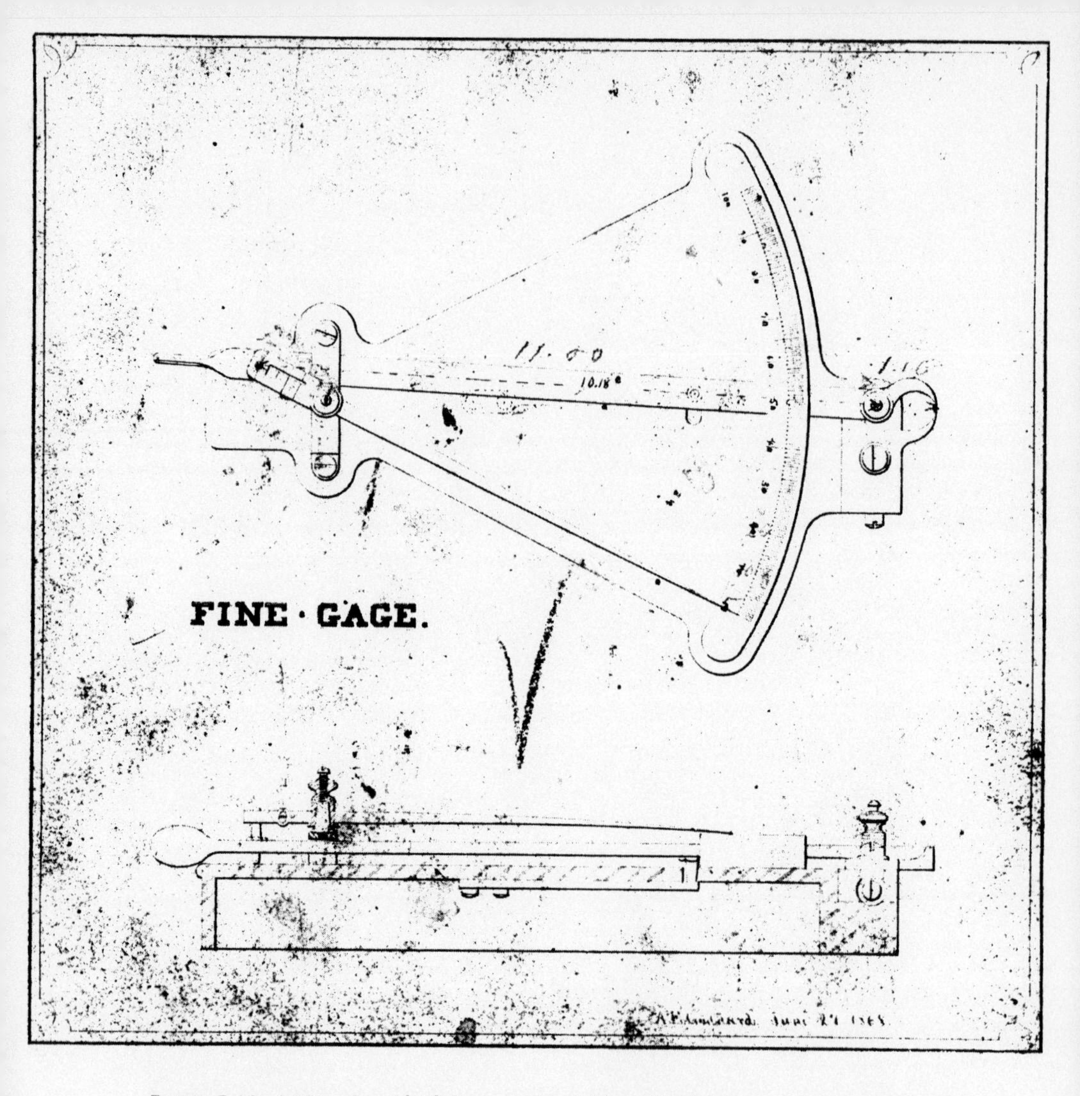

FIGURE 5.16. A "Fine Gage" [sic] drawn at the Waltham Watch Company on June 27, 1868. SOURCE: Waltham Watch Company Drawings, Smithsonian Institution Specimen. COURTESY: Smithsonian Institution, Museum of American History. Negative no. 83-12687.

duced at just about one-half the cost of imported movements of similar grade.[64]

Furthermore, automatism in tools is the coming necessity for cheapening labor.[65]

The manufacture of watches was a very complex process when it was started in the 1850s, and over the remaining decades of the nineteenth century it became increasingly complex with the introduction of more complicated watch designs and increasingly sophisticated machinery. It would be impossible to describe the process of watch manufacturing in

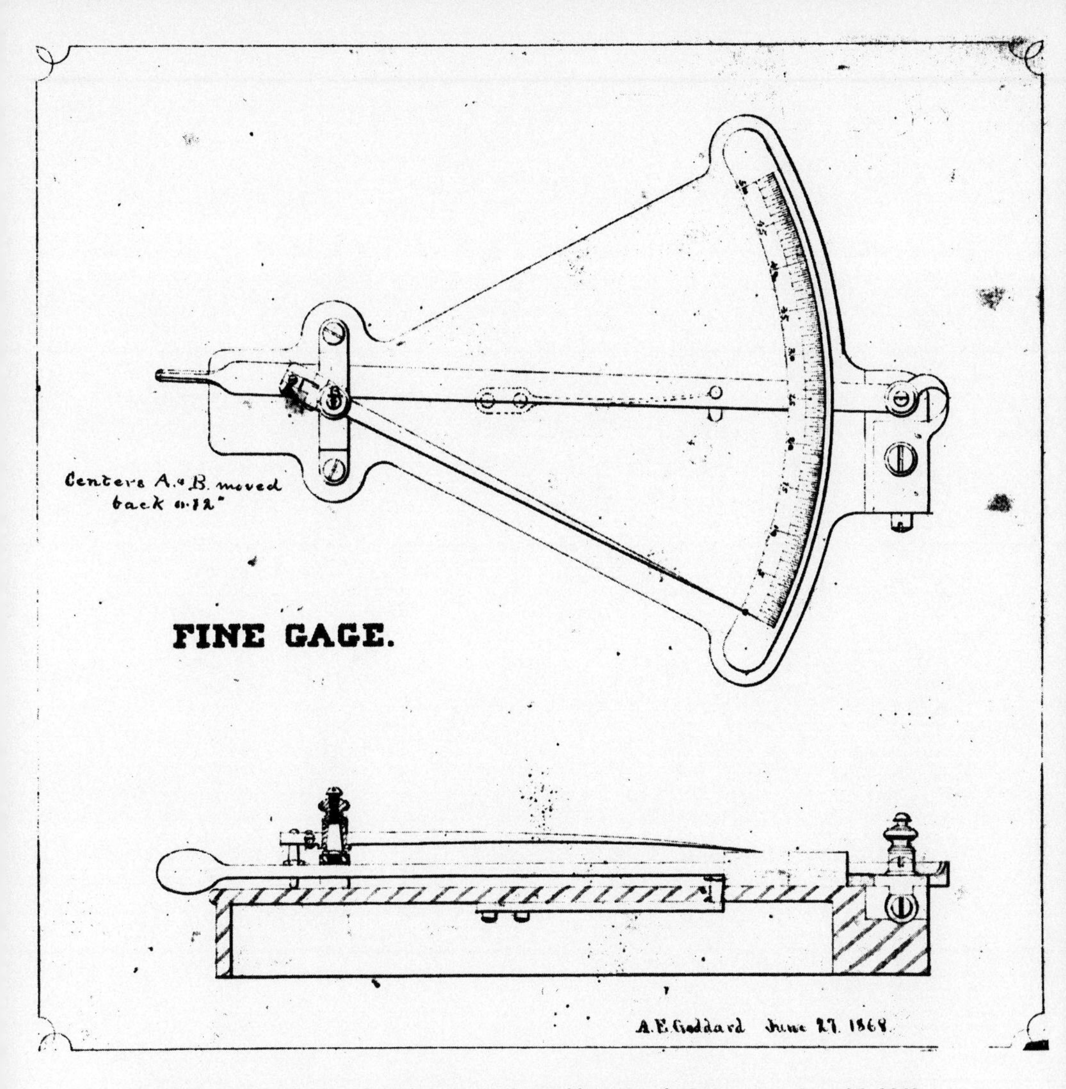

FIGURE 5.17. A "Fine Gage" [sic] drawn at the Waltham Watch Company on June 27, 1868. This drawing probably coincides with the change from the English system of measurement to the metric system in 1868. Notice how the gauge is designed for quick and easy use. The lever at the left opens and closes the jaws at the right. The needle sweeps over a large arc, indicating the distance the jaws are opened. The lower jaw is adjustable, so that the needle can easily be set to "0." SOURCE: Waltham Watch Company Drawings, Smithsonian Institution Specimen. COURTESY: Smithsonian Institution, Museum of American History. Negative no. 83-12697.

complete detail.[66] Because it is so well documented with surviving tools and manuscripts, watch screw production has been selected to illustrate how automation occurred at Waltham. The complicated process of producing the balance and escapement also illustrates the automation trend at Waltham, but also shows sophisticated techniques for selectively matching parts and adjusting the final product.

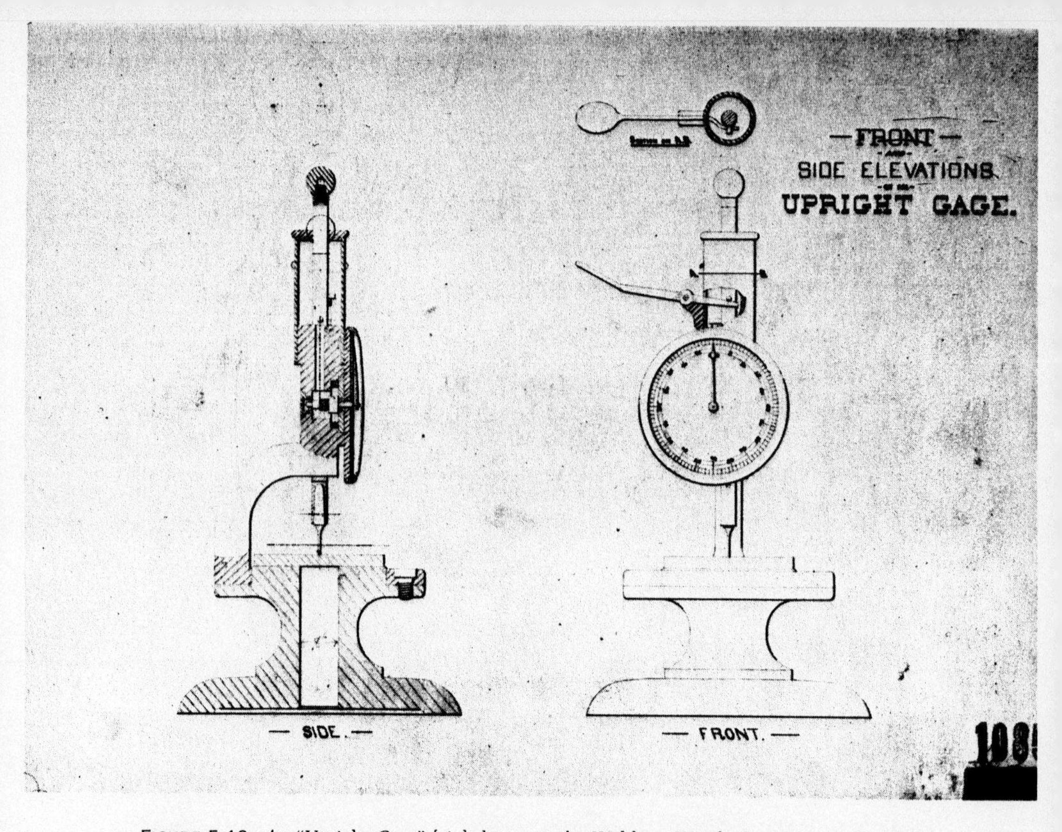

FIGURE 5.18. An "Upright Gage" [sic] drawn at the Waltham Watch Company on March 11, 186?. This drawing may also coincide with the change from the English system of measurement to the metric system in 1868. It is graduated in hundredths. Notice how the gauge is designed for quick and easy use. The lever at the left raises and lowers the spindle in the center of the table. The needle sweeps around the dial, indicating the distance the spindle is raised. The gauge is adjustable by rotating the dial or by adjusting the spindle. SOURCE: Waltham Watch Company Drawings, Smithsonian Institution Specimen. COURTESY: Smithsonian Institution, Museum of American History. Negative no. 83-12683.

Screw Manufacturing at Waltham[67]

The number of screws in a nineteenth-century pocket watch was roughly one-fourth the total number of parts in the watch. Screws were used to secure the top plate to the pillar plate, hold jewels in place, poise the balance wheel, and secure various parts such as the winding wheels, click, and escapement lever bridge. It is not surprising to find the mechanics at Waltham turning their attention to the development of automatic machinery to make screws at an early date.

The traditional technique of making watch screws prior to mechanization at the Waltham Watch Company was to cut each screw by hand with the use of a jam plate. Until the mid to late 1850s each screw was

THE SCIENTIFICALLY BUILT WATCH

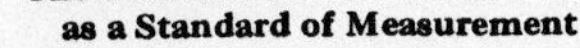

ACCURACY

Standing Gauge

Lower Plate

The Twelfth Part of a Human Hair as a Standard of Measurement

IMAGINE the twelfth part of a human hair being the difference between Waltham standardized accuracy and the variable guess work in foreign watches.

Waltham produced, by Waltham genius, methods of measurement and gauges to measure so infinitely accurate that the Waltham Watch *became* and *is* the most perfectly constructed watch in the world.

If in the lower plate (illustrated) there was a measurable difference between the location of one bearing from another, it would mean irregularity in the time-keeping performance of that watch. Such a minute error *cannot* occur in a Waltham Watch without discovery.

When you buy a Waltham you own a watch that can be readily, and what is most important to you, perfectly and economically repaired—at an upkeep cost at least 50 per cent lower than the repair of foreign made watches.

This is one more good reason why you should own a Waltham.

This story is continued in a beautiful booklet in which you will find a liberal watch education. Sent free upon request to the Waltham Watch Company, Waltham, Mass.

Pendant and Bow Patented

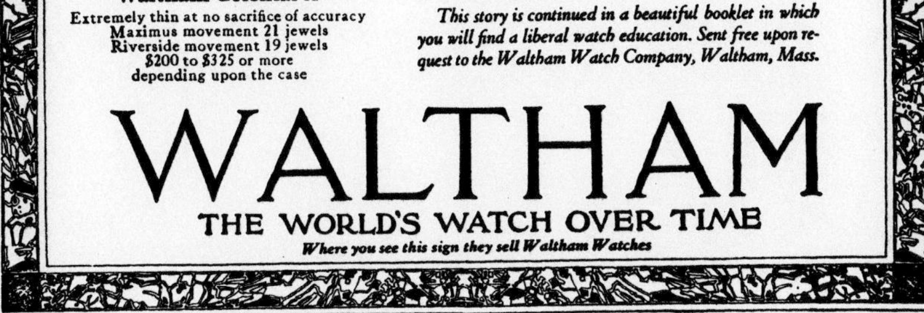

FIGURE 5.19. Aaron L. Dennison, the "Father of American Watchmaking and the Waltham System of Standardization," was featured in this April 1921 Waltham Watch Company advertisement from the *National Geographic*. Note the illustration of the "Standing Gauge," invented and used at Waltham nearly sixty years before. In November 1921 the upright gauge was again featured in a different Waltham advertisement; see MPM H 33550/25215. COURTESY: Milwaukee Public Museum. Negative no. H-668-A-49. MPM H 44999.4/27054.

FIGURE 5.20. This needle gauge almost certainly came from the Elgin National Watch Company, ca. 1920–1940 (perhaps much earlier). COURTESY: Milwaukee Public Museum. Negative no. H-627-19-L-3. MPM H 41377/26313.

forced into a threaded hole or die cut in a steel plate (figure 5.22).[68] Waltham's first screw cutting machinery was in use perhaps as early as 1855 and almost certainly by 1858.[69] It consisted of two machines, one to cut the screw threads and the second to slot the screws (figures 5.23, 5.24).[70]

In 1871 Charles Vander Woerd developed the first automatic screw machine (figure 5.25).[71] Its design remained in use at Waltham through the 1950s and at the Waltham Screw Company of Keene, New Hampshire, until 1981.[72] Vander Woerd's machine is based on the conventional three-bearing, self-closing, sliding spindle lathe and double slide rest (as had the hand operation), but it is activated by a series of cams mounted on a camshaft running parallel to the lathe spindle.[73] The Vander Woerd machines remained in production until the Waltham factory closed in 1954 (figures 5.26, 5.27, and 5.28).[74]

Edward A. Marsh invented an automatic screw machine that supplemented the Vander Woerd machine and patented it on October 27, 1885.[75] By 1886 Marsh's machine was in production. In design, Marsh's machine departed radically from the Vander Woerd machine in which a single work piece was acted on successively by different cutting tools brought to a single place. Marsh's screw machine employed the same tools working simultaneously on numerous workpieces (figure 5.29).[76] Duane Church subsequently produced an improved version of the Vander Woerd screw machine (figure 5.30).[77]

The evolution of screw making technology from jam plates to tumble tailstock lathe to Vander Woerd screw machine to Marsh screw machine to the Church screw machine over a thirty-five year period is typical of the continual evolution of machinery at Waltham. Between 1849 and the death of Duane Church in 1905, there was never a period of technological stagnation at Waltham.[78]

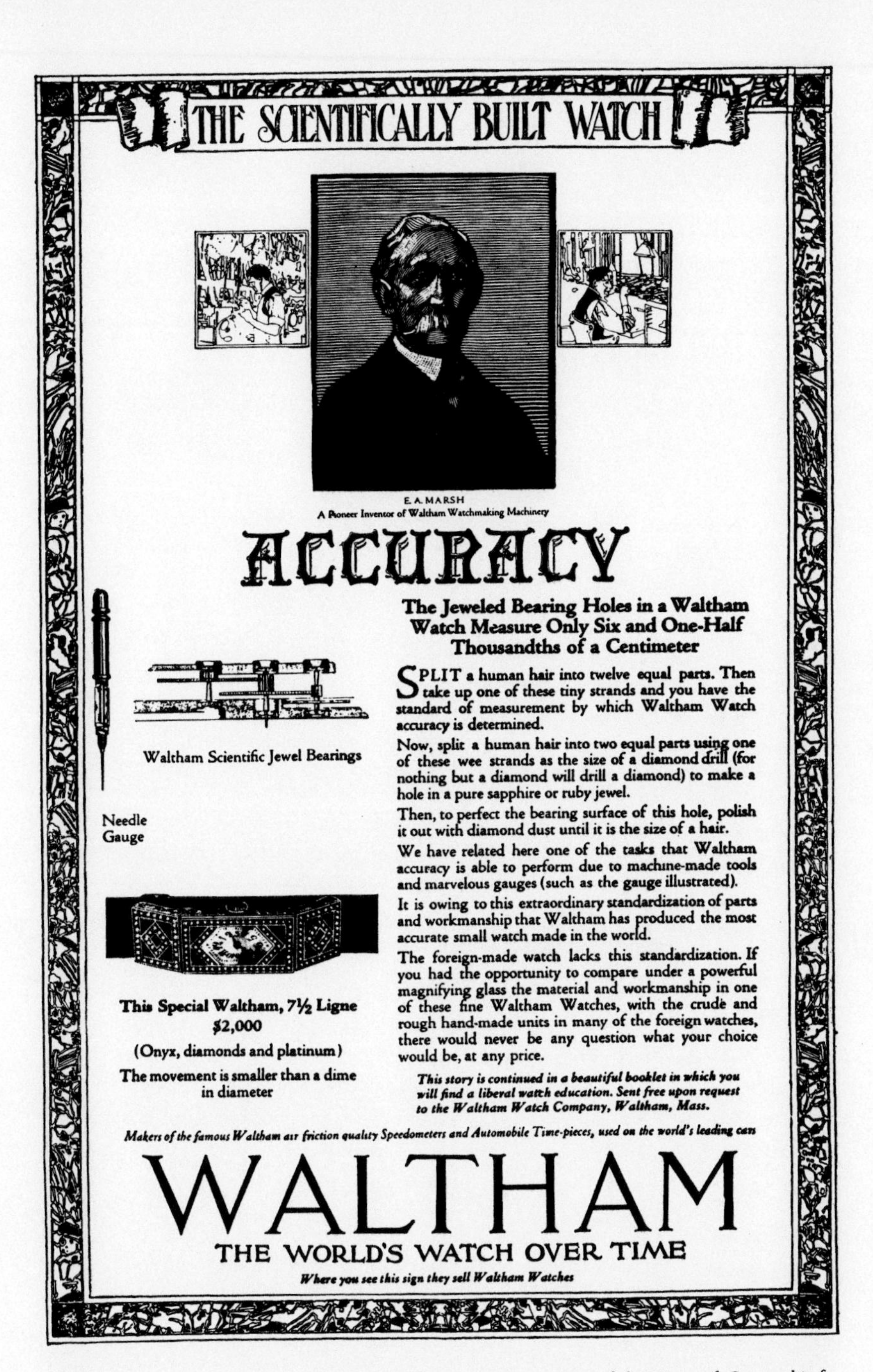

FIGURE 5.21. This magazine advertisement appeared on page 35 of the *National Geographic* for July 1921. Note the needle gauge illustrated at the left, as well as the portrait of Edward A. Marsh, who was still living at that time. COURTESY: Milwaukee Public Museum. Negative no. H-668-A-31. MPM H 33549/25215.

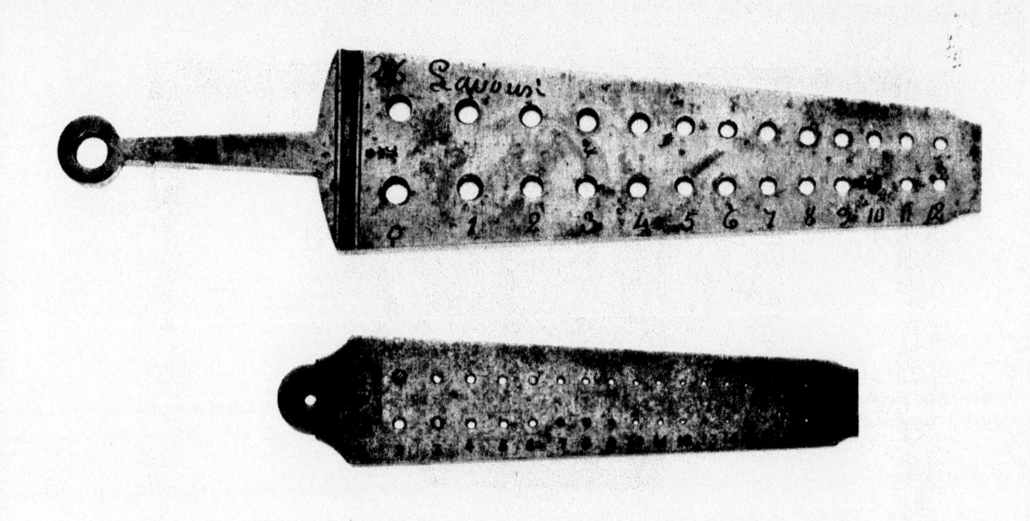

FIGURE 5.22. Two screw cutting jam plates. To produce a screw, the workman physically turned the wire into the threaded hole. COURTESY: Milwaukee Public Museum. Negative no. H-627-19-L-4. MPM H 44406/26877 and H 44397/26877.

Adjustability: The Watch Escapement and Balance

As was the case with typewriters, watches required adjusting as an integral part of their manufacture. This is particularly true with the escapement and balance. A discussion of this mechanism, its manufacture, its assembly, and its adjustment will illuminate the extent to which all American System manufacturers and watch manufacturers in particular were driven by the nature of their products to adopt particular technologies.

A nineteenth-century American pocket watch consisted of two round plates, between which ran a series of brass wheels and steel pinions driven by a steel mainspring. This train of gears was stopped five times each second by an escapement that transmitted the power of the mainspring to an oscillating balance wheel that stored this energy in a steel hairspring.[79]

The form of escapement generally used by Waltham and most American manufacturers was the detached straight line lever escapement with a club-tooth escape wheel and an escape lever with jeweled pallets (figure 5.31). Similarly, most manufacturers used a temperature compensating (cut), bimetallic balance with timing and poising screws in its rim and a blued steel (hardened and tempered) hairspring.

The escapement has two distinct actions, that between the escape wheel and the escape lever pallets and that between the escape lever fork and

FIGURE 5.23. Screw cutting lathe, Waltham Watch Company, ca. 1866–1870. This lathe is fitted with a foot-operated, self-closing, three-bearing, sliding spindle headstock, a double compound tool post with gimballed universal joint handle, and a three spindle tumble tail stock. It was used at Waltham as early as 1859 and is essentially identical to the lathe described by Marsh in *Automatic Machinery* and illustrated in Waltham drawings. It was superseded by Vander Woerd's automatic screw machine in 1871, although it was probably kept in service for some time after that. COURTESY: Smithsonian Institution, Museum of American History. Catalogue no. 336,478, Negative no. 77-14090.

the roller jewel on the balance wheel staff. "The function of the escapement is to impart to the balance, regularly, . . . the power which has been transmitted through the train from the mainspring to the escape [wheel]."[80] The balance wheel, which is controlled by a hairspring, receives an impulse from the escape lever that spins it. Most of this energy is stored in the hairspring, which arrests the rotation of the balance and subsequently begins to spin the balance wheel in the opposite direction. As the balance wheel passes the escape lever, its roller jewel unlocks the escapement and it receives an impulse in the opposite direction. This process occurs five times each second, 60 seconds per minute, 60 minutes

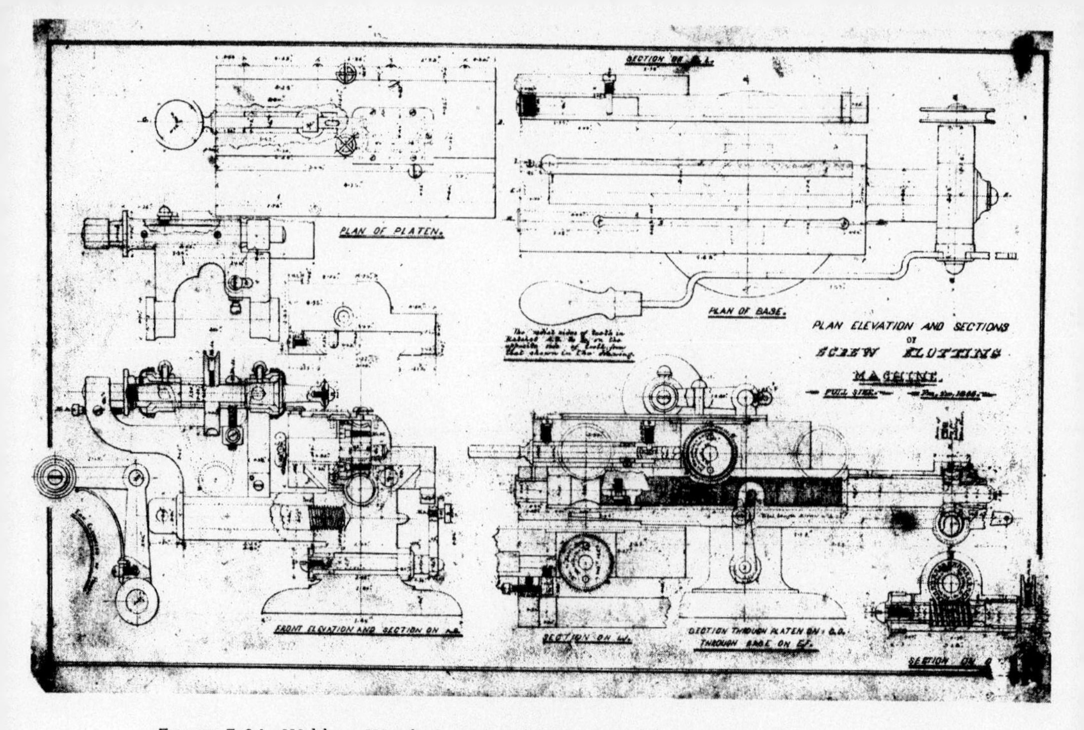

FIGURE 5.24. Waltham Watch Company drawings of a "Screw Slotting Machine," dated February 2, 1866. This machine completed the production of screws (except for polishing) which had been threaded on the screw cutting lathe. SOURCE: Waltham Watch Company Drawings, Smithsonian Institution Specimen. COURTESY: Smithsonian Institution, Museum of American History.

FIGURE 5.25. Charles Vander Woerd. SOURCE: Marsh, *Automatic Machinery*, p. 69. COURTESY: Smithsonian Institution, Museum of American History.

FIGURE 5.26. This is apparently the 1875 version of Charles Vander Woerd's automatic screw machine, invented in 1871. This machine, designed for larger screws, is nickel plated and was apparently exhibited at the American Centennial at Philadelphia in 1876. COURTESY: Smithsonian Institution, Museum of American History. Catalogue no. 316,564, negative no. P65727.

per hour, 24 hours per day, 365 1/4 days per year— or 157,788,000 times per year.

The entire mechanism consisted of the escape wheel and its arbor, the escape lever and its arbor, two pallet stones, a guard pin, the roller (and sometimes a double roller), roller jewel, balance staff, hairspring, hairspring collet, hairspring stud, hair spring stud screw, hairspring regulator, balance wheel, and up to twenty-two timing and poising screws, the cocks and screws to hold the escapement, and the end stones and hole jewels and jewel screws—as many as fifty-six individual parts depending on the model. Clearly, this is a complex precision mechanism (figure 5.32).[81]

Most manufacturers used brass escape wheels, although some higher grade watches had steel escape wheels. "The proper cutting of the teeth of escape wheels is certainly a matter of great importance and of no little

FIGURE 5.27. Screw Department at the Waltham Watch Company, 1884. The machines seen in operation are Vander Woerd automatic screw machines. Note the upright gauge on the bench at the lower right of the photograph. SOURCE: H. C. Hovey, "The American Watch Works," *Scientific American* (August 16, 1884), N.S., 51(7):102. COURTESY: Milwaukee Public Museum. Negative no. H-627-15-K-24.

difficulty; the peculiar form of the teeth demanding the utmost accuracy in workmanship, and requiring a succession of cuts by as many different shaped cutters."[82] These wheels were first stamped from strips of sheet brass, the center hole and the spokes being punched out simultaneously (figure 5.33). These blanks were then stacked on an arbor threaded at both ends and held tightly with two nuts. The arbor was placed in a special escape wheel cutting machine, which apparently featured in its earliest version a series of cutter spindles mounted "in a single rotatable block or head."[83]

Waltham's escape wheel cutting machines, like its screw cutting machines, underwent an evolution between the late 1850s and the 1880s. The earliest identifiable version featured four cutters in its "rotatable block" which was moved by hand using a long lever (figure 5.34). Ambrose Webster improved upon this machine with a semi-automatic version, probably in the mid 1860s. It was "automatic to the extent of moving the carriage and operating the index, and also stopping itself on the com-

FIGURE 5.28. "The Smallest Screw in the World." As late as 1919 Waltham featured Vander Woerd screw machines in its advertising. SOURCE: Waltham Watch Company, *Waltham and the European Made Watch*, p. 28. COURTESY: Milwaukee Public Museum. Negative no. H-668-A-16. MPM H 45866/27141.

pletion of the work of each of the six cutters required."[84] Following the completion of each cutter's work, the operative manually brought the next cutter into place and started the machine (figure 5.35).

Charles Vander Woerd invented the third generation escape wheel cutter about 1870.[85] It carried thirty wheel blanks and used six sapphire cutters. It was "automatic to about the same extent, and in the same features as the Webster machine, but omitted one motion which the Webster machine obtained, viz., the lifting of the cutter to avoid contact with the work during its return movement" (figures 5.36 and 5.37).[86]

In 1883–84, Edward A. Marsh designed Waltham's fourth generation automatic escape wheel cutting machine. It carried fifty wheel blanks, automatically moved the carriage into the cutters, lifted the cutters during the return pass of the carriage, indexed the carriage, successively changed the six cutters (three steel and three sapphire), and stopped itself after the ninety cuts required to cut fifteen escape wheel teeth (figures 5.38 and 5.39).[87]

The use of sapphire cutters eliminated the need to polish the faces of each individual club tooth, which would otherwise have been necessary since this was a point of critical wear.[88] The completed escape wheel was then press fitted onto its arbor and pinion.

FIGURE 5.29. Edward A. Marsh patented this automatic screw-making machine on October 27, 1885, U.S. Patent No. 329,182. This screw machine was featured in Waltham advertising as late as 1907. SOURCE: Marsh, *Automatic Machinery*, p. 106, figure 42. COURTESY: Smithsonian Institution, Museum of American History. Negative on file in the Division of Mechanisms.

The escape levers were stamped from flat steel with precision punches and dies. They were probably tumble polished, being too small to be polished by hand.[89] The slots holding the jewels were then ground out with diamond dust ready to receive the pallet jewels or stones as they were also called, which were held in place with shellac. Waltham originally purchased its jewels in England, but soon began to manufacture its own. They were usually made from natural garnet, although sapphires, rubies, and diamonds were also used in the higher grades.[90]

"Matching the escapement" was a particularly skilled job at any watch factory, because the performance of the watch depended greatly on the relationship between the escape wheel and the escape lever (also called a fork) with its jewels. Matching the escapement consisted of cementing the pallet jewels into the pallet fork using heated shellac. Special hand tools for holding and heating the fork and the jewels were used to heat the parts in an alcohol lamp. Excess shellac was scraped off using a brass tool. Despite the production of wheels and forks and jewels to precise standards, each had to be matched independently during the process of

FIGURE 5.30. Duane Church invented this machine, an "Automatic Screw-Making Machine," which was an improvement on Vander Woerd's screw machine. This machine probably dates from 1893, as Church received U.S. patent no. 512,156, on January 2, 1894, for a screw making machine. It was said to be simpler and stronger, thus capable of doing work that the "older" machine could not do. SOURCE: Marsh, *Automatic Machinery,* p. 104, figure 41. COURTESY: The Time Museum, Rockford, Ill.

assembly. The pallet fork and its jewels were designed to be adjusted as an integral part of that assembly.[91]

The balance wheel started as a steel disk, drilled and turned to size, onto which a ring of brass was melted. The brass encased steel disk was again turned to the proper thickness and "rolled to condense the brass." The back of the disk was then ground smooth and the interior of the front recessed, leaving a steel and brass rim. The thin steel surface was then punched out to create the arms of the balance. The rim was then drilled, tapped, and cut, ready to receive the poising and timing screws, which are chosen to conform to the strength of the hairspring (figure 5.40).[92]

Waltham mechanics developed a series of automatic and semi-automatic machines to produce the balance wheel. These included machines

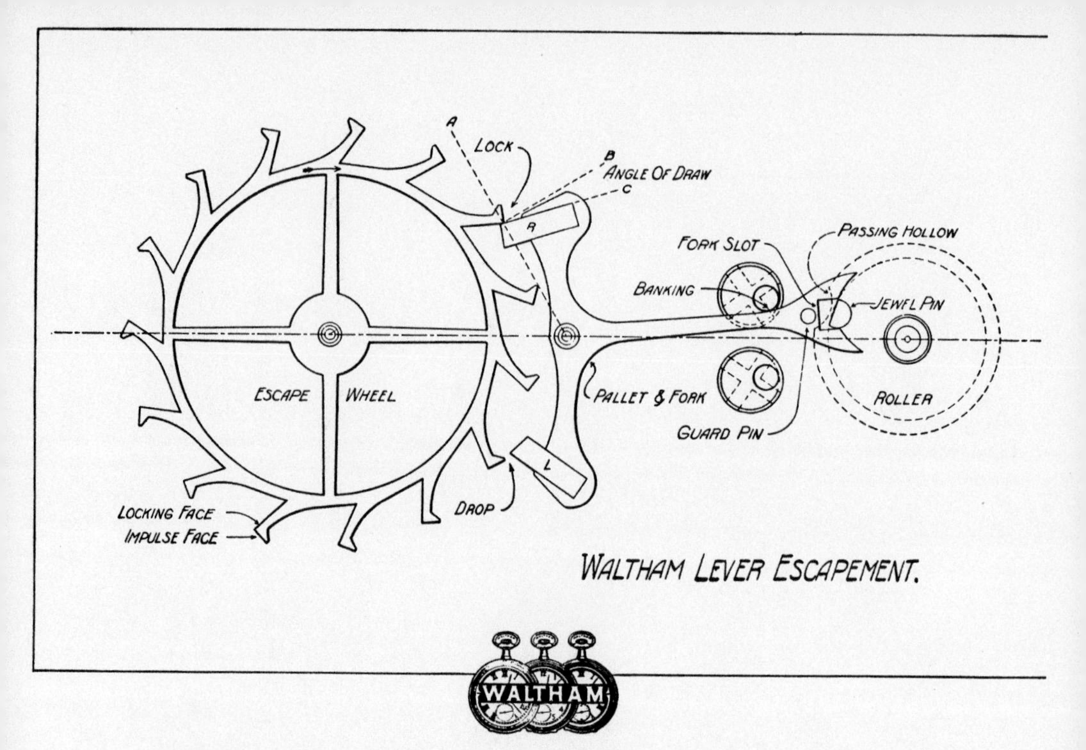

FIGURE 5.31. This is Waltham's late nineteenth-century version of the straight line lever escapement. SOURCE: Olof Ohlson. Helpful Information for Watchmakers, p. 23. COURTESY: Milwaukee Public Museum. Negative no. H-688-A-15. MPM H 45890/27141.

to compress the brass, drill and tap the screw holes, and polish the rim (figures 5.41, 5.42, 5.43, 5.44, and 5.45).

Originally Waltham purchased its hairsprings, apparently from English suppliers. However, as with other parts, they soon began manufacturing their own. This was an evolutionary process, as the company first purchased wire, then about 1881 began producing its own wire.[93] By the mid 1890s the company had a series of automatic machines in place requiring only a single operator (figure 5.46).

The hairsprings were made from steel wire, which was drawn down from .022 cm to .018 cm in ruby and sapphire dies. It was then rolled between hardened steel rollers and drawn through diamond dies to a width of .027 cm and a thickness of .908 mm. It was then cut into fourteen-inch lengths, wound, hardened and tempered, cleaned with acid, and blued. The winding or "coiling" process remained a skilled hand procedure and resisted automation.[94] The inner terminal of the spring was attached to a brass collet, and its outer end was attached to a hairspring stud.

Each hairspring was then gauged to determine its strength. Since the quality of the steel wire varied, the hairsprings—despite their identical

FIGURE 5.32. This watch is Waltham's *Model KW-16,* also known as the *Nashua Model.* It was made sometime between 1864 and 1872 in the so called "Nashua Department." It is Waltham's highest grade, *American Watch Co.,* serial no. 125,489. This 18-size, 3/4 plate, gilt, key wind back, key set back movement has Geneva stopwork and maintaining power. It features seventeen jewels in settings, with an exquisitely engraved balance cock and escapement lever bridge. The escapement and balance feature a straight line lever with brass club-tooth escape wheel and a cut, bimetallic, compensation balance with gold screws regulated by a free sprung, blued steel, Breguet hairspring set in a vibrating hairspring stud pivoted and jewelled between the balance cock and pillar plate. The movement is inscribed "American Watch Co. Waltham, Mass. Stratton's Patent No. 125,489"—on top plate, "Fogg's Pat. Feb. 2, 1864"—on balance cock. COURTESY: The Time Museum, Rockford, Ill. Inventory no. 2006.

FIGURE 5.33. "Punching Blank Wheels." SOURCE: American Watch Company, "II. The Watch as a Growth of Industry," *Appleton's Journal of Literature, Science, and Art*, (July 9, 1870), 9(67):33. COURTESY: Milwaukee Public Museum. Negative no. H-627-15-K-43.

FIGURE 5.34. The "Old Escape Wheel Cutter," which Edward A. Marsh believed was used at an early date. This is perhaps Waltham's second generation escape wheel cutter, as it seems a bit too sophisticated for the 1850s. SOURCE: Marsh, *Automatic Machinery*, p. 80, figure 28. COURTESY: The Time Museum, Rockford, Ill.

FIGURE 5.35. This is Ambrose Webster's semi-automatic "Escape Wheel Tooth Cutter," which Webster claimed to have been "the first wheel tooth-cutting machine with automatic motions which was ever used in American watchmaking." It moved the carriage, operated the index, and stopped itself after a cut. SOURCE: Marsh, *Automatic Machinery*, p. 81, figure 29. COURTESY: The Time Museum, Rockford, Ill.

size—varied in strength. The strength of the hairspring and the mass of the balance had to be perfectly matched in order to obtain the proper rate for each watch. Originally this was accomplished by the "cut and try" method, in which springs and balances were chosen at random until a pair matched correctly. This was necessarily time consuming and expensive, particularly for the higher grade Breguet hairsprings, while the flat hairsprings could be easily lengthened or shortened.

John Logan, who was originally a watch springer, rose to become an inside contracting department head at Waltham, evidently on the strength of his inventions in both hairspring and mainspring production. "He invented a system of testing all hairsprings by a standard balance, and all balances by a standard spring, and grading the springs according to their relative strength, and, by means of a long studied and carefully prepared schedule, or table, selecting the springs adapted to the various balances" (figures 5.47 and 5.48).[95]

FIGURE 5.36. Charles Vander Woerd's semi-automatic escape wheel cutting machine, ca. 1869. SOURCE: Marsh, *Automatic Machinery*, p. 82, figure 30. COURTESY: The Time Museum, Rockford, Ill.

FIGURE 5.37. "Escapement-Wheel Machine." This machine was designed by Charles Vander Woerd sometime prior to 1870. Compare it with the photograph from Marsh, *Automatic Machinery*, in figure 5.36. SOURCE: American Watch Company, "II. The Watch as a Growth of Industry," *Appleton's Journal of Literature, Science, and Art*, (July 9, 1870), 9(67):33. COURTESY: Milwaukee Public Museum, Negative no. H-627-15-K-32.

FIGURE 5.38. The "Marsh Escape Wheel Tooth Cutter" was designed about 1883. It was a fully automatic machine, requiring only that the operative load the wheel blanks and start it. SOURCE: Marsh, *Automatic Machinery*, p. 83, figure 31. COURTESY: The Time Museum, Rockford, Ill.

Thomas Gill, the inside contractor who ran the hairspring department in 1884, also invented gauges to measure hairsprings and balances. Gill's "scale for weighing balances" measured the "avoirdupois" of the complete balance (sans hairspring). He also invented a "gauge for determining the relative strengths of hairsprings."[96] The average weight of a balance was eight grains (figure 5.49).[97]

Gill's hairspring strength gauge consisted of a dial plate about seven inches in diameter graduated into 2,000 divisions, each .01 inches wide, in the center of which was a staff connected to a spring of known strength. Each hairspring was mounted on the center staff with its collet while its stud was attached to an arm extending from an outer ring revolving about the stationary dial. The ring was revolved once in each direction and its strength was determined by a hand attached to the center staff. The average strength of a hairspring on Gill's gauge was 1,000′, while the vari-

FIGURE 5.39. "The American Watch Co.—Making Escapements." This cut clearly illustrates two Marsh automatic escape wheel cutting machines. Compare these machines with figure 5.38. SOURCE: H. C. Hovey, "The American Watch Works," *Scientific American* (August 16, 1884), N.S., 51(7):103. COURTESY: Milwaukee Public Museum, Negative no. H-627-15-K-39.

ation between the directions was usually from 5° to 10° (figures 5.50, 5.51, and 5.52).[98]

The variation of .01 grains in the balance and 1° on Gill's guage were each equal to four seconds per hour. Thus by balancing the weight of the balance and the strength of the hairspring, Waltham watchmakers could match a balance wheel and hairspring accurately. Even after this matching procedure, each balance wheel and hairspring had to be properly "vibrated," the process of testing the balance wheel with its hairspring and making the necessary adjustments to the timing screws. As the company's advertising noted, "each hairspring belongs to its own balance and each balance to its own watch, but they do not meet each other till the watch is done" (figures 5.53 and 5.54).[99]

Once all the parts of the watch came together, it was finished and adjusted. Like typewriter manufacturers, watch manufacturers never automated the assembly and adjusting phases of watch production. Depending on the model and grade of watch, adjusting could take up to five months. In the higher grades, adjusting consisted of running the watch

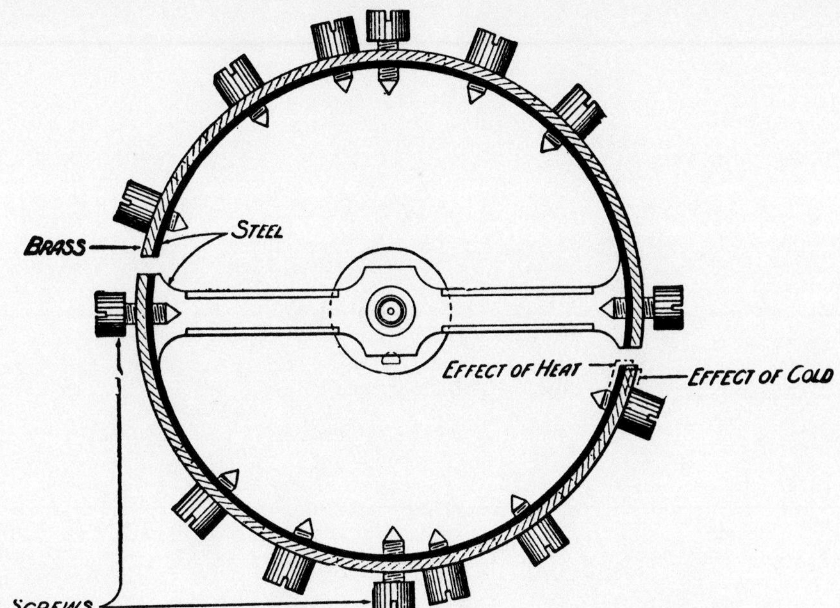

FIGURE 5.40. A typical Waltham Balance Wheel. SOURCE: Olof Ohlson. *Helpful Information for Watchmakers*, p. 6. COURTESY: Milwaukee Public Museum, Negative no. H-627-19-L-84. MPM H 45884/27141.

in five positions—pendant up, pendant down, pendant side, dial up, dial down—in temperatures ranging from 95° to 38°, and adjusting for isochronism.

The adjusters could make as many as eleven adjustments in the process of manufacture. These include:

1. placement of the two pallet stones in the escape lever,
2. adjusting the two escape lever banking pins,
3. placement of the roller jewel in the roller table,
4. placement of the roller table on the balance staff,
5. placement of the hairspring collet on the balance staff,
6. adjusting the effective length of the hairspring with the regulator lever,[100]
7. adjusting the placement of the balance jewels (hole jewels and end stones),
8. adjusting the timing and poising screws (may include adding timing washers or undercutting screws),
9. bending the guard pin on the escape lever,
10. repositioning the hairspring in its stud, and
11. raising & lowering the hairspring stud.

Lower grade watches received no such adjusting.[101] Indeed, some watches were shipped with labels instructing the retailer to do the ad-

FIGURE 5.41. This "Early Balance Rim Drilling Machine," like much of Waltham's early machinery, is based on the simple lathe with a specially designed headstock and its work holding devices in combination with the tumble tail stock. Compare the tail stock of this machine with that of the screw machine in figure 5.23. SOURCE: Marsh, *Automatic Machinery*, p. 123, figure 53. COURTESY: The Time Museum, Rockford, Ill.

justing (figure 5.55). Retail watch repairmen could replace damaged hairsprings (indeed virtually any part) with original factory material. This was available in various assortments, in the case of hairsprings Waltham offered forty different hairsprings by watch size and screw count in its April 1909 material catalogue.[102] The catalogue also listed sixty-two different hairsprings for Waltham's various models beginning with the *Model 1877* and running through its *Models 1900, 1907,* and *Jewel*. Within these sixty-two different springs, there were one to five grade variations. In its hairsprings alone (not counting the models made during the twenty years before the *Model 1877*), Waltham had 162 hairspring variations available in its material catalogue. Even with this diversification, Waltham reminded potential watch repairmen that "close timing of hairsprings makes it necessary that the movement or the complete balance should be forwarded to the factory for vibration."[103]

FIGURE 5.42. Charles Vander "Woerd's Automatic Balance Drilling Machine." This machine also incorporated a tapping mechanism. It was automatic to the extent that the operative need only insert and remove the balance wheel. One operative tended up to twelve machines. SOURCE: Marsh, *Automatic Machinery*, p. 125, figure 55. COURTESY: The Time Museum, Rockford, Ill.

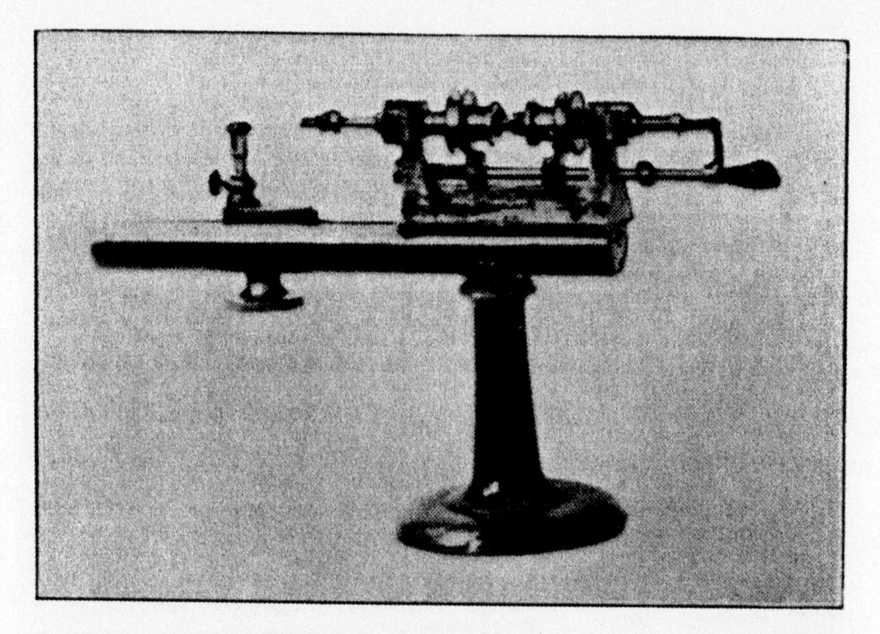

FIGURE 5.43. An "Early Balance Tapping Machine" which, like the "Early Balance Rim Drilling Machine" in figure 5.41, is based on a simple lathe. SOURCE: Marsh, *Automatic Machinery*, p. 124, figure 54. COURTESY: The Time Museum, Rockford, Ill.

The watch escapement and balance was undoubtedly the most complex assembly produced by American System manufacturers in the nineteenth century. It was only part of the watch.

DUANE HERBERT CHURCH AND THE END OF INNOVATION AT WALTHAM

On August 1, 1905, Duane Herbert Church died in his home in West Newton, Massachusetts.[104] He was born in 1849 in Madison County, New York, but by the age of sixteen was an apprentice to a St. Paul, Minnesota, watchmaker named J. E. Gridley. Following his apprenticeship, he worked for seventeen years in St. Paul and for Matson & Co. of Chicago as a "watchmaker at the bench." In 1882 he went to work for the Waltham Watch Company, first as a traveling "missionary" expounding the virtues of machine made watches, then in the Boston office on "experimental work," and finally became master watchmaker at the factory that year. Church spent eight years as master watchmaker before taking the position of superintendent of toolmakers, apparently in 1890. There he remained until his death in 1905.

FIGURE 5.44. The "Marsh Modern Balance Tapping Machine" performed the final tapping of the balance wheel. This machine tapped all the holes simultaneously. It could also be set up to drill all the holes. Marsh patented this machine on February 8, 1887, U.S. Patent No. 357,398, "Machine for Drilling and Tapping Watch Balances." SOURCE: Marsh, *Automatic Machinery*, p. 126, figure 56. COURTESY: The Time Museum, Rockford, Ill.

As master watchmaker, Church first improved Waltham's existing full plate model. In 1884 he completed the design of a new movement featuring stem (or pendant) winding and setting. This eliminated the lever setting mechanism and its related cams and levers. More important, it allowed the fitting of watches into cases without having to file the lever setting notch in the case ring. This concept was patented and remained in force through the turn of the century, when, upon its expiration, it was quickly adopted by all existing manufacturers.[105]

Church quickly turned his attention to manufacturing technology. As early as March 1887, he, John Logan, and Edward A. Marsh jointly patented a machine for "Testing and Grading Watch Balances and Hair Springs.[106] In July of that year, he and Marsh patented an "Automatic Blank Feeding Device."[107] Evidence of the close and harmonious working

FIGURE 5.45. The "Balance Making Room" at the Waltham Watch Company. The operative on the right is tending a bank of four semi-automatic machines. While the details are unclear, these strongly resemble "Woerd's Automatic Balance Drilling Machines." Compare with figure 5.42. SOURCE: H. C. Hovey, "The American Watch Works," *Scientific American* (August 16, 1884), N.S., 51(7):101. COURTESY: Milwaukee Public Museum. Negative no. H-627-15-K-28.

FIGURE 5.46. "Modern Automatic Hair Spring Wire Forming Machine" at the Waltham Watch Company, ca. 1895. SOURCE: Marsh, *Automatic Machinery*, p. 134, figure 59. COURTESY: The Time Museum, Rockford, Ill.

FIGURE 5.47. A "Device for Testing Balances and Springs" invented by John Logan. SOURCE: Marsh, *Automatic Machinery*, p. 141, figure 63. COURTESY: The Time Museum, Rockford, Ill.

FIGURE 5.48. "Logan Device for Testing Balances and Springs." SOURCE: Marsh, *Automatic Machinery*, p. 142, figure 64. COURTESY: The Time Museum, Rockford, Ill.

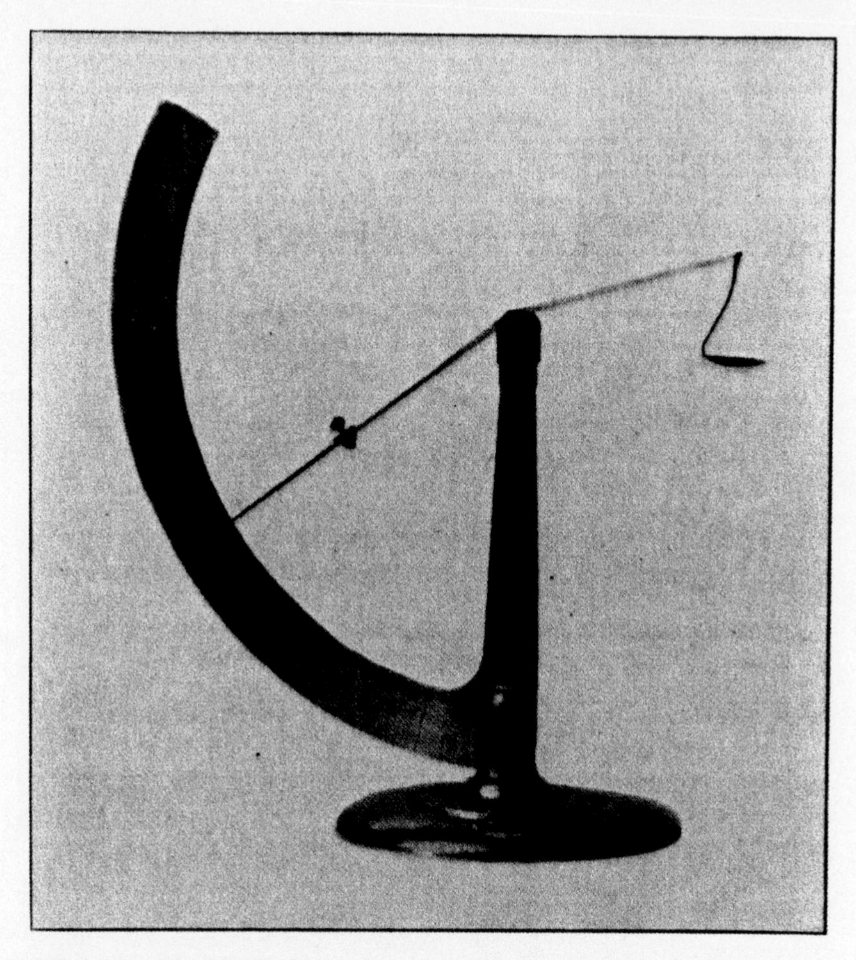

FIGURE 5.49. "The Gill Scale for Weighing Balances." SOURCE: Marsh, *Automatic Machinery,* p. 145, figure 67. COURTESY: The Time Museum, Rockford, Ill.

relationship between Marsh and Church is found in the writings of Marsh, whose compositions refer to Church in a saintly manner.

What inspired Marsh to canonize his colleague was Marsh's understanding of what Church accomplished. Seen in the perspective of watch manufacturing at Waltham, Church's work represents a pivotal change in theory. Church did nothing less than do away with jigs and fixtures.

To share Marsh's appreciation of the Church revolution in theory and technique, consider the evolution of machinery at Waltham from 1850–1890. The first mechanics began with, but quickly discarded, European ideas and devices. The development in the mid 1850s of the basic American watchmaker's lathe was the basis of most of Waltham's first generation

FIGURE 5.50. "The Gill Gauge for Determining the Relative Strength of Hair Springs." Compare with figures 5.51 and 5.52. SOURCE: Marsh, *Automatic Machinery*, p. 146, figure 68. COURTESY: The Time Museum, Rockford, Ill.

FIGURE 5.51. "Gauging Hair Springs" in 1884. The operative is using Gill's "Gauge for Determining the Relative Strength of Hair Springs." Behind Gill's gauge on the work table is a grid, each square of which is probably designated for a particular strength hair spring. Compare with figures 5.50 and 5.52. SOURCE: H. C. Hovey, "The American Watch Works," *Scientific American* (August 16, 1884), N.S., 51(7):101. COURTESY: Milwaukee Public Museum. Negative no. H-627-15-K-28.

FIGURE 5.52. This is the Waltham factory drawing of Gill's hair spring gauge, dated May 20, 1879. Compare with figures 5.50 and 5.51. SOURCE: Waltham Watch Company Drawings, Smithsonian Institution Specimen. COURTESY: Smithsonian Institution, Museum of American History, Negative no. 83-12693.

production machinery. The first screw cutting machine, the early balance drilling and tapping machines, and a host of other machines were based on this simple lathe.

A second generation of machines appeared in the late 1860s and early 1870s, the Vander Woerd automatic screw machine being typical of this new technology. It generally used existing ideas (such as turrets, slide rests, and self-closing collets) combined with screws and cams to produce automatic motions. These kinds of machines were particularly adaptable to screws, balance staffs, wheels, pinions, and other regularly shaped parts. Such parts as plates and cocks remained to be automated. By the late

FIGURE 5.53. "Balance and Hair Spring Vibrating Machines" in the Waltham watch factory in 1896. These are probably variations of the machine jointly patented by Edward A. Marsh, John Logan, and Duane H. Church on March 29, 1887, U.S. patent no. 360,234. Compare with figure 5.54. SOURCE: Marsh, *Automatic Machinery*, p. 147, figure 69. COURTESY: The Time Museum, Rockford, Ill.

FIGURE 5.54. "The Hairspring and Balance" illustrates the vibrating machine in Marsh. Compare with figure 5.53. SOURCE: Waltham Watch Company, *Waltham and the European Made Watch*, p. 36. COURTESY: Milwaukee Public Museum. Negative no. H-627-19-L-80. MPM H 45866/27141.

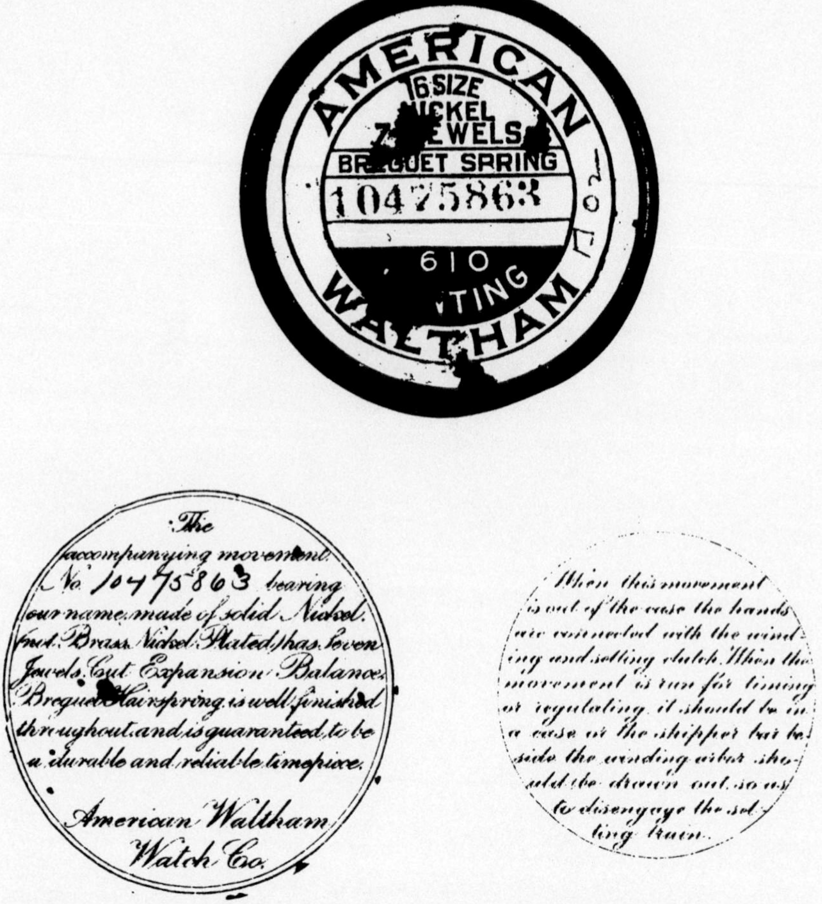

FIGURE 5.55. These labels came from the Waltham factory with its watch movement in a small tin. The lower right label reads "When this movement is out of the case the hands are connected with the winding and setting clutch. When the movement is run for timing or regulation, it should be in a case or the shipper bar beside the winding arbor should be drawn out so as to disengage the setting train." The movement was made about 1901. It was a 16-size, *Model 1899*, nickel, seven jewel, hunting movement, grade 610, serial no. 10,475,863. COURTESY: Milwaukee Public Museum. Negative no. H-627-19-L-83. MPM H 42636/26656.

1880s this second generation of machinery was fairly well developed—and then Mr. Church came to Waltham (figure 5.56).

Church believed that once a part was in a machine "it should not be released until fully completed." During his tenure as Waltham's chief tool maker, he built this theory into the manufacture of most watch parts. His theory is best known in the practice of watch plate manufacturing. Upon his arrival at Waltham, watch plates arrived as punched blanks

FIGURE 5.56. Duane H. Church, ca. 1896. This portrait appears as the frontispiece in Marsh's book *Automatic Machinery*. On the last page, Marsh writes that Church's "fertility and originality in the field of invention, achieved so much in the embodiment of automatic features as to render his recent machines wonders of mechanism." COURTESY: Smithsonian Institution, Museum of American History. Negative no. 17969a.

from several suppliers in Waterbury, Connecticut. These were turned to size and had the three dial feet holes drilled in them. The dial feet holes served as the reference points throughout the fabrication process. Drilling was done through jigs, and each operative had a three spindle drill press with three different sized drills. She placed the plate in her fixture and drilled all the holes of the sizes of her three drills. She then took the plate out of its jig and passed it along to the next operative who drilled all the holes of the sizes of her three drills. This step-by-step process necessarily introduced some errors and caused some problems.

Church's watch parts bed lathe changed plate manufacturing. The machine featured a series of machines mounted on a long cast iron bed. At one end, blank watch plates were automatically taken from a magazine and carried to the first machine by a special transfer arm. The first machine gripped the blank and proceeded to drill a series of holes and recesses, while the transfer arm returned to the magazine for another blank. When the first process was completed on the first blank, it was removed by a second transfer arm that carried it to the second machine. The second machine grabbed the partially completed blank and performed the next series of drillings and millings, while the second transfer arm returned to the first machine to remove the second blank that had been

FIGURE 5.57. Front view of a watch plate drilling machine, Waltham Watch Company, ca. 1895, designed by Duane H. Church. COURTESY: Smithsonian Institution, Museum of American History. Catalogue no. 316,561, negative no. 46767-B.

placed in the first machine by the first arm while the second arm was moving the first blank. This process continued unabated through seven steps before the completed plates were deposited in a bin at the opposite end of the machine (figures 5.57, 5.58, 5.59, and 5.60).

This last generation of Waltham manufacturing machines used no jigs to locate the work with respect to the drills and mills. This was done by the machine, specifically the mechanism holding the work piece. Church accomplished this with the use of compressed air in cylinders and pistons with cam operated valves to prevent shock in starting and stopping against the hardened steel stops. He also developed a series of mechanical transfer elements that could move a piece, rotate it 180°, and place it precisely in a self activating holding mechanism. This process was adapted for staff turning as well as plate drilling and many other processes. Like the generations of machinery that preceded them, these machines improved the quality of Waltham watches and lowered the cost of their production.

Within ten years after Church died in 1905, the Waltham Watch Com-

FIGURE 5.58. Rear view of Church's automatic plate drilling machine. Notice particularly the air lines running to the cylinders and pistons located between each of the transfer arm housings. COURTESY: Smithsonian Institution, Museum of American History. Catalogue no. 316,561, Negative no. 46767-C.

FIGURE 5.59. Detail view of Church's automatic plate drilling machine. The transfer arms are missing, as is the plate magazine that was carried in the V-shaped device located above the catalogue number. The air lines in this view are apparently designed to blow away chips, rather than control the mechanism. COURTESY: Smithsonian Institution, Museum of American History. Catalogue no. 316,561, Negative no. 46767-D.

FIGURE 5.60. "The Lower Plate" illustrates Marsh's plate drilling machine in 1919. Compare with figures 5.57, 5.58, and 5.59. SOURCE: Waltham Watch Company, *Waltham and the European Made Watch,* p. 22. COURTESY: Milwaukee Public Museum. Negative no. H-627-19-L-82. MPM H 45866/27141.

pany began the gradual decline that ended in its demise in 1954. Many of Church's machines were still in use when much of the factory machinery was sold at auction that year. By then, it was worn out and sadly out of date, yet still testified to the last of a series of dazzling American watch factory mechanics whose inventions brought watch manufacturing hegemony to the United States in the nineteenth century. Their prowess in machine design was not, unhappily, matched by the managerial skills of their employers in the twentieth century. By the 1920s Waltham was in severe trouble, but managed to struggle on for another twenty-five years on its reputation alone, a reputation earned by Webster, Vander Woerd, Marsh, and Church (figure 5.61).

Factory Organization

> The plan of manufacture . . . extends to every part of the watch, commencing with the rolled plates of brass, steel and silver, the wires used for pinions, pins and screws, and the gems for jewels; and by punching, swaging, cutting, turning, polishing, burnishing, drilling, enameling, gilding, etc., brings out the perfect living mechanism. All is done by machinery, each machine doing its peculiar work to a gauge [sic] or pattern . . . With the exception of the jewels and the pivots that run in them, every watch is in every part exactly like every other, so that a thousand might be taken to pieces and then re-constructed with pieces taken indiscriminately.[108]

A Campaign to Protect You in Buying Your Watch

RALPH WALDO EMERSON, speaking in one of his essays of a distinguished man, said: "He is put together like a Waltham Watch."

This remarkable tribute to Waltham greatness is the result of the genius of men whose inventive faculties have been concentrated for nearly three-quarters of a century to make it the wonderful time-keeping device it is.

The buying of a watch is an investment in time-keeping. And time is the most valuable possession of man.

You purchase a watch for one thing — to keep correct time for you — to tell it to you with dependability at any moment of the day or night.

A good watch, therefore, must have something more than good looks — it must have good "works."

Millions of people imagine that the "best" watch is made abroad — or, at any rate, that its works are imported from there.

Yet, in competitive horological tests at the world's great Expositions, Waltham has not only defeated these watches of foreign origin, but all other watches as well.

In a series of advertisements we are going to show Americans that there is a watch built in the United States whose time-keeping mechanism is more trustworthy than those of foreign make,—

A watch that is easily and reasonably repaired because its parts are standardized,—

A watch that represents American leadership in mechanical skill,—

Duane H. Church, famous inventor who filled the great shops at Waltham, Massachusetts, with exclusive watch-making machinery that performs miracles of accurate and delicate work which the human hand could never equal.

A watch that has revolutionized the art of watch making and assured accurate and dependable time-keeping.

We are going to take you through the "works" of a Waltham — lay bare those hidden superiorities which have led the horological experts of the greatest nations to choose Waltham as *the* watch for the use of their government railroads.

When you have finished reading these advertisements, which will appear regularly in the leading magazines, you will walk up to your jeweler's counter and demand the watch you want — because you will know how it is built and why it is superior to the foreign watch.

"Mention the Geographic—It identifies you."

FIGURE 5.61. This January 1919 *National Geographic* advertisement is apparently the first in a series. In it, Waltham featured Duane H. Church, who had been dead for over a decade, and quoted Ralph Waldo Emerson. By the 1920s Waltham was surviving on its reputation, having lost the best of its inventive mechanics and being slowly bled to death by its new owner. COURTESY: Milwaukee Public Museum. Negative no. H-668-A-23. MPM H 44999.7/27054.

The mass production, assembly, and adjustment of interchangeable watches at Waltham required an effective and efficiently organized factory. At least as early as 1863, and probably by 1857, the Waltham Watch Company developed functionally differentiated departments staffed by a highly skilled and specialized work force coordinated by the superintendent.

The superintendent prepared a monthly production card for each of twenty-five department foremen "stating the number and kinds of watches to be made."[109] Each department foreman reported daily to the superintendent on the work completed and the parts transferred to other departments. The company provided each department with a bookkeeper to assist the foreman in keeping track of his materials, tools, workers, etc., but who answered to the superintendent, not the foreman. The superintendent was helped by an assistant superintendent in overseeing the 3,746 different operations required on the 150 parts of each watch. The company also employed a master mechanic and a master watchmaker: the former helped design new machinery and the latter worked on new watch movement designs.[110]

The superintendent usually decided to produce watches in lots of 1,000 of a particular model and grade.[111] He communicated this decision to his foremen via the production card, and each foreman drew his material from the company. As an inside contractor, each foreman parceled out this material to his department's employees whom he hired. These operatives generally worked by the piece (although there were some people working on day rate and some on an annual salary). Once they finished their allotted tasks, they returned the work to the foreman who then passed the parts to the storeroom (if they were finished) or on to an other department for further work.

The company did not usually finish all 1,000 movements of a particular lot at one time. After fabrication, most parts were stored "in the flat" (unassembled and unfinished) and were passed out for final assembly and adjustment in lots of ten as watches were demanded.[112] Some highly finished or specially adjusted movements were probably finished individually.[113] There is some evidence that some watches remained unfinished for decades at Waltham. Unfinished movements were stored in rectangular wooden trays, with ten circular compartments, each of which held the parts of one particular watch separate from the parts of the other watches in the tray (figures 5.62, 5.63a, 5.63b, and 5.64).

Many other parts were simply stored in jars, such as "10,000 seconds hands in one jar" (one day's work), and "19,979 center wheels in one

FIGURE 5.62. Watch subassemblies such as balances were manufactured and assembled prior to final assembly in a finished watch. This ten hole balance tray is especially interesting, as it contains some political graffiti. It comes from the Elgin National Watch Company, Elgin, Ill., ca. 1880-1890. COURTESY: Milwaukee Public Museum. Negative no. H-627-19-L-5. MPM H 36646/25622.

box."[114] Interestingly, Waltham inventoried much of its stock by weighing the total production and dividing by the average weight of each part, presumably determined by weighing a random sample of those parts.

The wooden, ten-compartment trays served several purposes in the watch factory. First, they were used to carry partially finished watches from department to department. Second, smaller versions were employed for sub-assemblies such as hairsprings and balance wheels, which, when finished, were united with frame parts and train parts in the larger trays to complete the watch. Third, the ubiquitous wooden trays kept the parts

FIGURES 5.63A AND 5.63B. A watch movement tray from the Howard factory, containing unfinished Keystone Howard watch movements. The label attached to this tray indicates that these watches were in the Finishing Room at the Howard factory. There are at least sixteen different operations per movement, ranging from matching the escapement to adjusting for temperature and position. COURTESY: American Clock & Watch Museum, Bristol, Conn.

of each watch together during the manufacturing process, which often lasted four to six months but which could last years.

The segregation of partially finished watches was critically important, because, at certain points in the manufacturing operation, some of the parts of each watch were machined with respect to each other and had to be kept together. Although the vast majority of watch parts—screws, wheels, hands, dials, mainsprings, winding wheels, click springs, dial washers—were fully interchangeable within a model, several parts were somewhat less than fully interchangeable. These were the frame parts,

FIGURE 5.64. "The Pinion Room" at the Waltham Watch Company in 1884. Notice the stacks of wooden boxes behind each machine. These boxes contain partially finished watches, segregated into separate compartments, and were carried throughout the factory during the entire process of manufacture and adjustment. SOURCE: H. C. Hovey, "The American Watch Works," *Scientific American* (August 16, 1884), N.S., 51(7):102. COURTESY: Milwaukee Public Museum. Negative no. H-627-15-K-49.

such as the pillar plate, the top plate, the balance cock, the potence, the barrel bridge, the balance wheel and its hairspring, and the escape wheel and its pallet fork with stones. This lack of perfect interchangeability was not due to any method of manufacturing, but rather to the very nature of the watch itself.

The frame parts illustrate the technical necessity of keeping certain parts together after particular stages of manufacture. Each pillar plate required eighty operations after it was received from either the Scoville Manufacturing Company or the Benedict & Burnham Manufacturing Company, both of Waterbury, Connecticut.[115] One of the first operations was punching the dial feet holes that served as the "single reference point" for all subsequent operations. Another of the eighty operations was stamping a serial number on each pillar plate and placing the plate in one of the compartments of the wooden trays. Each of the other frame parts passed through a similar process, including serial numbering, and was added to its particular tray compartment. When all the frame parts for a particular watch were collected in their compartment, they were assembled for the first time and had their train holes drilled. After this operation, the plates had to be kept together to ensure that the train holes were properly aligned.[116]

Later in the manufacturing process, these parts were boiled several times in soapsuds to clean them of dirt and grease, fully assembled and run "in the gray" (before gilding), and then disassembled, cleaned, stoned, and gilded before being assembled again and sent to the adjusting department. At each step, there was the possibility of confusing the (apparently) identical parts of several watches, hence the use of serial numbers on those parts.

Balance wheels and hairsprings were matched at one point in their manufacture due to minute differences in balance wheel weights and spring strengths. Edward Howard noted the "distinct individuality" of each watch.

> Some parts are so minute that, although you suppose you have them all alike, the fact is that no two have been made without some little variation, having an appreciable effect on its action as a timekeeper. That is where the individuality comes in. . . . We must make the best of such a condition.[117]

Waltham enjoyed two major advantages of manufacturing interchangeable parts outside the manufacturing process itself. First was the sale of repair parts. By 1885 sales of replacement parts for the million-plus Waltham watches in use justified the company's publishing its first parts catalogue. The catalogue illustrated all the parts available and listed their prices, with the exception of frame parts which, the company noted, had to be made to order.[118] This catalogue listed some nineteen different series of watch movements with as many as six grades per series. For each series there were some one hundred parts.[119]

The second advantage of manufacturing interchangeable parts was found in the company's advertising. Fixing watches was no longer a problem, since parts were readily available. As early as 1858 the company advertised that by sending the serial number and describing the part, replacements could be had by mail.[120] In 1884 the company advertised in the *Scientific American* that it kept accurate records of all its watches and that the "work . . . is so nearly perfect that should any part of a watch fail in actual use the owner need only send on the number of the movement to enable the factory to supply an exact duplicate . . ."[121]

There are two implications of Waltham's ability to supply parts given only the watch's serial number. First, the company had a sufficiently sophisticated record keeping system to provide the needed data. Second, the identified part was either in stock or could be made quickly and easily. Both are indications of a well organized and well managed factory.

The functional production department structure of the factory changed little after 1857, although the machinery in each department changed several times over the course of the nineteenth century. In addition to these production departments and the assembly and adjustment departments, there were several departments serving the entire factory in general. The two most important were the carpenter's department and the machine department. The carpenters performed a wide variety of jobs from building new benches to constructing the models for industrial fairs. The machine department designed and built the new machinery for the various production departments and served as an in-house research and development laboratory for testing new mechanical ideas.[122]

The Waltham Watch Company not only developed the most sophisticated production technology in the nineteenth century but, along with its automatic machinery, developed the most highly organized form of factory organization. Marsh, Waltham's master mechanic and historian, opined that "there is no manufacturing industry . . . in the world which is so complex, and which demands so much brain-work and careful management, as does the manufacture of watches by modern methods."[123]

THE ECONOMICS OF WATCH PRODUCTION

> That party who can make watches the cheapest will have them to make. We have the lead so far in all the world in this particular—that we are able to give the best watch for the money asked in all grades, and that money is now little enough in the case of the cheapest watch to command every market.[124]

> [I]n 1857, [Waltham] employed about ninety people and produced five watches per day,—one watch for each eighteen workmen. At that rate of production there would be required 56,394 people to produce the 3,133 movements per day, which were made in 1907 by 4,300 people, being one movement for each 1.37 people.[125]

The mechanization of watch manufacturing at Waltham was very profitable. The direct economic returns to the development of increasingly sophisticated machinery and eventually to fully automatic machinery were crystal clear to R. E. Robbins, the Waltham management, and the mechanics on the shop floor who designed and built the machinery. From the early 1860s Robbins believed strongly in the new technology and its capability to lower watch prices through lower manufacturing costs. In 1867 Robbins assured the stockholders that "our manufacturing . . . has gone on most encouragingly—the gain in number of watches made . . . [being] . . . forty percent; while the average number of hands has . . . increased . . . about 12 1/2%."[126] Similarly, in 1876 Robbins told the stockholders, "It is a fact that this year's selling price is the cost of last year and so it has been these three years. And *these reductions have been largely due to improvements in mechanical means and methods, the end of which we have not even approached.*" (Italics added.)[127]

Robbins further reported that the industry had overproduced for the market but that Waltham continued to produce.

> You will naturally ask why then do we propose to increase our production. The hard and ungenerous answer is that as it is impracticable to bring all the makers to an agreement to curtail, the weaker ones must be forced to it by those who have the power. The purpose is by large production to effect cheap production and thus to be able to undersell.[128]

In 1885, after a major expansion of the factory, Robbins again repeated his message to the stockholders.

> Being fully equipped at last with a completed and magnificent factory and ample machinery, we set out at the beginning of the year with a determination to make the most of them. . . . We never made so good watches nor made them so cheaply as now—nor were they ever so popular . . .[129]

Robbins' statements to the stockholders were reflected in the Waltham's investment in machinery (table 5.1). This increased use of machinery and especially automatic machinery did lead to a decline in the prices of watches, particularly the lower grades. In 1864 Waltham's *WM. ELLERY* grade (the so-called *Soldier's Watch*) sold for $13.[130] By 1907 the price of Waltham's lowest grade was $1.75.[131] From 1890 to 1899, a period of intense technological innovation in highly automatic machinery, the average price of a Waltham movement (all grades) fell from $10.09 to $4.49 (table 5.2).[132]

The constantly increasing productivity of labor at Waltham was probably the single most significant contribution to the decline in watch prices. At Waltham, throughout the late nineteenth and early twentieth centuries the number of man-days required to produce a watch constantly fell. The decline was most dramatic between 1854 and 1859, hence Robbins' strong belief in increasing investment in machinery (table 5.3).

Waltham's management and its mechanics had learned early that they would succeed primarily in manufacturing inexpensive watches for a large market, not by producing high grade watches. Marsh correctly analyzed

TABLE 5.1
Waltham Watch Company—Investment in Machinery 1859–1879

Year	*$ Invested in Machinery*
1859	$ 59,744.81
1860	66,668.48
1861	72,646.78
1862	74,255.97
1863	88,074.61
1864	86,318.41
1865	114,310.47
1866	141,233.69
1867	164,373.92
1868	174,867.54
1869	175,000.00
1870	185,000.00
1871	200,000.00
1872	212,000.00
1873	225,000.00
1874	253,813.51
1875	238,126.95
1876	245,000.00
1877	255,000.00
1878	255,000.00
1879	255,000.00

SOURCE: Waltham Watch Papers, R. E. Robbins, MSS. AD-1 to AD-2.

NOTE: These figures include "furniture and fixtures" as well as some other miscellaneous costs, but generally reflect the level of investment in tools and machinery at Waltham. They also reflect the inflation resulting from the Federal Government's financing the Civil War with paper currency, the so called Greenbacks. The figures for the years 1869–1873 and 1876–1879 appear to be budgeted or estimated numbers rather than actual expenditures—called to my attention by David Landes.

the Nashua Watch Company's failure when he described the finishing of the Nashua material at Waltham following its purchase in 1862.

> These were of three-quarter plate model . . . and finished and adjusted with special care, they proved to be superior timekeepers. The 16-size movements of this model were manufactured for several years and mar-

keted at high prices, which, however, did not secure a corresponding profit, owing to the small number produced, as well as the relatively high cost of their production.[133]

TABLE 5.2
Average Price per Watch Movement, 1890–1899

Year	*Average Price*
1890	$10.09
1891	9.86
1892	7.28
1893	missing
1894	5.10
1895	5.01
1896	4.84
1897	4.80
1898	5.12
1899	4.49

SOURCE: Moore, *Timing A Century*, pp. 67–68 and 87.

Another indication of the economic success of rapid technological change at Waltham, and particularly the automatic machinery, is Waltham's profitability. Table 5.4 provides data proving that Waltham was a very profitable firm.

TABLE 5.3
Increasing Labor Productivity at Waltham, 1854–1905

Year	*Man Days per Watch*
1854	21
1859	4
1862	3
1883	2.2
1905	1.5

SOURCE: Moore, *Timing A Century*, p. 233.

TABLE 5.4
Dividends of the Waltham Watch Company, 1860–1884

Year	*Dividends*
1860	3 1/8% cash
1861	13% cash, 10% scrip
1862	Passed
1863	4% cash
1864	11% cash
1865	22% cash
1866	60% cash, 150% stock
1867	28% cash
1868	5% cash
1869	11% cash
1870	20% cash
1871	20% cash, 66 2/3% stock
1872	20% cash
1873	12% cash
1874	12% cash
1875	5% cash
1876	6% cash
1877	3% cash
1878	7 1/2% cash
1879	4% cash
1880	5% cash
1881	9% cash
1882	8% cash
1883	8% cash
1884	9% cash

SOURCE: Moore, *Timing A Century*, p. 63.

By all indications—investment in machinery, watch prices, company profits, and the active market in watch parts—the Waltham Watch Company competed quite successfully in a highly competitive market by producing large quantities of lower grade interchangeable watches with increasingly sophisticated automatic machinery. There is no doubt that the American System was very profitable at Waltham, as both R. E. Robbins and the inventive mechanics realized.

So profitable was the Waltham Watch Company in the early 1860s that other firms soon began to enter the market. The most interesting aspect of this invasion was the degree to which the new companies relied on Waltham mechanics for technical expertise. The Nashua Watch Company was the first firm to compete with Waltham. In 1859 Nelson P. Stratton, with the help of Belding D. Bingham (who came to Waltham from Nashua, New Hampshire) led a group of Waltham mechanics, including Charles Vander Woerd, Charles S. Moseley, and others, to start this firm. It failed in 1862 and was purchased by Waltham, which hired back all the employees who had left.[134]

The pattern of new watch companies hiring experienced mechanics from established firms, especially from Waltham, continued throughout the nineteenth century. In 1864 midwest investors began the National Watch Company (soon to become the Elgin National Watch Company) and at least seven important mechanical positions at Elgin were first filled by men from Waltham. Charles S. Moseley became the Elgin superintendent, while George Hunter, Otis Hoyt, John K. Bigelow, Charles E. Mason, D. R. Hartwell, and P. S. Bartlett (for whom a Waltham grade was named) all assumed responsible jobs. In addition to the usual financial inducements, each man was offered an acre of land in Elgin. How many others came to Elgin from Waltham to fill nonsupervisory positions is unknown.[135]

Even in the smaller watch factories, men from Waltham played important roles. The Aurora Watch Company of Aurora, Illinois (1883–1892), was a short-lived failure and eventually became part of the Hamilton Watch Company.[136] In the process of starting their factory, the Aurora entrepreneurs hired Charles C. Hinckley. Hinckley worked at one time in Samuel Colt's Hartford armory before working for the Waltham Watch Company. In 1872 Hinckley went west to Grand Crossings, Illinois (a suburb of Chicago), to work for the Cornell Watch Company. When the Cornell Watch Company failed, many operatives, including Hinckley, went to Rockford, Illinois. There they joined the Rockford Watch Company, formed from the debris of the Cornell concern. Subsequently, Hinckley was hired away by the Illinois Watch Company in Springfield, Illinois. There, he worked as the model maker until lured to Aurora in 1883. At Aurora, Hinckley designed the new Aurora watch and built some of the Aurora machinery. He left Aurora in 1885.[137]

While Hinckley's case may seem extreme, he is typical of a sub-culture of watch factory mechanics who had worked at Waltham at one time or another. They worked not only for other watch factories but also for a host of small precision machine tool companies that helped to spread the new technology. For example, Ambrose Webster left the Waltham Watch

Company to help found the American Watch Tool Company. This company sold some tools to Waltham, but sold also to such firms as the Aurora Watch Company. The American Watch Tool Company also sold machinery to both the Swiss and the English and by 1884 was supplying domestic manufacturers of other precision products as well.[138]

In sum, the Waltham Watch Company not only developed watch manufacturing, but also set the economic and technological terms on which other watch manufacturers were forced to compete. Watch manufacturers, perhaps more than other American System industries, such as wooden clock makers and typewriter manufacturers, required a large market in which to sell their goods because they were so highly capitalized and paid their skilled labor such relatively high wages. Expensive automatic machinery was profitable only above a certain level of production.[139] For the Waltham Watch Company and other private sector American Systems industries, the American economy, combined with exports primarily to Western Europe, provided the market necessary to absorb the production of automatic machinery.

CONCLUSION

> Under the administration of such great establishments there have grown up completely new methods of work. The claim that the machinery, processes, [and] methods of organization of a great American watch . . . company are not original, is as ridiculous as to say that the Falls of Niagara are not original.[140]

In the interchangeable manufacture of watches at the Waltham Watch Company between 1849 and 1910, the American System reached its zenith. Watch factory mechanics developed the most sophisticated automatic machinery in the nineteenth century. They enforced a series of standards unknown in other American System factories based on the metric system and extremely precise gauges designed for highly specialized purposes. They developed a system of factory organization far beyond any other industry at the time.

Like other American System industries, watch manufacturers were constrained by the technical requirements of their product. They produced a fully interchangeable watch, subject to the necessary assembly and adjustment requirements of the watch itself.

Watch factory mechanics and entrepreneurs were initially enthralled with armory practice, and several mechanics employed at Waltham had previously worked at the Springfield Armory. However, these mechanics borrowed very little from the armory directly. The most important transfer of technology was the concept of a system of manufacturing with strictly imposed standards. Once the idea of a system was accepted at Waltham, the watch factory mechanics quickly broke what few ties they had with armory practice and developed their own version of the American System to suit the specialized needs of watch manufacturing.

In this process, they drew heavily on English watchmaking technology, Connecticut clockmakers, American watchmakers and repairers, and native American mechanics with no horological experience. This mix of traditional techniques, armory practice, and new technology made the Waltham Watch Company the most innovative firm in the late nineteenth century.

The Waltham Watch Company found it innovations to be highly profitable, as did other American System industries who preceded and followed it, such as typewriter and bicycle manufacturers. The new machinery of watch manufacturing increased output and productivity, cut labor costs dramatically, improved the quality of watches, lowered the costs of assembly and adjustment, and created a new market for watches and watch material. The automatic machinery was especially profitable and produced parts in quantities considered huge by nineteenth-century standards. Waltham management believed strongly that only more sophisticated machinery would allow the company to compete successfully in the increasingly competitive watch market. The company's profits substantiate Royal Robins' belief in the new technology.

The Waltham Watch Company made innovations in other areas besides production. It pioneered the full product line as early as 1864. Later in the century it expanded its line and offered over 200 different styles of dials (including special order dials) and numerous styles of hands.[141] Mass production in the watch industry did not force uniformity on the public. Indeed, quite the opposite is the case. Mass production offered the buying public a greater diversity of watch styles, grades, and prices.

The watch industry illustrates clearly the importance of the working relationship between the mechanic and the entrepreneur. Both R. E. Robbins and the many men who worked for the Waltham Watch Company understood precisely what they were doing. They were lowering the cost of watch manufacturing through automatic mechanized production of fully interchangeable parts on a scale unheard of in the nineteenth century, as evidenced by one visitor's astonishment at seeing 10,000 seconds hands in one jar and 19,000 wheels in one box.[142]

Robbins understood the necessity of giving his most skilled and creative mechanics a large measure of freedom. For example, in 1862, when the Waltham Watch Company bought the bankrupt Nashua Watch Company, it made a major effort to keep the renegade mechanics happy, despite their having left Waltham to start a competitor in 1859. This effort took form in the *Nashua Department*, a separate wing built onto the Waltham factory for the Nashua machinery and run from 1862 to 1876 by the very men who had started the Nashua concern. The *Nashua Department* was eventually integrated into the factory in 1876, but remained semi-independent, producing the company's high grade 3/4 plate movements. Its superintendent, Charles Vander Woerd, invented the first automatic screw machine.[143] Clearly, Robbins understood Vander Woerd's need to experiment. Indeed Vander Woerd had experimented with automatic machinery as early as 1859, and it is tempting to speculate that Robbins may have held the reins a bit too tightly then. Perhaps, once he got the ingenious Vander Woerd back in Waltham, Robbins had learned from his earlier mistake and left the former sailor unbridled to devise new machines.

The Waltham Watch Company served as an inspiration to others in the manufacturing field. Its techniques were copied by every other American watch factory, each of whom hired men and women directly from Waltham or men and women who had once worked there. Mechanics also left Waltham to start precision tool manufacturing firms, the American Watch Tool Company being the best known and most important, which sold the Waltham technology domestically and internationally. The Waltham measuring techniques were spread throughout American industry by many firms that manufactured Waltham-like tools. Today, dial micrometers based on the original Waltham upright gauge are the world wide standard. So great was the Waltham reputation for precision manufacturing that an entire advertising campaign was built upon it in the 1920s.

Finally, as Aaron L. Dennison was inspired to manufacture watches by his trips through the Springfield Armory in the 1840s, Henry Ford was reportedly inspired to manufacture automobiles by his trip through the Waltham Watch Company. While Dennison can hardly be entitled the Father of American Automobile Manufacturing, the company he founded had a broad impact not only on Henry Ford (for whom it later made speedometers) but on American industry generally and the economic growth and development of the United States. Each Waltham watch should teach us a little about the interwoven fabric of watch manufacturing in particular and the American System of Manufactures in general. Each Waltham watch should serve as a reminder of our rich American industrial heritage (figure 5.65).

FIGURE 5.65. This Waltham advertisement ca. 1905-1910 (based on the text which mentions the Russian/Japanese War of 1904-1905) draws nostalgically on the American Civil War. SOURCE: Inside back cover of an unidentified magazine. COURTESY: Milwaukee Public Museum Negative. MPM H 33468.a/24844.

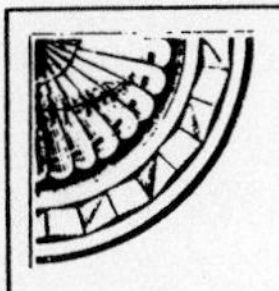

CHAPTER

The American System in Perspective: The Public and the Private *Sectors*

INGENIOUS YANKEES examined four American System industries in detail: wooden movement clock manufacturing, axe manufacturing, typewriter manufacturing, and watch manufacturing. The data provide evidence for a significant reinterpretation of the rise of the American System in the private sector. First, armory practice was not the significant factor previous interpretations have suggested. Second, the American System developed quite independently in the private sector, particularly in the antebellum period. Neither the federal nor state governments played any role in financing or subsidizing these industries. Their sources of capital were invariably private. Third, major product design innovation occurred with the new production technology, notably changes which simplified the product and provided for adjustability during assembly. Fourth, given a realistic definition of interchangeability, the private sector American System industries did manufacture on the "interchangeable system" and did achieve functional, indeed "perfect," interchangeability. Fifth, mass production did lower unit costs and consumer prices which substantially increased the size of the market. Sixth, technological change originating within each industry was often a function of the product being manufactured. Seventh, private sector American System industries introduced new materials and new techniques related to those new materials.

In sum, the rise of the American System as a production technology in the private sector of the American economy was propelled by economic factors but controlled by technical considerations. The American System of Manufactures was not merely the product of idealistic engineers and "technological enthusiasts," nor was it simply a response to economic incentive and relative factor costs. The American System appeared in the nineteenth-century American economy as the result of the crucial interaction between the engineers and the entrepreneurs.

In two antebellum industries, the American System appears without the influence of armory practice. In the wooden movement clock industry, there was no armory practice influence. There is no evidence of any transfer of technology from the infant federal armories. Eli Terry could have learned about interchangeable parts from Eli Whitney, but despite the fact that both worked in the Connecticut River Valley, there is no evidence of contact between them. Terry received the first American horological patent in 1797 at a time when Thomas Jefferson was actively promoting manufacturing. Perhaps during the patent process Jefferson communicated the ideas of LeBlanc on the interchangeable system to Terry who applied the idea to clocks.

Wooden movement clock manufacturers used a model-based system in manufacturing, which antedates the introduction of this concept in the federal armories. Arms makers probably learned from the clockmakers in this respect. Thus, there is no data to indicate any transfer of armory practice to the wooden movement clock industry.

Similarly, there are no data to indicate that the axe industry benefited from developments in the federal armories. Axe manufacturers at Collins & Co. used a form of die forging developed in its own shops. The techniques of shaving, grinding, tempering, and hardening axes were also developed in the axe factory by Collins' gifted superintendent, Elisha K. Root. Root had no previous experience in arms manufacturing. The case of Root and Collins & Co. offers an interesting case of the transfer of technology within the private sector from axes to arms. Although none of Root's inventions were transferred to the factory of his new employer, Samuel Colt, the general idea of creating a system of manufacturing was transferable. Rather than the armories influencing the axe makers, exactly the opposite occurred.

In the watch industry, there was an indirect armory practice influence. Aaron L. Dennison, the "Father of the American Watch Industry," was inspired to manufacture watches after a visit to the Springfield armory. Several early watch factory mechanics came from armories. Beginning with the Waltham *Model 1857*, these mechanics introduced the concept of a model-based system, although this technique had been practiced in and may have been transferred from the clock industry. Watch manufacturers quickly developed their own manufacturing techniques, notably a new gauging system. By the late 1860s they had begun to make semi-automatic machinery. This trend culminated in the early 1900s with fully automatic machinery and important efforts toward mechanical self gauging. In the watch industry, technical developments after the Civil War far surpassed those in the arms industry. Any transfer of technology from the arms industry must have been quite insignificant after about 1860, as virtually all watch factory developments took place in house. Aside from Dennison's inspirational tour in the 1840s and perhaps the concept of a model based system, the watch industry developed independently of the arms industry.

Like the watch industry, the typewriter industry was little influenced by armory practice. The typewriter industry is an interesting case study because it allows historians to explore how an entirely new product is produced when introduced in a technically sophisticated environment. The typewriter manufacturers used a model-based system, a common practice in 1873 both inside and outside the arms industry.

Were the typewriter manufacturers copying armory practice in using the model-based system or were they copying the sewing machine manufacturers or were they simply adopting the general technique oblivious to its origin? The first typewriter was manufactured in an armory/sewing machine/farm implement factory, which insures ambiguity. The first typewriter manufacturer, E. Remington & Sons, gathered techniques from both their armory and their sewing machine operation. Clearly they employed some armory practice, but equally clearly, they employed many new techniques and materials. Typewriters had cast iron, fancy painting, wire, steel type, rubber, delicate wood, glass, and sheet metal, none of which were found in firearms. Typewriters also had many small forged parts and screws, similar to firearm parts produced on automatic screw machines. The typewriter manufacturers also used machines to "exercise" or break-in their parts, a technique quite foreign to armory practice. In sum, they chose the techniques and materials required to produce typewriters, where appropriate those techniques included armory practice.

The private sector American System industries used relatively little armory practice. However, they developed a host of new technologies and employed a wide range of new materials. Closely related to these new technologies and new materials was product design.

Three aspects of product design strongly influenced nineteenth-century manufacturers: 1) end use, 2) simplicity, and 3) ease of manufacture. An analysis of these design features with respect to the federal arms industry and the private sector will do much to illustrate why the most significant technological developments occurred in the private sector outside armory practice. The military firearm was designed to perform well in battle. Interchangeable parts were important so that battle-damaged arms could be easily repaired. Only within the most general constraints were the federal armories concerned with the economics of production with interchangeable parts, since their product must fit the specifications of war, not commerce.

In sharp contrast, the private sector manufacturers faced the need to sell goods in the market, often in competition with similar goods, hence price was critical. Simultaneously, the manufacturers of goods for the general public were concerned only that their goods were made well enough to attract a market. The difference in quality demanded by the federal government and the private sector is perhaps nowhere more graphically illustrated than in the case of Samuel Colt and an early batch of his patented revolvers. These particular guns were rejected by the federal government as inferior and not up to standards, yet were sold quickly to the general public. Colt could afford to make inferior arms (by federal government standards) because he had only to worry that they would sell to the public.

Likewise, Eli Terry was unconcerned with making a masterpiece clock. He was interested in making a cheap clock that would keep reasonably accurate time for a few years. Thus, the federal government and private sector firms differed fundamentally in their approach to manufacturing methods and product design.

The second aspect of product design was simplicity. Again, a sharp contrast existed between the federal government's arms requirements and the needs of the private sector. The government wanted a weapon that worked under all conditions and was easily repaired by cannibalizing damaged arms. They were relatively unconcerned with simplicity as a design concept. While simplicity may have been desirable, it was hardly an imperative. In the private sector, there was great concern with simplicity, especially as it concerned the number of parts to be manufactured. The fewer the parts, the lower the cost. This movement toward simplicity in the watch industry resulted in the Waltham *Model 1857*, a greatly simplified watch design, which was widely copied in the American watch industry. In the 1880s Americans produced an even cheaper watch, the "Dollar Watch," an excellent example of product development in the private sector. Private sector American System industries designed and manufactured products to sell in all parts of the market, products greatly simplified from their higher priced counterparts, but which performed adequately for the intended user.

The third important aspect of product design dealt with the cost of the manufacturing process. This design aspect must have been of little interest to the federal armories, which were not overly concerned with cost but rather with producing a weapon with strategic and tactical characteristics. In the private sector, however, the manufacturers were concerned with production costs. They designed their products for easy and inexpensive manufacturing. Eli Terry's wooden movement shelf clock is perhaps the earliest product designed for ease of manufacture. Terry designed the outside escapement to simplify the process of depthing the escapement during manufacturing. Terry's invention allowed the manufacture of interchangeable parts which were only interchangeable at assembly due to their adjustability.

The typewriter industry also designed its products for ease of manufacture. The typewriter was the most complex American System product judging by the number of parts and necessary adjustments. To ease their assembly problems, typewriter manufacturers designed their machines to be adjustable during the assembly process. The Sholes & Glidden (Remington) and the Caligraph both had numerous adjustments to be made during assembly. Nor was this adjustability simply an aspect of early typewriter manufacturing. The standard Remington typewriters through the early 1920s (Models 10 and 11) also had many built-in adjustments. As late as 1954 the Royal Typewriter Company still employed a final adjustment and inspection process during which one employee finished two to four machines daily.

The design of the product as a function of the manufacturing process is an aspect of the American System which developed in the private sector and was unknown in the federal armories.

The use of interchangeable parts depends on the definition of interchangeability which is a function of the product being manufactured. Historians cannot establish an arbitrary standard of perfect interchangeability based on small arms production, but must consider the industry and the product in determining if the parts were interchangeable.

The data developed from two collections of unused wooden movement clock parts prove conclusively that wooden movement clocks were fully interchangeable. Clock parts could be chosen and assembled at random. These clocks were not, however, highly precise scientific instruments. They had a tolerance of about .01″ and were designed to have their most critical parts, the escapement wheel and the verge, adjusted at the time of assembly.

Watches were also interchangeable within the confines of this new definition of interchangeable. Most parts, such as screws, wheels, and dials, were completely and fully interchangeable, while some parts were interchangeable until assembly, such as escapement pallet jewels, hairsprings, and balance screws. While each was interchangeable, the delicacy of the watch required that final adjustments be made in the assembly process. Prior to this step in assembly, most parts could be chosen at random for assembly. After this step—matching the escapement, poising the balance, and setting the hairspring—the parts were not interchangeable without considerable readjustment.

Typewriters were also fully interchangeable within the confines of their adjustments, a feature designed into the machine itself and fully allowed for in the manufacturing process. The typewriter was so complex that its manufacturers took adjustment for granted. The machine-made watch may have been a more precise instrument, but the typewriter was also a precision instrument requiring precise multiple adjustment and alignment of every typebar during its manufacture and assembly. Neither the process of assembly nor the aligning and adjusting steps could be eliminated with improved parts manufacture. With the numerous connections, levers, and movements, the typewriter simply needed adjustment. It was, nevertheless, fully interchangeable.

The four case studies examined here demonstrate dramatically that the use of American System technology—the mass production of interchangeable parts on specialized machine tools—was largely, although not exclusively, responsible for lowering unit costs in manufacturing and hence the purchase price to the consumer. In the case of wooden movement clocks, prices fell drastically between 1807 and 1840, as did the cost of making axes at Collins.

Based on their production technology and especially their automatic machinery, the Waltham Watch Company was able to compete very successfully in a highly competitive market and was able to reduce the labor cost of its lower priced watches greatly over the second half of the nineteenth century. Together, Waltham and its "competitor" in Elgin successfully drove most of the smaller watch manufacturers out of business by 1910, based on their ability to produce cheap watches. Royal Robbins, the treasurer of the Waltham Watch Company, believed strongly in the development of automatic machinery as the most important technique for reducing costs and lowering prices. The extension of this concept to the "dollar watch" firms, notably the Waterbury Watch Company, is also strong evidence that mass produced, cheap, reliable goods made with special machinery and fully interchangeable, were very profitable.

The typewriter manufacturers with good typewriter designs were highly successful and profitable, using interchangeable parts produced on specialized machine tools. As prices began to fall, the largest firm, the

Remington Typewriter Company, formed a typewriter trust, the Union Typewriter Company, to maintain prices at about $100 per machine. Firms outside the trust advertised equally good machines with improved designs at lower prices and broke the trust. As prices throughout the economy rose, those of typewriters remained fairly constant for nearly six decades, thus the real price of typewriters fell.

In all the private sector industries using the new technology of mass production, the cost of producing goods fell and the consumer's price fell as well. The entrepreneurs employing the American System understood fully the economies of scale they achieved by producing large quantities of interchangeable parts and the money they saved in assembling and adjusting those parts. The American System of Manufactures was not "as much a political as technical" system. In the private sector of the American economy, the American System did not abound "in 'irrationalities,' such as the insistence upon interchangeability even when it raised rather than lowered costs."[1] The firms which adopted and adapted the American System in the private sector acted very "rationally." There was nothing "irrational" about lowering the cost of manufacturing and increasing the size of the market through mass production.

In choosing and developing technologies to lower their costs, the private sector American System industries generated most of their technological changes internally. Relatively little technology came from federal armories. Wooden movement clock manufacturers based their work on the earlier technology of wooden movement clock making, but developed most of the new technology themselves. There is no evidence to suggest that the new technology of wheel sawing, wire pointing, plate drilling, and wire bending and gauging came from anywhere but the clock manufacturers themselves. Developments in the axe industry occurred within the firms of that industry, made by people working in the business.

The watch industry drew heavily on the European technology of watchmaking, particularly the English methods. Watch manufacturers not only copied the British technology but purchased parts and supplies. Once past this initial phase of technological transfer, the American watch manufacturers independently developed new methods and new machinery to suit their particular needs. While there was some borrowing of general concepts from the arms industry, the product and the materials with which it was made were so strikingly different that the transfer of technology could only have been of the most general kind.

The typewriter industry borrowed technology from the sewing machine industry and the arms industry, but also developed new technology independently. The early machine decorations and the use of cast iron in early typewriters were influenced by the Remington sewing machine factory whose chief mechanic, William K. Jenne, did much of the development work. The typewriter's many identical screws and small forged parts influenced its manufacturers to use some armory techniques. Jefferson Clough, the superintendent of the Remington armory, contributed to the early typewriter manufacturing efforts. The early typewriter manufacturers developed techniques outside the arms and sewing machine industries. The use of light wood, steel type, vulcanized rubber, glass, and light sheet iron have no history of use in the arms industry or the sewing machine industry. Some manufacturers adopted sheet steel stamping and bending techniques from the bicycle makers.

The manufacture of typewriters and watches demonstrates most clearly that the development of the American System in the private sector after mid century did not rely on the armories.

In conclusion, the American System of Manufactures had deep roots in both the public sector and the private sector prior to the Civil War. That which occurred in the private sector was of far greater importance.

The antebellum economy featured rapid, economically viable technological change in both the axe and wooden movement clock industries, independent of the work being done in the federal armories. These manufacturers found the mass production of interchangeable parts mechanisms more profitable than continuing the craft traditions from which they emerged. Simultaneously, under vastly different economic and technological conditions, the American System of Manufactures was developing independently in the federal armories, but not to such a degree in the private armories.

After 1850, and especially after the Civil War, the federal armories lost the technological lead in the American System of Manufactures to the private sector. Driven by the technical needs of the products they manufactured, the watch and typewriter industries developed new techniques and employed new materials to mass produce their goods.

Driven by economic forces from which the federal armories were sheltered, the watch and typewriter manufacturers found profit in the mass production of interchangeable parts mechanisms. The conclusion is inescapable, the American System of Manufactures—the mass production of consumer durable goods composed of interchangeable parts made with specialized machine tools—was highly profitable.

The American System was especially profitable for companies in the early phase of an industry's development, before competitors copied the technology. This is clear in the cases of wooden movement clocks, axes, typewriters, and watches. Even after the market was flooded with competitors, the successfully established firms competed effectively with increasingly advanced technology, as did the Waltham Watch Company and the Remington Typewriter Company. The only way competitors were able to capture any market share was by introducing a newly designed product and manufacturing it with American System technology. Technological enthusiasm notwithstanding, the American System did pay, and paid very well.

The American System returned a profit to its investors because it was the product of two very talented but very different groups of men, entrepreneurs and engineers. The entrepreneurs provided capital and marketing skills while the engineers supplied the technical know-how and the new ideas. Neither group was capable of succeeding independently. Each required the skills of the other. The "enthusiastic" engineers had no outlet for their ideas without the economic opportunity offered by the entrepreneurs.

Similarly, the entrepreneurs could not exploit the opportunities of the market without the necessary products of manufacturing technology. Independently, each was incomplete, yet together, they loosed the creative forces which shaped the American economy and provided the mechanism for economic growth.

Notes

Introduction

1. David F. Noble, "Command Performance: A Perspective on the Social and Economic Consequences of Military Enterprise," ch. 8 in Merritt Roe Smith, ed., *Military Enterprise and Technological Change: Perspectives on the American Experience* (Cambridge: MIT Press, 1985), p. 337.

2. The exception being axes.

3. Edwin A. Battison, "Eli Whitney and the Milling Machine," *Smithsonian Journal of History* (1966), 1:9–34

4. David A. Hounshell, *From the American System to Mass Production, 1800–1932,* (Baltimore, Md.: Johns Hopkins University Press, 1984); Merritt Roe Smith, ed., *Military Enterprise and Technological Change.*

5. Merritt Roe Smith, "Army Ordnance and the "American System" of Manufacturing, 1815–1861," ch. 1 in Merritt Roe Smith, ed., *Military Enterprise and Technological Change,* pp. 63–64.

6. Hounshell, *From the American System to Mass Production,* p. 25.

7. Barton C. and Sally L. Hacker, "Smith, Merritt Roe, ed., *Military Enterprise and Technological Change; Perspectives on the American Experience,*" *Technology & Culture* (July 1987), 28(3):709–712.

Throughout the literature on the American System runs the theme that its adoption deskilled the American craftsman and reduced him to the level of a machine operative. Lost in the debate is the fact that the new technology required many new skills. These included invention, production, maintenance, and repair. The new technology so lowered the price of many consumer goods that it usually created many new jobs. Nineteenth-century factory operatives running automatic or semi-automatic machine tools developed different skills than their craftsman fathers, but they were still very skilled people. See, for example, such works of John R. Harris as: *Liverpool and Merseyside* (London: Frank Cass, Ltd., 1969); "Skills, Coal and British Industry in the Eighteenth Century," *History* (1976), pp. 167–182; *Industry and Technology in the Eighteenth Century: Britain and France* (published lecture, University of Birmingham, 1971); and "SaintGobain and Ravenhead," *Great Britain and Her World* (Manchester: Manchester University Press, 1975), pp. 27–70.

8. Noble, "Command Performance," p. 338. Having studied clock, watch, axe, and typewriter manufacturing, I find virtually no industrial strife connected with the adoption of the American System in the private sector. Indeed, in watch and typewriter manufacturing just the opposite seems to be the case. The problem experienced by the craftsmen at Harpers Ferry has no counterpart in watch manufacturing, where

extremely rapid technological change took place. See, for example, Donald R. Hoke, "No Luddites Here! The Technology and Culture of American Watch Factories," paper delivered at the Social Science History Association meeting St. Louis, Mo., October 1986.

There is a useful comparison to be made between the Harpers Ferry experience and that of Collins & Co. Collins' problem was not merely with a competitor seventy miles west, but with the agricultural cycle of life. His workers wanted to return to their farms to plant and reap, while Collins required them to continue full time employment with the axe company. The problems at Harpers Ferry and Collins & Company both occurred in large, water powered, capital intensive settings. Contrast this with the smooth adoption of the new technology by the wooden movement clock contractors who apparently worked out of their homes primarily in the winter. There appears to be a technological imperative at work here, the large capital intensive establishments requiring more industrial discipline than the putting out system in private homes.

9. Hounshell, *From the American System to Mass Production*, pp. 48–49.

10. Hounshell, *From the American System to Mass Production*, p. 4.

11. Hounshell, *From the American System to Mass Production*, p. 120.

12. Hounshell, *From the American System to Mass Production*, p. 33.

13. Hounshell, *From the American System to Mass Production*, pp. 20–23.

14. Hounshell, *From the American System to Mass Production*. With the single exception of two sewing machines—that are distinctly different models—all the objects used in Hounshell's work are employed for illustrative purposes only.

15. I apologize for this untimely pun.

16. Terry's Porter Contract clocks were sold complete with dial, hands, weights, and pendulum. They were intended to be cased and were not intended to be "hang-up" clocks as were some of the earlier American wooden movement clocks or the Black Forest "Wag-on-the-Wall" clocks.

Chris Bailey, former curator and director of the American Clock & Watch Museum in Bristol, Connecticut and an expert on early American clocks, was of great help in teaching me the differences between Terry's early shelf clocks. He was particulary helpful in stressing the different movements and their features. A trip to the American Clock & Watch Museum should be part of every historian's education.

17. Hounshell, *From the American System to Mass Production*, p. 54.

18. Hounshell, *From the American System to Mass Production*, p. 56.

19. Hounshell, *From the American System to Mass Production*, p. 56.

20. Hounshell, *From the American System to Mass Production*, p. 56.

21. Hounshell, *From the American System to Mass Production*, p. 57.

22. The concept appears once with respect to sewing machines, p. 112, and once with respect to Ford production, pp. 283–286.

23. Hounshell, *From the American System to Mass Production*, p. 83.

24. Hounshell, *From the American System to Mass Production*, pp. 68 and 75.

25. Roger Moore to Donald Hoke, David Hounshell et al., Dear Park Inn, Newark, Delaware—Spring 1975, "over a tall frothy one."

26. At least as much respect as one can have for the ideas of a Dallas Cowboy's fan.

27. Hounshell, *From the American System to Mass Production*, p. 8.

28. The Eagle Bicycle Manufacturing Company of Torrington, Connecticut, also used a combination of new and existing technologies. In 1895, its hubs and bearings were "turned out of the solid bar," as were Pope's. It cold swaged its spokes and also used special tools to assemble its machine. "The pieces being assembled in a heavy metal frame, called the "jig," are drilled and pinned together and are brazed, being subjected to operations of alignment between the brazing operations." The Eagle featured aluminum wheels, but apparently riveted rather than welded them together. See, "The Eagle Bicycle," *Scientific American* (January 11, 1896), 74(2):17 and 20.

29. Hounshell, *From the American System to Mass Production,* p. 205.

30. Hounshell, *From the American System to Mass Production,* p. 212.

31. No Hagley Fellow passes through the University of Delaware without being influenced by Eugene Ferguson (some of us more than others). One of the debates between graduate students in economic and technological history was Ferguson's notion of the fervent, indeed rhapsodic, engineer and "technological enthusiasm." Ferguson had clearly identified an important phenomenon; none of us could quite figure out how to interpret it and put it in perspective.

32. It is no longer a ballistics range, one shot was fired in 1963 and the round ricocheted so badly that it was never used again.

33. See Milwaukee Public Museum accession record 26819. This typewriter was donated by Mrs. Wilson E. Mayer, Carl P. Dietz Typewriter Collection, H 43403/26819.

34. The exception being John Hall's breach loading rifle.

35. Hounshell, *From the American System to Mass Production,* p. 92.

36. Lynn White, Jr., "The Discipline of the History of Technology," *Journal of Engineering Education* (January 1964), 54(10):349–351.

1. *Meanwhile, Over in the Private Sector . . .*

1. Edward A. Marsh, "History," manuscript RC-2 in the Waltham Watch Company Papers, Baker Library, Harvard University. The internal evidence suggests that the manuscript was written about 1921 by Edward A. Marsh, the Master Mechanic who worked through and participated in the automation of the Waltham Watch Company. Marsh wrote a number of articles and a book on the technical history of Waltham. He died in 1934 and was probably the only employee at Waltham in the early 1920s who could have written this "History."

Throughout this work I have used the terms "mechanic" and "engineer" more or less interchangeably. If pressed for a sharper definition, I would define the mechanic as an antebellum individual lacking formal training. The engineer was a postbellum individual with formal training of some sort. The transition from mechanic to engineer in the nineteenth century is unclear and no definition can do more than suggest the difference, if there is indeed a difference.

2. This definition is taken from my class notes of lectures given by Eugene S. Ferguson between 1973 and 1975. I don't recall having seen it published.

3. This work seeks to complete the study of the industries traditionally known as the American System industries and noted by Roe in 1916, typewriters, and watches in particular. See Joseph Wickham Roe, *English and American Tool Builders* (1916),

and Joseph Wickham Roe, "Interchangeable Manufacture in American Industry." See also David A. Hounshell, *From the American System to Mass Production*.

4. L. T. C. Rolt, *A Short History of Machine Tools* (Cambridge: MIT Press, 1965), p. 147, "It is not the inception of the interchangeable system in America but its rapid application to industry generally that requires explanation."

5. Adjustment is often designed into the product itself. Consider the adjustment phase of American watch manufacturing in the late nineteenth century. The pallet stones were cemented in the pallet fork matched to a particular escape wheel, the hairspring was made to reflect the strength and length of the particular piece of steel in hand as well as the particular balance wheel it was to oscillate for the next billion oscillations, and the balance and poising screws which provide the adjustment for temperature and position were properly positioned in the balance wheel rim. All of these manufacturing operations were performed with highly specialized tools which required a great deal of highly skilled hand work. In addition, watches were built with a final adjusting feature, the hairspring regulator lever which lengthened or shortened the effective length of the hairspring and hence changing the beat of the watch. Given the product, it was the only method to produce inexpensive watches. It was certainly possible to make watches without this final, user adjustment feature. Waltham and Howard both had extremely high grades of free-sprung watches which were made without micrometer hairspring regulator levers, but they are the exceptions and were extremely expensive to make and sell.

In typewriters, the critical adjusting problem was alignment of the type. The American Writing Machine Company, makers of the Caligraph, designed adjustability into their type bars so that the owner could adjust the machine after it had begun to wear after hard use and "get out of alignment."

6. That Pope, an armory practice bicycle maker would produce such a slow, cheap typewriter as the World is puzzling. There is very little if any armory practice found in the manufacture of this machine. Perhaps Pope wanted into the typewriter market so badly that it took anything it could find. It is not entirely clear that Pope manufactured these machines, as many are marked "Typewriter Improvement Company."

It is worth noting that Pope evidently attempted to enter the high quality typewriter market in June 1889, when it contracted with Walter J. Barron, a well-known typewriter inventor, to produce a machine on the "Wagner" model. Wagner's model eventually became the Underwood Typewriter, the machine that sparked the transition from "blind" to "visible" typewriters. See Densmore Papers, Box 7, H 40100/25593, "W. J. Barron With A. Densmore Agreement," in the Milwaukee Public Museum.

7. Merritt Roe Smith, *Harpers Ferry Armory and the New Technology* (Ithaca, N.Y,: Cornell University Press, 1977), and Merritt Roe Smith, "John H. Hall, Simeon North, and the Milling Machine: The Nature of Innovation among Antebellum Arms Makers," *Technology & Culture* (October 1973), 14(4).

8. David A. Hounshell, "From the American System to Mass Production," (Ph.D. dissertation, University of Delaware, 1977). Hounshell has the best discussion of the literature on the American System.

9. Robert A. Howard, "Interchangeable Parts Reexamined: The Private Sector of the American Arms Industry on the Eve of the Civil War," *Technology & Culture* (October 1978), 19(4):633–649.

10. Donald Hoke, "Conference Report: A Symposium on the Rise of the American System of Manufactures," *Technology & Culture* (January 1980), 21(1):67–70.

11. Bruce Bliven, *The Wonderful Writing Machine* (New York: Random House, 1954), ch. 11, "Adjuster at Work."

12. David A. Hounshell, "From the American System to Mass Production."

13. As a youngster, I recall riding my sister's bicycle (without permission) and having the misfortune of smashing into my brother's bike, ripping out several spokes of my sister's front wheel. I attempted to repair the wheel by removing every other spoke, hoping the wheel would run true and the accident go unnoticed. In my futile effort to repair the wheel, I came to understand the importance of wheel truing. (The repair cost was eventually garnished from my allowance).

14. Henry F. Piaget, *"The Watch": Its Construction, Merits and Defects; How to Choose it and How to Use it* (New York: Published by the Author, 1868), 88 pp. The title page contains an overpasted label indicating that the 1868 edition is the "Third Edition, Improved and Enlarged, Suitable to the Present Style of Watches," but the 1877 edition is clearly printed on the cover "Third Edition," leading to the belief that the 1868 edition is the second edition. Piaget mentions an edition of 1860 which was presumably the first edition. See also Henry F. Piaget, *"The Watch": Its Construction, Merits and Defects Explained and Compared. History of Watch Making by Both Systems* (New York: Published by the Author, A. N. Whithorne, 1877), 36 pp.

15. Henry Ford stated that "in mass production there are no fitters." He was referring to the assembly of his Model T. But Henry Ford was simply another in a long line of entrepreneurs who claimed full interchangeability without ever having achieved it. I distinctly recall a visit to a (General Motors?) automobile assembly plant in Baltimore, Maryland, with a Boy Scout troop about 1964. My most vivid memory of the tour was watching a worker "fit" rear doors on station wagons with a large rubber hammer. When the door didn't fit correctly, he gave it a good hard smash with his hammer or leaned on the door or violently pulled it up into position. These parts were no more "perfectly interchangeable" than were Henry Ford's, and this worker's hammer illustrates graphically one of the most important aspects of the American System and mass production, assembly and adjustment. In mass production, Henry Ford notwithstanding, there are always fitters and adjusters.

16. W. F. Durfee, "The History and Modern Development of the Art of Interchangeable Construction in Mechanism," *ASME Transactions* (1893), 14:1225–1257. See also W. F. Durfee, "The First Systematic Attempt at Interchangeability in Firearms," *Cassier's Magazine* (April 1894), 5(30):469–477.

17. Contrary to Marxian historiography, the labor history of the private sector American System industries in the nineteenth century was remarkably free of labor strife. This is especially true of the more highly paid mechanics. Technological change in the American System industries did not lead to labor problems.

18. Eugene S. Ferguson, "The Americanness of American Technology," notes that: "As a group, economic historians have grappled more seriously with explanations of technological development in America than have other historians, yet their explanations are mechanical, elegant perhaps, but useless when we begin to probe for meaning and significance." He continues, "To suppose that those Americans calculated the subtleties of factor endowment and incremental gains on alternative in-

vestment possibilities is to miss the strong romantic and emotional strain in the narrative of the American involvement with its technology." See also Eugene S. Ferguson, "Enthusiasm and Objectivity in Technological Development," paper read Dec. 29, 1970, at AAAS symposium, "Technology: Nuts and Bolts or Social Process;" Ferguson, "Toward A Discipline of the History of Technology," *Technology & Culture* (January 1974), 15(1):13–30; and "On the Origin and Development of American Mechanical "Know-How,"" *Midcontinent American Studies Journal* (Fall 1962), 3(2):2–16; H. J. Habakkuk, "The Economic Effects of Labor Scarcity," in Saul, ed., *Technological Change*, pp. 23–76; "Second Thoughts on American and British Technology in the Nineteenth Century," *Business Archives & History* (August 1963), 3(2):187–194.

19. Merritt Roe Smith, "From Craftsman to Mechanic: The Harpers Ferry Experience, 1789–1854," in Ian M. G. Quimby and Polly Anne Earl, eds., *Technological Innovation in the Decorative Arts* (Charlottesville: University Press of Virginia, 1974).

20. Antiquarians who collect watches, clocks, and typewriters take great pride in identifying and accumulating the unusual designs and products of small peripheral companies. They often spurn the common, but vastly more important machines. Thus watches made by the Freeport Watch Company are historically unimportant, yet highly prized by collectors, while the products of the Waltham Watch Company (with some exceptions) are not actively sought.

21. Eugene S. Ferguson, "The Critical Period of American Technology," (paper, ca. 1961; copy in Eleutherian Mills Historical Library, Greenville, Wilmington, Del., 19807). "The fundamental decisions to make the most, if not the best, were reached . . . before 1853. Stated another way, I think we got the way we are today during the first half of the nineteenth century."

22. David A. Hounshell, "The Bicycle and the "American System" of Manufactures" (May 1975).

23. H. J. Habakkuk, see note 18 above.

24. Paul Uselding, "Elisha K. Root, Forging, and the "American System,"" *Technology & Culture* (October 1974), 15(4):543–568. Uselding's article is an excellent example of the theoretical economist's attempt to explain technological change without understanding the technology itself.

25. Eugene S. Ferguson, see note 18 above; David A. Hounshell, *From the American System to Mass Production*, especially ch. 2, "The Sewing Machine and the American System of Manufactures." My good friend David Hounshell, (a student of Ferguson's), goes to great lengths to avoid seemingly obvious economic forces influencing the sewing machine manufacturers. See also Russell I. Fries, "British Response to the American System: Small Arms Industry after 1850," *Technology & Culture* (July 1975), 16(3):377–403, for a discussion of the noneconomic response to technological change.

2. *Wooden Movement Clock Manufacturing in Connecticut, 1807–1850*

1. Eugene S. Ferguson, "History and Historiography," in Otto Mayr and Robert C. Post, eds., *Yankee Enterprise: The Rise of the American System of Manufactures*, pp. 1–23, n.14.

2. John Joseph Murphy, "Entrepreneurship and the Establishment of the American Clock Industry" (Ph.D. dissertation, Yale University, 1961). Hereafter cited as: Murphy, "Entrepreneurship." See also John Joseph Murphy, "Entrepreneurship in the Establishment of the American Clock Industry," *Journal of Economic History*.

3. Murphy, "Entrepreneurship," pp. 61–62.

4. David A. Hounshell, "From the American System to Mass Production, The Development of Manufacturing Technology in the United States, 1850–1920" (Ph.D. dissertation, University of Delaware, 1978), ch. 1, p. 58. Hereafter cited as Hounshell, "Mass Production."

5. See, for example, Brooke Hindle, "A Bibliography of Early American Technology," pp. 29–94, in *Technology in Early America, Needs and Opportunities for Study*. "The manufacture of clocks has been properly assigned a special place in the rise of machine technology. Early, in the handicraft period, a standard of measured precision and of machine production became a feature of clockmaking; it was not attained in other crafts until much later," p. 71. See also Eugene S. Ferguson, "History and Historiography," especially p. 4, and *n*14.

6. Notably absent from recent historical work are the efforts of the antiquarians in the National Association of Watch and Clock Collectors, Inc., whose headquarters are in Columbia, Penn. Within that group there is an informal subgroup known as the Cog Counters who specialize in wooden movement clocks. Their literature is cited often in the notes below.

7. For a superb review of the literature on the technology of antebellum America, see Hounshell, "The American System of Manufactures in the Antebellum Period," pp. 7–85, in "Mass Production." Hounshell has cited virtually all the published sources.

8. Merritt Roe Smith, *Harpers Ferry Armory and the New Technology*.

9. Brooke Hindle, *Technology in Early America: Needs and Opportunities for Study*, p. 71.

10. Edwin A. Battison and Patricia E. Kane, *The American Clock, 1725–1865: The Mabel Brady Garvan and Other Collections at Yale University*, p. 16. Hereafter cited as Battison and Kane, *American Clock*.

11. Penrose R. Hoopes, *Shop Records of Daniel Burnap, Clockmaker*, especially part three, "Shop Methods and Equipment," pp. 95–104, and part four, "Memorandum Book," pp. 107–132, including Burnap's description of clockmaking.

12. Penrose R. Hoopes, *Connecticut Clockmakers of the Eighteenth Century*, p. 33. Hereafter cited as: Hoopes, *Connecticut Clockmakers*.

13. David Todd, the Smithsonian clockmaker, has worked on thousands of clocks. He assures me that clockmaking is an art in many instances and not a science or business. No one, he argues, would go to the trouble to create such an item as a clock with such beauty and finish if it were merely intended to indicate the time. Having discussed this with Mr. Todd and having looked at many clocks myself, I am inclined to agree. Careful observers can still see the marks of the depthing tool on many clocks, indicating that each clock was an individual effort.

14. John Wyke, *A Catalogue of Tools for Watch and Clock Makers by John Wyke of Liverpool*. Wyke illustrates some 492 tools, mostly for watchmakers and clockmakers although some are cabinetmaker's tools.

15. Murphy, "Entrepreneurship," p. 17. See also Battison and Kane, *American Clock*, p. 17.

16. For example, in the collections of the Milwaukee Public Museum there is a brass movement tall clock by Jacob Gorgas, Jr., of Bucks County, Penn., ca. 1795–1800. This clock has its original dial which is clearly marked "Wilson, Birmingham" on the back of the calendar wheel. Milwaukee Public Museum catalogue no. N 991. Numerous examples of the same phenomenon can be found in museums and private collections today. See, for example, Stacy B. C. Wood and Stephen E. Kramer, III, *Clockmakers of Lancaster County and Their Clocks, 1750–1850,* pp. 86, 101, 111, 117, 120, 122, and 129.

17. Battison and Kane, *American Clock,* p. 16, Cheney, "native school." These American clockmakers were not the first to make wooden movement clocks. European clockmakers made wooden clocks long before they were manufactured in New England. Many examples of these European clocks survive, both English and Continental. For example, the famous English chronometer maker John Harrison began life as a carpenter and used wood extensively in his early clocks. However, none of these makers seem to have influenced the American clockmakers who began working in wood. Their clocks are strikingly different both in design and the handling of the wood itself. See, for example, Albert L. Partridge, "Wood Clocks, The Art or Mystery of their Manufacture." Partridge cites a number of European wooden works clocks and concludes, on the basis of clock design, the location of manufacture, and the lack of any connection with Connecticut and/or Terry, that there was no connection at all with the Connecticut wooden works makers, and that Terry et al must have developed their wooden works clocks and their technology with no knowledge of their European antecedents. See also Albert L. Partridge, "Connecticut Enters the Eight-Day Field," and "Wood Clocks, Connecticut Carries On." See also an unentitled note on Harrison in *The Cog Counter's Journal* (May and August 1977), 14: 56. See also, Hoopes, *Connecticut Clockmakers,* "Wood Clockmaking, pp. 32–36. Hoopes believes there was a connection between the European wooden works makers and the American makers through Gideon Roberts of Bristol who spent some time in the Wyoming Valley in Pennsylvania among the Germans. Presumably these Germans had a tradition of wooden clock making and influenced the design of the wooden clock Roberts made which differed from the Cheney type. I find this difficult to accept in light of the design of the clocks in question.

18. Laurence Luther Barber, "The Clockmakers of Ashby, Massachusetts." See, for example, a clock by Alex Willard of Ashby, Mass., in Battison and Kane, *American Clock,* pp. 74–77.

19. Hoopes, *Connecticut Clockmakers,* p. 33. For a typical nineteenth-century example, see Abram English Brown, "By Sun Dial and Noon Mark" (1899), in an unknown newspaper. ":598 Photograph File Watches," Baker Library, Harvard Business School. Brown's article carries a lengthy subtitle, "Primitive Ways of Telling the Time Were Good Enough for Our Ancestors—Clocks Became a Necessity in Every House as Soon as Yankee Genius Made this Possible—Some Pioneer Manufacturers of Tall Clocks—Wonderful Growth of the Industry During the Present Century."

20. Hoopes, *Connecticut Clockmakers,* p. 34.

21. Chauncey Jerome, *History of the American Clock Business for the Past Sixty Years and Life of Chauncey Jerome Written by Himself,* p. 36. Hereafter cited as: Jerome, *American Clock Business.* Jerome is an important source of data on the early American

clock industry. He lived through the wooden movement era and well into the brass era as an active participant. His recollections are generally quite good.

22. Hoopes, *Connecticut Clockmakers*, pp. 35–36, and 38. See also Chris H. Bailey, *Two Hundred Years of American Clocks and Watches*, pp. 69–70, where the "crude clocks" and "hand sawn" teeth myth is repeated.

23. Ward Francillon to Donald Hoke, February 27, 1983.

24. Henry Terry, "A Review of Dr. Alcott's History of Clock-Making: By A Clock-Maker." Hereafter cited as: Henry Terry, "A Review." Henry Terry was the son of Eli Terry and a clockmaker himself. Henry was fully familiar with the wooden movement manufacturing process and his account is quite accurate. Since Terry's article is reprinted in full in Kenneth D. Roberts, *Eli Terry and the Connecticut Shelf Clock*, I have used Roberts' page numbers in citing this article.

25. Connecticut Historical Society wheel cutting engine, accession no. 1981,117.0, gift of Kenneth D. Roberts.

26. Hoopes, *Connecticut Clockmakers*, pp. 34-35.

27. "Pull-up" refers to the method of winding the clock. Rather than using a key, each winding drum carried two cords wound in opposite directions. The owner wound the clock daily by pulling down on the cord with the light weight which had been wound onto the drum during the previous day. In the process, the driving weight of the clock was "pulled-up" and the clock wound. Many wooden movement makers painted false winding holes on their clocks' dials to imitate the more expensive brass clocks.

28. Jerome, *American Clock Business*, p. 38, and Henry Terry, "A Review," p. 36, provided data for the initial list; additional data came from William H. Distin and Robert Bishop, *The American Clock: A Comprehensive Pictorial Survey, 1723–1900, With A Listing of 6,153 Clockmakers*, as well as various issues of *The Cog Counter's Journal* and *The Bulletin of the N.A.W.C.C., Inc.*

29. Daniel Burnap is a good example. He retired from clockmaking and went into farming after the onslaught of the cheap clock. It is ironic that he was put out of the clock business by one of his own apprentices, Eli Terry.

30. Murphy, "Entrepreneurship," p. 3. The invention and development of the 30-hour, wooden movement shelf clock and the evolution of a mass production technology for its manufacture have been credited to Eli Terry. Unlike his contemporary, Eli Whitney, Eli Terry's accomplishments are well documented, and there are no pretenders to claim credit for Terry's accomplishments. The heroic American inventor has been out of favor with American historians for the past two decades, but Eli Terry seems to qualify for just such a title.

31. Henry Terry, "A Review," p. 36.

32. I have drawn on six basic sources for this short biography of Terry: Murphy, "Entrepreneurship," and "Entrepreneurship in the Establishment of the American Clock Industry"; Chris H. Bailey, *Two Hundred Years of American Clocks and Watches*, pp. 103–134; Roberts, *Eli Terry;* and Virginia and Howard Sloane, "4,000 Clocks: The Story of Eli Terry and His Mysterious Financiers." "Mr. Cheeney" probably refers to Benjamin or Timothy Cheney, the wooden movement clock makers working as early as 1759 in East Hartford.

33. Penrose R. Hoopes, *Shop Records of Daniel Burnap, Clockmaker.*

34. Henry Terry, "A Review," in Roberts, p. 35, plate V.

35. Hoopes, *Connecticut Clockmakers*, pp. 32–36. See also Partridge, n17 above.

36. Ward Francillon to Donald Hoke, February 27, 1983.

37. Battison and Kane, *American Clock*, p. 17. Battison cites no evidence here for Terry's decision to quit the 30-hour, pull up, tall clock business, but he was a craftsman and the explanation that he was disturbed at the low quality of work is not inconceivable. See also John Joseph Murphy, "Entrepreneurship in the Establishment of the American Clock Industry," for craftsmanship of Terry.

38. Henry Terry, "A Review," p. 36.

39. Jerome, *American Clock Business*, p. 41.

40. Battison and Kane, *American Clock*, p. 17. Battison cites no data to support his hypothesis about house size. This explanation is not found in the contemporary accounts nor is it mentioned by other historians.

41. Brooks Palmer, *The Book of American Clocks*, p. 3.

42. Jerome, *American Clock Business*, pp. 37, 39, 42, and 44.

43. Snowden Taylor, "Daniel Pratt, Jr., Boardman & Wells, Levi Smith—The End of the Wood Clock Era."

44. The wooden movement shelf clock was probably the first mechanical, self-acting product to find its way into the average American home. It stood apart from the firearm, the rotating meat spit, the spinning wheel, and the linen press, in that it was a luxury item and was far more complex, especially those clocks with alarm mechanisms. See Hoopes, *Connecticut Clockmakers*, and Palmer, *The Book of American Clocks*, p. 4.

45. Henry Terry, "A Review," p. 36; Murphy, "Entrepreneurship," p. 114; and Jerome, *American Clock Business*, on the four-cent clock case.

46. This clock is known among collectors as the Standard Terry-Type, five-arbor, thirty-hour, wooden movement shelf clock. See, for example, Snowden Taylor, "Characteristics of Standard Terry-Type 30-Hour Wooden Movements as a Guide to Identification of Movement Makers."

47. These clocks are usually identifiable through the term "Patent Clock," which appears on the clock label. Eli Terry & Sons, Eli Terry, Jr., Henry Terry, and S. B. Terry were some of the firm names under which the five-arbor, thirty-hour, wooden movement clocks were made by members of the Terry family.

48. One of the more interesting sidelights on Terry, and one outside the scope of this study of mass production, was his other work with clocks. Eli Terry was a highly skilled craftsman and, as Jerome noted, a "natural philosopher" as well. His first patent, the first American horological patent, was issued on November 17, 1797. It covered an "equation clock," which had two dials, one showing the apparent time (solar time) and the other showing mean (or true) time. He built one of these clocks for the City of New Haven, Connecticut, in 1826, but the dual dials so confused the townspeople that they demanded one of the dials be removed. It is not known if Terry developed this clock independently. Such clocks had existed for some time in England. Terry also made a fine regulator and many other clocks. His son Henry wrote of him that, after he retired from business in 1833, Eli Terry "for many years before his death, . . . never abandoned his workshop. No year elapsed up to the time of his last sickness without some new design in clockwork, specimens of which

are now abundant." See Bailey, *Two Hundred Years,* pp. 104 and 120; Henry Terry, "A Review"; and Elmer C. Korten, "An Eli Terry Regulator."

49. For example, see Chris H. Bailey, "Notes on 'Torrington' Clocks."

50. Bailey, *Two Hundred Years,* p. 121; and Bailey, "Mr. Terry's Waterbury Competitors, The Leavenworths and Their Associates." This phenomenon, the immediate imitation of Terry's design, was not new in the clock industry. Simon Willard's "Patent Timepiece," the so-called "banjo" clock, had also been widely and quickly copied in the first decade of the century. Subsequently, other clock and case designs were widely and quickly copied as well. It was (and remains to this day) a feature of the horological market. Walter A. Dyer, *Early American Craftsmen,* pp. 104–130, and pp. 133–161. See also Allan Dodds Frank, "Genuine Phonies." Frank relates the story of Rolex's problem with imitation products. See the *Washington Post* classified advertisement section (January and February 1983) for examples of such advertisements.

51. Snowden Taylor, "Characteristics of Standard Terry-Type 30-Hour Wooden Movements as a Guide to Identification of Movement Makers." Some of these "makers" may have been assemblers only or purchasers of other maker's movements. Kenneth D. Roberts, "Documented Listing of Connecticut Firms Manufacturing or Marketing Wooden Movement Shelf Clocks, 1816–1850." Roberts does not distinguish between the makers (i.e., manufacturers) and the dealers, and his listing is undocumented.

52. Elisha Manross account book, Connecticut Historical Society, Ms. 80834.

53. Murphy, "Entrepreneurship," p. 197, Jerome, *American Clock Business,* p. 7.

54. Murphy, "Entrepreneurship," p. 83.

55. Murphy, "Entrepreneurship," p. 92.

56. Jerome, *American Clock Business,* p. 43.

57. Felicia J. Deyrup, *Arms Makers of the Connecticut Valley: A Regional Study of the Economic Development of the Small Arms Industry, 1798–1879.*

58. John A. Diehl, "Luman Watson, Cincinnati Clockmaker." All the Downs and Watson data are from Diehl.

59. Murphy, "Entrepreneurship," p. 83.

60. Chris H. Bailey, "George B. Seymour's Accounts." This account book is in the collections of the American Clock & Watch Museum in Bristol, Conn.

61. "From the Archives—Connecticut State Library, Probate Files"; and Samuel Terry to Norman Olmstead, April 29, 1829, reprinted in Roberts, *Eli Terry,* p. 224, table 45, where Samuel Terry notes that he has wheels in stock. Both Seth Thomas and Hopkins & Alfred carried inventories of parts, as their surviving inventories indicate.

62. SAS Institute, Inc., *SAS User's Guide: Basics.*

63. It is possible, indeed likely, that defective parts were discarded and that this collection represents only those parts that were made to gauge. There is no way to guess what percentage of total production was acceptable.

64. Samuel Eliot Morrison, *Admiral of the Ocean Sea: A Life of Christopher Columbus,* p. 2.

65. Brooke Hindle, ed., *America's Wooden Age: Aspects of its Technology,* p. 4.

66. Bruce A. Burns, "Terry Standard Thirty-Hour Wood Movement Contemporaries."

67. Ward Francillon, "Some Wood Movement Alarms"; and Kenneth D. Roberts, "Documented Listing of Connecticut Firms Manufacturing or Marketing Wooden

Movement Shelf Clocks, 1816–1850." These movements were made by about a dozen manufacturers, "while there were hundreds of case makers, jobbers and distributors who bought movements and cased, labelled and sold clocks under their own names. Even among the major producers, exchanges of cases for movements was (sic) common."

68. Snowden Taylor, "Characteristics of Standard Terry-Type 30-Hour Wooden Movements as a Guide to Identification of Movement Makers."

69. By 1822 Terry had developed his five arbor, thirty-hour, wooden movement cased in his famous "Pillar and Scroll Top" case. Terry attempted to cover this and other designs with United States patents, but like his contemporaries (Eli Whitney and Oliver Evans, for example), Terry was unable to protect his inventions through the patent office. Imitators and infringers sprang up quickly, and although some makers paid Terry a license fee for using Terry's patents, most did not. Some competitors redesigned the wooden movement shelf clock to get around Terry's patent, the most notable movement designs were the so called "groaner movements," the "Torrington" (or "East-West" movements made by Norris North and others in and around Torrington, Connecticut), and the Hoadley "upside-down" movement which was strikingly similar to Terry's movement but was mounted upside down in the case. All of these movements and numerous others incorporated all the Terry improvements, but in an arrangement that freed them from infringing on Terry's patent. Although there is great external variety in these clocks, they are actually more the same than different. See Roberts, *Eli Terry;* Lockwood Barr, *Eli Terry Pillar and Scroll Shelf Clocks;* Ward Francillon, "Some Wood Movement Alarms"; and Chris H. Bailey, "Notes on 'Torrington' Clocks." For early American patent problems and patent laws, see, for example, Constance Green, *Eli Whitney and the Birth of American Technology;* Bathe and Bathe, *Oliver Evans;* and Eugene S. Ferguson, *Oliver Evans.*

70. Ward Francillon, "Some Wood Movement Alarms."

71. George Bruno, during long conversations at his home and later at the N.A.W.C.C. National Meeting in Philadelphia in July 1983, noted the common sizes found on most wooden movement clocks, 5/8", .358", .410". The holes vary in size, but the sizes are standard. George Bruno exemplifies the spirit of the N.A.W.C.C. in sharing his knowledge freely. I thoroughly enjoyed learning from George, but I simply enjoyed his company as well.

72. Bailey, *Two Hundred Years,* p. 111, illus. 111.

73. Both Seth Thomas and Hopkins & Alfred carried inventories of parts, as the surviving inventories indicate. In addition, Samuel Terry notes that he has wheels in stock. See Samuel Terry to Norman Olmstead, April 29, 1829, in Roberts, *Eli Terry.*

74. Some individuals believe the plate fixture was used to mark holes on each plate. These hole would then have been drilled individually using a drill press.

75. Brooke Hindle, ed., *America's Wooden Age: Aspects of its Technology,* p. 7.

76. Paul Richard, "National Gallery Names Chief Curator, Harvard's S. J. Freedberg Assumes Post Sept. 1."

77. Smithsonian Institution specimen, Museum of American History catalogue no. 334,012, accession no. 306,558. See Edwin Pugsley to Silvio Bedini, April 17, 1973, accession file 306558.007: "I shall be interested if you are able to learn more about its history. It has been in my cellar shop for over 40 years and that an old resident of Derby, Ct. took me to one of the oldest houses in Derby to get it." See also Edwin Pugsley to Silvio Bedini, March 28, 1973, accession file 306558.005: "Any-

way, Mr. Burr took me one day to one of the oldest houses in Derby and in the half basement was this device. Neither the old lady owner nor Mr. Burr knew what it was, but Burr confirmed my guess that you could rack up several wooden gear blanks on the arbor and, using the indexing device, could cut any number of teeth probably using a shaped fly cutter. Mr. Burr died probably 25 years ago."

78. Smithsonian Institution specimen, Museum of American History catalogue no. 326,469, accession no. 260,025, collected by Edwin Battison in the spring of 1965, a gift of Mr. Newton L. Lockwood and came "from Mr. Lockwood's former residence in Connecticut." The machine is described as a "large wooden framework of machine for cutting gears used in wooden clocks," weighs 400 lbs., and measures 5′ x 3′ x 4′.

79. Hounshell, "Mass Production," pp. 56–65.

80. This adjustment actually occurs through an arc rather than vertically. However, this is of no concern to the clockmaker, since the workpiece is moved into the cutter and only the distance between the center of the gear and the edge of the cutter is critical. As long as the tooth is cut properly and the face of each tooth is parallel to the centerline of the arbor, the position of the cutter is of little or no importance.

81. The three rails are extremely important because three rails are required to provide a perfectly straight cut into the tool. Two rails would not provide the necessary alignment. This principal was developed in a screw cutting lathe by David Wilkinson of Rhode Island, patented in 1789. Wilkinson's three bearing lathe carriage is significant, because it allows for the accurate machining of an item such as a lead screw with a relatively inaccurate machine. See Edwin A. Battison, "Screw-Thread Cutting by the Master-Screw Method Since 1480," especially illustration 17, the third wheel of Wilkinson's carriage is torn away in the drawing. See also Battison, *Muskets to Mass Production*.

82. This machine could also have been used for a variety of turning jobs as well by using the powered arbor to hold a workpiece instead of a cutter, for example, turning pinion blanks. Given Hiram Camp's description of pinion blank cutting, this engine could easily be set up to perform that job. Note the two square holes behind the missing arbor. These could have held the two centers into which the drop gauge fit. Camp described cutting the pinions to length by mounting two circular saws on a single arbor the proper distance apart and then making both the end cuts on the pinion simultaneously. This machine could also perform that operation.

Indeed, with a variety of fixtures, this machine could easily be set up to perform numerous jobs, including turning wheel blanks to size (diameter), wheel tooth cutting, cutting pinion blanks to size (length), pinion roughing, turning pinions to size (diameter), plate pillar turning, and cutting cam sections.

Suppose that manufacturers first performed one operation on all of one particular part on this machine, e.g., cutting pinion blanks to length, then refitted the machine with a different set of fixtures, and performed the second operation, e.g., roughing the pinion blank and then turning it to size, then this relatively simple but very versatile machine assumes a significant place in the factory. Subsequently, he could refit the engine again to turn wheel blanks to size and then cut the wheel teeth. There is evidence to suggest this sort of management may have occurred. Eli Terry reportedly used a single arbor machine to cut his wheels, and the arbors in such a machine could easily be changed just as they could be on this machine.

There is also evidence to the contrary—that more specialized machinery was used. The inventory of Asaph Hall's shop (1843) lists "1 Wheel engine, 1 pinion engine, 1 pinion lathe, 1 large lathe, 1 small lathe with apparatus, 1 mandrill, 2 shafts for turning, 1 pinion lathe (a second machine), 2 old circular saws," along with hand tools, parts, and stock. Such an inventory suggests that many parts, especially the wheels and pinions, were produced on specialized machine tools of some sort. The precise configurations of these machines remains unknown. See "From the Archives—Connecticut State Library, Probate Files"; Lawrence P. Hall, "Asaph Hall of Goshen and Clingon"; and Snowden Taylor, "Research Activities and News."

83. Smithsonian Institution specimen, Museum of American History catalogue no. 315,806, accession no. 224,779, gift of Mr. & Mrs. Newton Lockwood, collected in 1959 by Edwin Battison

84. Smithsonian Institution, Museum of American History accession file 226,926, gift of Mr. & Mrs. Newton L. Lockwood. This building was collected through the efforts of Eugene S. Ferguson with the help of Mr. George Watson and Edwin Battison in November 1959 after lengthy negotiations with the Army Corps of Engineers. See Forrest Palmer, "Harwinton Clock Shop Earns Niche in U.S. History: Headed For Exhibit At Smithsonian"; "Old Clockshop To Be Exhibited In Smithsonian." For a published reference to the Hopkins & Alfred account book, see Ward Francillon, "Some Wood Movement Alarms." Francillon reports that some 4,246 clocks and movements were sold by Hopkins & Alfred according to the "Company Book" which begins on January 1, 1834, and continues to February 26, 1842, when the factory was evidently closed. The original account book has been lost but survives in two Xerox copies, one in the possession of Mr. Francillon and the other in the American Clock & Watch Museum, Bristol, Conn. This account book does not record all the business of the factory, only the local trading business, exchanges of clocks for local products (notably tobacco, probably snuff). The factory certainly had a larger output than 4,246 clocks in this eight-year period.

85. Newton L. Lockwood to Mr. George Watson, September 22, 1959: "There are a few small old common brick in a pile, just north of the shop, where I piled them when we were cleaning up, about twelve years ago. I will try to get these brick out of your way, as I have wanted to use them for a fireplace at some future date. I think they may have come from an old dry kiln, a few rods north of the present building." Smithsonian Institution, Museum of American History accession file 226,926.026.

86. Newton L. Lockwood to Eugene S. Ferguson, May 23, 1960. Original letter in the files of the Division of Mechanisms, Museum of American History.

87. Gauges at the Connecticut Historical Society from Seth Thomas Factory, accession number 1978.44.1+ (the cataloging process at the Society has not yet been completed). See also Hounshell, "Mass Production," p. 60, and his note 124. The gauges were in the hands of William B. Kinter, unsorted and unidentified, at the time Hounshell saw them. They were subsequently donated to the Connecticut Historical Society. Hounshell's description: "Two types of "gauges" exist, marking and verifying gauges. Both types are extremely crude devices—sheet iron hastily cut and filed—which would give only the roughest kind of accuracy. The jigs that survive are timing jigs (for the correct location of striking cams) and jigs for correctly bending the rods of the striker. A templet also survives which was used for locating the bearings on the wooden plates of the Thomas clock. Although very ingenious, these devices have

little in common with the jigs, fixtures, and gauges used at the Springfield Armory. The demands for precision are worlds apart between wooden clock making and small arms production and, in truth, cannot be compared."

88. I spent five days measuring, describing, and photographing these tools at the Connecticut Historical Society in February 1983, and wish to thank Mr. Philip Dunbar for his hospitality and help in this endeavor. Several members of the N.A.W.C.C., notably from the Cog Counters, had already done some sorting and identifying of the tools in November 1982. I benefited from their work and wish to express my appreciation to them. Slides are in the files of the Milwaukee Public Museum.

89. The provenance is also rather remarkable. Related fully in a notebook kept by Winton, the story of their survival is almost as interesting as the tools themselves. In the late nineteenth century a Seth Thomas workman named Morse retired from the company after many years. Morse handled all the wooden movement clock repairs and inquiries. Shortly after his retirement the company decided to clear an old building. Someone noted that there were many old wooden movement parts and tools in the building and suggested passing them on to Morse. This was done and Morse continued his repair work after his retirement from Seth Thomas. Winton, an early collector, came to know Morse at that time. Winton knew of the material, then stored in the loft of Morse's barn, but never saw it. About 1920 Morse became senile and was sent to an old soldier's home by his daughter. The daughter called Winton offering to sell him some clocks, having found a note from Winton in one of Morse' clocks to that effect. Winton went to see Morse's daughter, bought the clocks, then asked about the material in the barn. That, he was told by the daughter, was to be hauled to the dump by a local boy. Winton managed to persuade Morse's daughter to let him clean the loft in the barn, which he did. He took several years to completely sort the material. Upon his death the material passed to Kinter and was subsequently donated to the Connecticut Historical Society.

90. There is an alternative interpretation of the use of this tool. The plate may have been marked with a punch driven through each iron bushing. Then the plate would be drilled individually. This process would tend to introduce errors.

91. Connecticut Historical Society specimen.

92. Roberts, *Eli Terry*, p. 300, plate 36, "1830 Bill to Samuel Terry for Cast Iron Clock Weights." From the Terry Papers, American Clock & Watch Museum, Bristol, Conn.

93. Hiram Camp, "A Sketch of the Clockmaking Business (1792–1892)," reprinted in part in Roberts, *Eli Terry*, p. 145. See also Murphy, "Entrepreneurship," p. 65.

95. Roberts, *Eli Terry*, p. 145.

95. Murphy, "Entrepreneurship," p. 90, cites "Address of General Joseph R. Hawley," in J. J. Jennings, ed., *Centennial Celebration of the Incorporation of the Town of Bristol*. Also cited in somewhat greater detail in Roberts, *Eli Terry*, p. 185, table 31: "Excerpts from address by General Joseph R. Hawley, Bristol's Centennial Celebration, June 17, 1885, *Bristol Centennial Celebration* (Hartford, Conn., 1885)," p. 72.

96. Norman Olmstead to Samuel Terry, April 24, 1829. Reprinted in Roberts, *Eli Terry*, p. 224, table 45.

97. Samuel Terry to Norman Olmstead, April 29, 1829. Reprinted in Roberts, *Eli Terry*, p. 224, table 45.

98. J. A. Wells, Esq., to Mr. (Daniel) Pratt, June 22, 1847. Reprinted in *The Cog*

Counter's Journal (February 1975), no. 5, p. 3. "From the papers of Daniel Pratt, Jr., courtesy of the Reading Antiquarian Society, Reading, Mass."

99. "Random notes from Samuel Terry accounts and loose papers," courtesy American Clock & Watch Museum, Bristol, Conn. *The Cog Counter's Journal* (May 1874), no. 2, p. 4.

100. Jerome, *American Clock Business,* p. 41.

101. Albert L. Partridge, "Wood Clocks: The Art or Mystery of their Manufacture." Partridge cites "Atwater, in his History of Plymouth published in 1895 . . ."

102. Jerome, *American Clock Business,* pp. 41 and 68.

103. Murphy, "Entrepreneurship," p. 65. Murphy concludes that Camp must have been correct in this assessment, since Terry could not have completed his contract with the Porters for 4,000 clocks without such an engine.

104. Murphy, "Entrepreneurship," p. 63.

105. Murphy, "Entrepreneurship," p. 64, cites Henry Terry, "A Review."

106. Hoopes, *Connecticut Clockmakers,* p. 35.

107. Jerome, *American Clock Business,* p. 68.

108. Murphy, "Entrepreneurship," p. 63.

109. The raw data and programs can be found in my dissertation, "Ingenious Yankees: The Rise of the American System of Manufactures in the Private Sector" (University of Wisconsin, Madison, May 1984).

3. *Elisha K. Root and Axe Manufacturing at the Collins Company, 1830–1849*

1. H. N. Brinsmade, D.D., *An Address at the Funeral of Samuel W. Collins, May 2, 1871.* This pamphlet is bound in MS. 72166, *The Collins Scrapbook, The Collins Company,* Connecticut Historical Society.

2. Paul Uselding, "Elisha K. Root, Forging, and the 'American System.'" See also Robert B. Gordon, "Material Evidence of the Development of Metalworking Technology at the Collins Axe Factory," *Industrial Archaeology* (1983), 9(1):19–28.

3. *Hartford Daily Post,* Saturday morning, September 2, 1865, 8(123[whole no. 2318]):2, cols. 2 and 3. See also *Hartford Daily Times, Hartford Evening Press,* and *Hartford Daily Courant,* of approximately the same date.

4. It is unclear why Root left Colt's employment in the mid 1850s, but that he did so is virtually certain. The Connecticut State Library Collection of Firearms contains a set of Colt firearms presented to Root by Colt on May 16, 1857. See *Samuel Colt Presents, A Loan Exhibition of Presentation Percussion Colt Firearms,* pp. 249–253. "The rifle, pistol, detachable stock, case, and accessories form part of a presentation of arms made to E. K. Root on the occasion of his resignation as superintendent of the Colt factory. According to a copy (Connecticut State Library, Colt Collection of Firearms) of a currently unlocated Colt factory record: "May 16th 1857 . . . Mr. Root having tendered his resignation as Superintendent of the Armory, the same was accepted," p. 253.

When Samuel Colt hired Root to be Master Armorer at his Hartford armory, Colt was purchasing Root's skill to conceptualize, design, and build a manufacturing system. Root built Colt's new pistol manufactory, the factory that was the showplace of the American System. His improvements in particular aspects of the manufacturing process, such as drop-forging, were relatively insignificant compared with his general design for the Hartford factory, and the sequential arrangement of the tools.

5. Carl W. Mitman, "Elisha King Root," in *Dictionary of American Biography*. See also MS. 65907, *Root Genealogy Co.*, Connecticut Historical Society, Hartford, Conn. There is some confusion on the specifics of Root's origin. See William B. Edwards, *The Story of Colt's Revolver: The Biography of Samuel Colt*, p. 257, where Edwards gives Root's birthday as May 5, 1808 in Belchertown, Hampshire County, Mass. Henry Barnard, *Armsmear: The Home, the Arm, and the Armory of Samuel Colt: A Memorial*, also gives this same date and place of Root's birth.

6. Henry Barnard, *Armsmear*, p. 257. Barnard states Root's age when he entered the cotton mill as ten, while MS. 65907, "Root Genealogy Co.," Connecticut Historical Society, indicates that his age was twelve. Because Root died so shortly after Colt, he was included in *Armsmear*, an extremely important source of information which cannot be excluded from any study of Root and Colt.

7. Philip K. Lundeberg, *Samuel Colt's Submarine Battery—The Secret and the Enigma*, p. 8.

8. Jack Rohan, *Yankee Arms Maker: The Story of Sam Colt and his Six-Shot Peacemaker*, p. 14. In a conversation between Root and Colt about "The Creative Phases of Mechanics" that reportedly followed the rescue, Root "explained the practices of first making working drawings of mechanical conceptions and then wooden models." Rohan does not cite the source of his information.

9. Carl W. Mitman, "Elisha King Root," in *Dictionary of American Biography* (1935), 16:144–145; "Elisha King Root," in National Cyclopedia of American Biography, (1922), 18:313; Barnard, *Armsmear*, p. 258; and Edwards, *The Story of Colt's Revolver*, p. 267.

10. Canton, Connecticut, is the town in which the Collins axe manufactory was located. Later, due in part to the large volume of mail addressed to Collins & Co., a part of the town was named Collinsville. All references to Canton and Collinsville are to the same town.

11. Barnard, *Armsmear*, p. 258. This information was provided by "the originator of that successful company" (Collins & Co.), undoubtedly S. W. Collins, who was not only living when *Armsmear* was in production, but was writing his own recollections. See *Dictionary of American Biography*, and Samuel Watkinson Collins, "The Collins Company," Connecticut Historical Society. In addition to the original manuscript at the Connecticut Historical Society there also a typewritten transcript compiled in 1926. My page citations refer to this bound typescript, and is cited as: Collins, "Reminiscences," p. 11.

12. "Work Agreement between E. K. Root and the Collins Co.," MS. 71553, "The Collins Company," Connecticut Historical Society. Signed in Canton on August 18, 1832.

13. H. N. Brinsmade, D.D., *An Address at the Funeral of Samuel W. Collins, May 2, 1871*. This funeral oration (for excerpt, see epigraph at the beginning of this chapter) explains why the Collins Company was so innovative. The reason rests with the company president, S. W. Collins, who obviously believed in progress through technological change, and who had a close personal relationship with Root.

14. "Terms with Workmen, 1834–1843," MS. 71553, "The Collins Co.," Connecticut Historical Society, p. 1.

15. *Ibid.*, p. 1.

16. Roger Burlingame, *Machines That Built America*, pp. 79–80. Burlingame credits Root with developing this division of labor, but the evidence is inconclusive. It is unclear if the recording of job titles and names begun by Collins on December 1, 1834, was the result of an entirely new division of labor or simply the earliest surviving record of an existing arrangement. It is conceivable that an earlier record book, now lost, contained the same information and that the December 1, 1834, date is of little significance. S. W. Collins noted that in 1833 "workmen were dissatisfied with their attempts to organize the business more economically" (Collins, "Reminiscences," p. 4), which tends to suggest a pre-1834 division of labor.

17. It is not clear if these people were inside contractors or merely salaried or daily employees.

18. Paul Uselding, "Elisha K. Root." Dr. Uselding's analysis would have benefited greatly had his understanding of the technology been somewhat more thorough. He errs in writing that "the accuracy of die forging very often eliminates the need for finishing cuts on the milling machine and invariably does away with 'heavy' milling cuts . . . Up to a point, die forging is, in a sense, a substitute for milling" (p. 565). Earlier, (p. 545), Dr. Uselding defined milling "as a sort of shorthand, or prototypic designation, for metal cutting in general—'making chips,'"

Die-forging is often the first step in manufacturing interchangeable parts and *all* die-forged parts require machining before they are interchangeable, because forged parts are covered with scale and have a band of flashing. It is the machining cuts that reduce the part to acceptable dimensions. Die-forging rarely comes within 1/32 of an inch of the final dimensions of the part, because of the allowance for machining. Furthermore, the surfaces of die-forged parts are too rough to work smoothly in any kind of a mechanism. No interchangeable parts were assembled directly after die-forging.

19. Hinman's work was apparently carried out in the Collins' factory, although this point is far from clear. Hinman may have been a private contractor who worked in Collinsville for only a short time. In the 1840s, he patented a pump in Ohio, suggesting he had gone west.

20. N. S. B. Gras, *Business History of the United States About 1650 to 1950s*, ch. 10, "Pioneers of the Industrial Revolution in America: Petty Capitalists Become Industrial Capitalists," pp. 95–98.

21. Collins, "Reminiscences," p. 8.

22. "Terms with Workmen, 1834–1843," MS. 71553, "The Collins Company," Connecticut Historical Society. Dec. 1, 1834.

23. Paul Uselding, "Elisha K. Root," p. 550.

24. Note that the innovative attitude of Collins and the die-forging work at Collins were well under way when Root arrived. All earlier scholarship on E. K. Root fails to take into account this earlier work.

25. This machine contained two dies attached to two vibrating arms pivoted in a heavy, iron, oblong frame and brought together by an ordinary crank and pitman turned by water power. Hinman did not claim novelty for his use of dies. In 1831 die-forging was a well known, widely practiced, and therefore unpatentable technique. On the contrary, he claimed only the method of applying power "to effect the work usually done by hammering or swedging in the ordinary methods . . ." "Patent of David Hinman, 2 November 1832 [unnumbered] improvement in the machine for

making Axes, Hatchets, Chisels &c.," MS. 71890, "The Collins Company," Connecticut Historical Society.

26. This second die-forging machine consisted of a heavy, oblong, iron frame with one die set in each short end of the frame. A large shaft with a fly wheel rotated in bearings set in the center of the long sides of the frame. Two eccentrics were constructed on this shaft to move two sliding bars one to two inches toward and away from the dies set in the short sides of the oblong frame. A second set of dies were set in the ends of these bars, each of which slid through a slot in a reciprocating shaft set parallel to the larger shaft in the center of the frame. As the center shaft turned, it alternately forced the two pairs of dies together at each end of the frame, with a slight rolling motion. Hinman claimed "the application of powerful press . . . operating dies, moulds or swedges . . . to give heated iron and steel the several forms which are successively requisite in forging it into all kinds of axes."

One is tempted to speculate that Hinman's first machine was built at Collinsville but discarded as unworkable and replaced by the second patented machine. This might explain the long delay in putting the machinery into production, but there is no supporting evidence. Even more unfortunate is the lack of evidence about Hinman's welding machine. Only a reference to it survives.

27. "Patent of David Hinman, 29 June 1833 [unnumbered] Improvement In Machinery For Forging Axes." MS. 71890, "The Collins Company," Connecticut Historical Society. A restored patent drawing of Hinman's second die-forging machine was made in 1841.

28. "Patent of Erastus Shaw, 29 June 1833 [unnumbered] Improvement in Forging Axes." MS. 71890, "The Collins Company," Connecticut Historical Society. Same date as Hinman's patent!

29. "Patent of Benjamin Smith, 2 November 1832 [unnumbered] improvement in the manufacture of Axes." Ms. 71890, "The Collins Company," Connecticut Historical Society. Dr. Uselding's article focused in part on milling technology, but surprisingly, he chose not to use the single, most exciting piece of information on milling in the entire Collins & Co. file. Same date as Hinman's first patent!

30. Merritt Roe Smith, "John H. Hall, Simeon North, and the Milling Machine: The Nature of Innovation among Antebellum Arms Makers," pp. 573–591. If milling technology was finding its way into private industry, perhaps Hall's drop-forging innovations were also finding their way north with the same individuals who transferred the milling technology.

31. Patents of David Hinman, Erastus Shaw, and Benjamin Smith, Ms. 71890, "The Collins Company," Connecticut Historical Society.

32. Henry J. Kauffman, *American Axes: A Survey of Their Development and Their Makers,* p. 137. "Shaw & Burke: Canton, Connecticut. 1831" is Kauffman's entire entry in ch. 7, "Directory of American Axe Makers."

33. Benjamin Smith may not have been a member of this tiny technical community and Hinman may only have been there temporarily. Nevertheless, Hinman, Kellog, and Shaw were apparently there when Root arrived and Root was almost certainly aware of Smith's patent which may only have been purchased by Collins.

34. Root claimed the use of rollers with dies attached to each, which moved in a reciprocating motion instead of a rotary motion. The pattern was plated out by dies set in reciprocating rollers instead of in dies (swedges) under the trip hammer. The

pattern was doubled over mechanically by a punch, two rollers and a pair of jaws instead of by hand, and was welded in reciprocating dies set in rollers instead of in dies (swedges) under the trip hammer. Root also invented a method of connecting and disconnecting the crank and the rollers. "Patent of Elisha K. Root, 30 March 1836, #767," MS. 71890, "The Collins Company," Connecticut Historical Society. I differ with Uselding's interpretation that the significance of Root's pattern rolling machine "lay in the use of forming dies attached to rollers." Uselding, "Elisha K. Root," p. 551. Shaw's patent already covered dies set in rollers. Root's patent explicitly states that "The use of Rollers to form the iron, or their application to the forming Iron for Axes is *not claimed*, but this improvement consists in placing and fastening on Dies of suitable forms to Rollers which *instead of revolving in a rotary manner as has heretofore been done*, are made to revolve partly round & back again" (italics added). The device causing reciprocating motion is the patented feature of this part of the machine, not the rollers and dies.

35. Root may also have been inspired by Hinman's unknown welding machine.

36. "Copy of Specifications for Punching Machine, 1 October 1838," 7 pp., MS. 71890, "The Collins Company," Connecticut Historical Society. Also U.S. National Archives, Records of the Patent Office, Record Group 241, p. 5.

37. A careful reading of the patent petition for Root's combination machine provides more accurate dating of the document than Uselding gives on p. 557 of "Elisha K. Root." Root mentions his October 1, 1838 petition for his punching machine, which was patented on December 10, 1838. Hence, Root wrote the specifications for his combination machine after October 1, 1838, but before news reached him that his punching machine petition had been granted, probably before January 1839.

38. A penciled note on the back of this document in the Connecticut Historical Society eliminates the apparent confusion about its submission to the patent office (Uselding, "Elisha K. Root," p. 557). "No patent issued for this—nor application made it being the opinion of [illegible, a surname] that it was already covered by our previous patents."

39. Edwin A. Battison, "The Auburndale Watch Company," p. 55.

40. Collins, "Reminiscences," p. 14.

41. *Ibid.*, p. 16.

42. Such trips were considered necessary as part of an early engineer's training and education. See, for example, Eugene S. Ferguson, ed., *George Escol Sellers' Early Engineering Reminiscences, 1815–1840;* Darwin H. Stapleton, "Moncure Robinson: Railroad Engineer, 1828–1840"; and Paul Uselding, "Henry Burden and the Question of Anglo-American Technological Transfer in the Nineteenth Century."

43. Collins, "Reminiscences," p. 16.

44. "Patent of Elisha K. Root, 10 March 1843, #2995, improvement in tempering steel." MS. 71890, "The Collins Company," Connecticut Historical Society. Uselding ("Elisha K. Root," p. 547, n.8) states "the C.H.S. also has a design for a patent for tempering steel dated March 1843. There is no evidence that patents were granted for these last two petitions." The earlier petition was for the combination machine.

45. Collins, "Reminiscences," p. 18, and Samuel Colt, "Steel Making in Hartford—Most Valuable Improvement," *The Weekly State & Union,* May 23, 1860, in

Colt Company Scrapbook, vol. 2, RG103, Connecticut State Library, Hartford, Conn.

46. Collins, "Reminiscences," p. 18. See also Barnard, *Armsmear*, p. 258.

47. Collins, "Reminiscences," p. 18.

48. *Ibid.*, p. 19. This passage provides the real motivation behind Root's development of his shaving machine—a particular kind of labor shortage. Uselding ("Elisha K. Root," p. 561) states: "It is an example of the substitution of a drawing knife, much like the bit of a shaper, for the process of grinding and, hence, of an alternative technology replacing an earlier one where the payoff to inventive activity for a class of applications within that earlier technology (grinding axes) had become sufficiently small to cue the inventor to look for alternative ways to accomplish the same objective." Here, I think, much of the evidence has been ignored. Uselding makes no mention of the skilled labor shortage problem at Collinsville, a theme that practically dominates S. W. Collins' "Reminiscences." Furthermore, diminishing returns to inventive activity in grinding had not set in to grinding technology. Robert Woodbury's monograph, *The History of the Grinding Machine*, provides very strong evidence to the contrary. One chapter is entitled "The Grinding Machine Comes of Age, Beginnings of Precision Grinding, 1860–1905." Note the dates here, twelve years after diminishing returns had supposedly set into grinding technology.

49. Albert S. Bolles, *Industrial History of the United States*, "Axes and Saws," p. 272. Despite the fact that Bolles was writing forty years or so after Root's grinding related activity at Collins, the grinding process remained unchanged, except that less of it was required. The actual technique was the same in 1881 as it had been in 1840. The same materials were used and same hazards remained.

50. Collins, "Reminiscences," p. 20. It is unclear if silicosis was known as such in 1845, but operatives knew that the dust killed and that "no grinder died of old age." Kauffman, Axes, pp. 37–38.

51. Collins, "Reminiscences," p. 20.

52. "Patent of Elisha K. Root, 22 August 1848, #5731." MS. 71890, "The Collins Company," Connecticut Historical Society. This patent was not analyzed in any of the scholarship on Root and Collins. There is evidence in Collins' "Reminiscences" that Root's shaving machine was built as early as 1844 or 1845.

53. Collins, "Reminiscences," p. 18. Samuel W. Collins' memory is slightly hazy here. He states that Root became superintendent in 1845, but calls Root a superintendent in 1843. Information gathered by the Root Genealogy Company in the 1930s dates Root's superintendency from September 1845. See MS. 65907 in the Connecticut Historical Society.

54. Barnard, *Armsmear*, p. 258.

55. MS. 65907, Connecticut Historical Society. E. K. Root to Matilda Root, Collinsville, August 25, 1847. Here is still more evidence that Root's stature as a mechanic and his reputation had spread far beyond the local community. In the late 1840s Root seems to be a much sought after man. Root wrote to his wife, Matilda, telling her of an offer by "brothers H & C" to invest in water privileges in Massachusetts prior to building a factory of some sort. Although Root did not invest in the property, he did promise to "render them all the aid I can consistently (?)—they are anxious to have me take a *direct interest* in the project, but I think I shall not do so." Matilda

Colt Root became Elisha's second wife on October 7, 1845. She was born on January 3, 1820, in Hartford, Connecticut, and died on July 14, 1898, in Hartford. Although unconfirmed, it appears that she was a fairly close relation of Samuel Colt. It is entirely possible that Root's connection with Samuel Colt was more than a business connection. Root's first wife was Charlotte Chapin whom he married on October 16, 1932. MS. 65908, "Root Genealogy Co.," Connecticut Historical Society.

56. Collins, "Reminiscences," p. 19.

57. Roger Burlingame, *Machines That Built America*, p. 78. "Old-timers at the Collins plant—it is flourishing today [1953]—still tell of the rumpus Root caused after a few days in the forge room. No one, of course, can remember this, but Root has become a kind of legend in Collinsville. Root thought all the machinery ought to be scrapped. The man in charge, known as Uncle Ben [B. T. Wingate?], was a conservative and started an argument which is said to have lasted for years. It was the old artisan tradition in conflict with the new revolutionary Yankee mind which knew that the time had come when machines must replace men as far as possible. Samuel Collins, however, was a sensible man and, himself, an adventurer in new fields so Root won out." Burlingame's story is undocumented, but its general theme is well in keeping with the available evidence.

58. Collins, "Reminiscences," pp. 14 and 24.

57. *Ibid.*, p. 30.

59. Loyalty to Collins & Co. and to S. W. Collins in particular may also have played a part in Root's decision to remain at Collinsville in 1845. Several readers have objected to the term "mechanical virility," but I have insisted upon its inclusion. Matt Roth, whom I knew as a Hagley Fellow and who worked for Brown & Sharpe as machinist's apprentice, found the term most appropriate.

60. "Our Visit to the Collinsville Ax Works," *Scientific American*, (July 16, 1859), N.S., 1(3):36–37. I am puzzled by the phrase "the anvil which it strikes." Perhaps Root incorporated an improvement about which we know nothing. Perhaps the writer misunderstood what he saw. Perhaps someone after Root improved or changed the machine. The Collins factory is also described, but in far more general terms, in Horace Greeley, et al., *The Great Industries of the United States*, "Axes and Plows," pp. 122–147, and in Albert S. Bolles, *Industrial History of the United States*, "Axes and Saws," pp. 270–274.

62. *National Cyclopedia of American Biography* (1922), 18(313). I have not yet found this, and it was evidently not built at Collins & Co.

63. MS. 71890, "The Collins Company," Connecticut Historical Society. "The U. States Patent Office. To All Persons . . . [S]pecifications of Thomas Blanchard's Letters Patent, dated 20th January 1820—which Letters Patent were extended by Act of Congress for 14 years from the 20th January 1848. In testimony Whereof [T]his third day of March one thousand eight hundred and forty-nine . . ."

64. Collins, "Reminiscences," pp. 24–25.

65. Barnard, *Armsmear.*

66. Joseph W. Roe, *English and American Tool Builders.*

67. The economic history of Collins & Co. was anything but smooth. Collins' "Reminiscences" draw a picture of constantly changing conditions. At times, axes could be sold at healthy profits, at other times, they could hardly be given away.

Competition for skilled labor was sometimes fierce and sometimes nonexistent. Axe prices also seem to have fluctuated. Collins & Co. was clearly reacting to market fluctuations, but deciding which fluctuation exerted how much influence is not easy. I conclude from S. W. Collins' writings that the inability to supply the demand for axes, even though that demand fluctuated greatly, was the prime economic force at work.

68. Collins, "Reminiscences," pp. 3, 7, 20.

69. Merritt Roe Smith, *Harpers Ferry Armory and the New Technology.*

70. *Ibid.*, p. 22.

71. One of the vexing problems of the theoretical economic historian's approach to the rise of the American System is the erroneous assumption that labor is homogeneous. Only occasionally does an economist assume a heterogeneous labor supply. As the case of Root's shaving machine illustrates, there were shortages of particular kinds of labor based on particular skills and particular working conditions within a context of an abundance of generally unskilled labor.

72. Collins, "Reminiscences." See, for example, pp. 20 and 24-25.

73. "Terms with Workmen," MS. 71553, "The Collins Company," Connecticut Historical Society. "Dec. 1, 1834." "The strikers must pick over their own coal daily & use no more borax than is absolutely necessary."

74. William B. Edwards, *The Story of Colt's Revolver: The Biography of Samuel Colt*, p. 267. Edwards cites Rohan, *Yankee Arms Maker*, as his source.

75. Barnard, *Armsmear*, p. 259.

76. Collins, "Reminiscences," pp. 25 and 27.

77. *Ibid.*, p. 25. In his comments on a dissertation draft, Jeff Williamson explicitly raised the question that the entire chapter has asked implicitly. What were Collins' competitors doing technologically? What problems did they face? How does Collins & Company fit into the "edge tool industry?" Are they leading or lagging technologically?

78. Collins, "Reminiscences," p. 24.

79. Alfred D. Chandler, "Anthracite Coal and the Beginnings of the Industrial Revolution in the United States," *Business History Review* (Summer 1972), 46(2); Thomas R. Winpenny, "Hard Data on Hard Coal: Reflections on Chandler's Anthracite Thesis;" and "A Reply by Alfred D. Chandler, Jr.," *Business History Review* (Summer 1979), 53(2).

80. Barnard, *Armsmear*, p. 259.

81. A handwritten copy of Blanchard's patent survives in the Collins & Co. papers at the Connecticut Historical Society.

82. "Patent of D. C. Stone, Number 699, 21 April 1838," U.S. National Archives, Records of the Patent Office, Record Group 241. Little is known of the axe manufactory at Napanoch, N.Y., but several axe making patents are known to have been issued to residents of the town. In his *Reminiscences*, Collins continually complains that his Napanoch competitors wait until he has trained the workmen before hiring them away. Napanoch was located on the Hudson River, only seventy miles west of Collinsville.

83. Robert C. Post, " 'Liberalizer' versus 'Scientific Men' in the Antebellum Patent Office."

4. *Typewriter Manufacturing, 1853–1924*

1. Terry Abraham, "Charles Thurber: Typewriter Inventor," pp. 430–434. Abraham's note on the scarcity of sources for typewriter history is well taken, especially with respect to major manuscript collections of the major manufacturers. Early in the research phase of this chapter, I contacted the corporate descendants of the major makers and met with complete failure in my attempts to find business records. For example, my good friend and former fellow Hagley Fellow David Hounshell said "There aren't any [typewriter manuscripts]" during a telephone conversation in January 1983. Mr. Truex of Sperry Univac (a former corporate successor of Remington Rand which was a corporate successor of the Remington Typewriter Company) said there was "Nothing [you] would be interested in" and suggested that I "talk to Remington Rand." "I suspect they [business records] were disposed of." This information also came during a telephone conversation in January 1983. Mr. Andy Sackerman of Varityper (corporate successor to the Hammond Typewriter Company) during a telephone conversation on February 2, 1983, said "I doubt very much if any [business records] would have survived." "I am absolutely certain they have not survived." Mr. Robert Guerney of the current Remington Rand Corporation was formerly an employee of the old Remington Rand Corporation. He has Remington Typewriter Company and Remington Rand serial numbers and some product data, but no business records. He has only one filing cabinet drawer full of information.

Late in the fall of 1983 I learned that unknown to Mr. Truex at Sperry Univac, much Remington Rand material had survived. It had been donated to the Eleutherian Mills Historical Library, Greenville, Wilmington, Delaware, and is now being processed. Unhappily, it surfaced too late to be used in this chapter. I can only await its availability with great anticipation.

There are some published sources of data on typewriter manufacturing, notably in the *Scientific American* which, we now know, are really paid advertisements. Other articles in trade journals and the popular press are probably in the same vein. In addition, there are several known typewriter trade catalogues which focus on the manufacturing operations of various factories, notably the Oliver Typewriter Company and the L. C. Smith & Bros. Typewriter Company. Here, I have relied heavily on the use and interpretation of photographs to generate data. This is a somewhat unconventional methodology in economic history, but, given the lack of manuscript sources, it is one of the only methods available.

Much of the work in this chapter is drawn from surviving typewriters, from the machines manufactured in the late nineteenth and early twentieth centuries. Although this is a relatively new approach in the history of technology and economic history, there is a precedent for such a methodology in the work of Arnold Thackery. Thackery wished to study early scientific societies in England, but found that most of their records had been destroyed during bombing raids in World War II. Determined to study these societies anyway, Thackery devised a methodology of looking at the members and analyzing their educations, businesses, religions, associations, and memberships.

2. Statement of Charles H. Jones, son of John M. Jones, made about 1912, just prior to his death. Typescript copy in the files of the Division of Mechanisms,

Smithsonian Institution, National Museum of American History, stamped "Min. & Mech. Tech. C. W. Mitman" "Feb. 18, 1931." This information was evidently supplied to Mitman by Harry A. Jones in 1931. Harry A. Jones evidently offered to sell this Jones Typographer, which had been exhibited at the New York Crystal Palace in 1853, to the Smithsonian for $1,000. It subsequently came to the Milwaukee Public Museum (under circumstances unknown), became part of the Carl P. Dietz Typewriter Collection, and may be seen today on exhibit. Milwaukee Public Museum catalogue no. E 47377.B/13890. See George Herrl, *The Carl P. Dietz Collection of Typewriters*, p. 48. The Dietz Collection machine once had a glass case of some sort which has been lost.

3. The Cooper Hand Printing Machine, Milwaukee Public Museum, Carl P. Dietz Typewriter Collection, catalogue no. E 47380/13890. George Herrl, *The Carl P. Dietz Collection of Typewriters*, p. 23. Cooper's patent model survives at the Smithsonian Institution's Museum of American History, catalogue no. 251,211, but the Dietz Collection machine is evidently the only surviving production model. While it is conceivable that the Dietz Collection machine is another model or prototype, it is not likely. The casting is far too detailed and sophisticated for a simple model. Thus my interpretation of the specimen as a production model. The production model was illustrated in both the *Scientific American* and the *Journal of the Franklin Institute*. See "Report on J. H. Cooper's Hand Printing Machine," *Journal of the Franklin Institute* (September 1857), 3rd Series, 34(3):213–214, and "Cooper's Hand Printing Machine, *Scientific American* (December 6, 1856), 12(13):100.

4. John Pratt to G. Brown Goode, October 1, 1891. Letter in the files of the Division of Mechanisms, Smithsonian Institution, National Museum of American History. "The machines (Pratt's Pterotype) were sold in '67 & '68. The invention dates from '64." Pratt was writing from Brooklyn to the Assistant Secretary of the Smithsonian inquiring if Goode would like Pratt to purchase one of his machines in England. Pratt was at that time an employee of the Hammond Typewriter Company which had acquired his patents and used them in its typewriter. Pratt did eventually sell a machine to the Smithsonian on March 18, 1892. Its catalogue card states: "The first key wheel machine. Invented in 1864 by John Pratt. Made by E. B. Burge, London, England, 1868." See also Bill Plott, "Simple Grave Alone Marks John Pratt's Fame," *Anniston Star*, Wednesday, March 30, 1966, 84(194):8-A, and "Pratt Not First Man To Invent Typewriter Although He Claimed the First 'Practical' One," *Anniston Star*, Thursday, March 31, 1966, 84(195):2-B, and "Pratt's Work Still Unknown To Many," *Anniston Star*, Friday, April 1, 1966, 84(196).

5. Richard Current, *The Typewriter and the Men Who Made It*.

6. There are three sources of typewriter invention—machines for the blind, hand printing machines, and telegraphy. It was not, as advertised, invented to replace the pen. The original makers weren't sure who would buy it.

7. L. C. Smith & Bros. Typewriter Company, *A Visit to the Home of the L. C. Smith Typewriter*. A trade catalogue in the collection of SCM, through the courtesy of Charles Molloy, Consumer Affairs Department, DeWitt, N.Y. Negatives and slides in the files of the Milwaukee Public Museum. See also Oliver Typewriter Company, *The Oliver Typewriters, The No. 3 Models*, pp. 16–19, "A Visit to Our Factory." A trade catalogue in the files of the Division of Mechanisms, National Museum of American

History, Smithsonian Institution. See also American Writing Machine Company, "Caligraph Writing Machine," *Scientific American* (March 6, 1886), N.S., 54(10):150.

8. *The Story of the Typewriter, 1873–1923*. For information on Jones' other inventions, see *Report of the Commissioner of Patents for the Year 1851: Part 1, Senate Ex. Doc. No. 118* (Washington, D.C.: Robert Armstrong, Printer, 1852), pp. 62, 93, 116, 205–206, 309; patent no. 8,234, "Carriage Bodies, Hanging," and patent no. 8,236, "Carriage Bodies, Hanging," and Reissue No. 191, "Improvements in Carriages," *Report of the Commissioner of Patents for the Year 1853: Part 1, House of Representatives, Ex. Doc. No. 39* (Washington, D.C.: Beverly Tucker, Printer to the Senate, 1854), pp. 41 and 410; patent no. 10,207, "Improvements in Machine for Cutting Straw," "John Jones and Alexander Lyle, of Rochester, N.Y.," *Report of the Commissioner of Patents for the Year 1857: Vol. 1, Senate Ex. Doc. No. 30* (Washington, D.C.: William A. Harris, Printer, 1858), pp. 115, 443, and 835; patent no. 18,907, "Hand Printing Press," "J. M. Jones, Palmyra, N.Y." This is not an exhaustive listing of Jones' patents, merely a sample from the 1850s.

9. "More Improvements Wanted," *Scientific American* (October 18, 1851), 7(5):37.

10. John Jones, "Printing Letters," *Scientific American* (February 7, 1852), 7(21):166; and Thomas L. Cook, *Palmyra and Vicinity* (Palmyra, N.Y.: 1930), p. 152.

11. "Letter Printing Press—The Typographer," *Scientific American* (March 6, 1852), 7(25):196.

12. *Report of the Commissioner of Patents for the Year 1852: Part 1, Senate Ex. Doc. No. 55* (Washington, D.C.: Robert Armstrong, Printer, 1853), pp. 103, 213–214; patent no. 8,980, "Copying Manuscripts," *Report of the Commissioner of Patents for the Year 1856: Vol. 1, House of Representatives, Ex. Doc. No. 65* (Washington, D.C.: Cornelius Wendell, Printer, 1857), pp. 96, 366, and 499; patent no. 14,919, "Printing Machine"; and "List of Patent Claims," *Scientific American* (June 12, 1852), 7(39):310.

13. "Crystal Palace Awards," *Rochester Daily Democrat*, January 24, 1854, 22(19):2, col. 4.

14. John Jones & Co., *Jones' Typographer: Patented June 1, 1852*. A broadside, found in the drawer of the Jones Typographer, Milwaukee Public Museum, Carl P. Dietz Typewriter Collection, catalogue no. E 47377.B/13890. The fact that Jones had other interests in patented products, notably agricultural implements and carriages, suggests that he may have formed a company for their production and sale, even if not for their manufacture.

This broadside is published without citation in Bruce Bliven, *The Wonderful Writing Machine*, p. 34. Curiously, Bliven virtually ignored the Milwaukee Public Museum collection, which was, at the time Bliven wrote, already the finest in existence.

15. Thomas L. Cook, *Palmyra and Vicinity*, pp. 152–153. Cook's production estimate is clearly based on information received from Harry A. Jones, who provided data to Carl Mitman at the Smithsonian Institution in 1931.

16. "Destructive Fire—Burning of the Rochester Novelty Works," *Rochester Daily Union*, April 25, 1856, 4(215):3, col. 1; and R. G. Dun & Co. Collection, Baker Library, Harvard University Graduate School of Business Administration, vol. 162, New York, Monroe County-Rochester, 1:262. "The firm has done a good business since it has been started, & was abt. to, or did, pay a dividend, when the building was destroyed by fire this spring." "July 16/56."

In the mid-1850s, the Rochester Novelty Works was precisely the kind of establishment to which the inventor of a "patent good" requiring "small castings" and subsequent "lathe work" might turn, as John Jones and Company did sometime before the New York Crystal Palace opened in 1853 to manufacture his Typographer. The Novelty Works was evidently quite a large establishment. Local Rochester newspapers lamented its fire loss in 1856, as well as its temporary financial failure in 1857 which "must certainly be regretted as the establishment was a credit to the manufacturing reputation of this city while it afforded employment to many hands who are now idle." "Suspension of the Novelty Works," *Rochester Union & Advertiser* September 30, 1857, 30(232):3, col. 2. The company featured a separate foundry and steam engine house in addition to the five-story building. A thermometer maker, George Taylor, rented several upper floors. According to its advertisements, the company manufactured a wide variety of castings and offered to produce "any description of small machinery, or small castings," as well as "patent goods made at short notice." *Daily American Directory of the City of Rochester, for 1851–2* (Rochester, N.Y.: Lee, Mann, 1851), p. 76, an advertisement for the Rochester Novelty Works. The company also advertised its capability to produce "Castings, Lathe Work, Geer Cutting, furnished to order." *Dewey's Rochester City Directory, Professional & Business Cards and Advertisements, 1857*, p. 64; bound with *Dewey's Rochester City Directory, 1857–58* (Rochester, N.Y.: D. M. Dewey, 1857), p. 302. On January 1, 1868, a complete inventory of the Rochester Novelty Works was taken, showing the company to have a wide range of machine tools, many files, and numerous items in the process of manufacture, including 400 sewing machines. See "*Inventory Book, 1868: Inventory Rochester Novelty Works, No. 4 & 6 Hill St.*" "Taken Jan. 1, 1868" in the Rochester Public Library, Local History Room.

17. Milwaukee Public Museum, Carl P. Dietz Typewriter Collection, catalog no. E 47477/13890. A second machine, almost certainly a standard production model, survives in a private collection in Norwich, Conn. Its owner kindly allowed me to examine it. A third Jones machine, his 1852 patent model, was photographed in the 1930s, but its whereabouts is unknown. Milwaukee Public Museum negative no. 421916, the Jones Typographer Patent Model. The MPM Photo Department date is January 30, 1939, and the source quoted is the Free Lance Photographers' Guild, New York City.

18. An index to the *Journal of the Franklin Institute* lists a long series of reports by a John H. Cooper, presumably the same Cooper. See also Michael McMahon and Stephanie Morris, *Technology in Industrial America: The Committee on Science & The Arts of the Franklin Institute, 1824–1900.*

19. *Report of the Commissioner of Patents for the Year 1856: Vol. 1, House of Representatives Ex. Doc. No. 65* (Washington, D.C.: Cornelius Wendell, Printer, 1857), pp. 68, 367, and 500; patent no. 14,907, "Printing Machine." See also "American Patents which Issued in May, 1856," *Journal of the Franklin Institute,* (July 1856), 32:33. Cooper's 1856 patent model survives at the Smithsonian Institution and what is probably a production model may be found in the Carl P. Dietz Typewriter Collection at the Milwaukee Public Museum. See George Herrl, *The Carl P. Dietz Collection of Typewriters*, p. 23. The Cooper Hand Printing Machine is Milwaukee Public Museum catalogue no. E 47380/13890.

20. "John H. Cooper to Committee on Science & Art, Philadelphia, April 8, 1857," Franklin Institute Archives.

21"No. 716, Report on J. H. Cooper's Hand Printing Press," manuscript report of the Committee on Science and the Arts, Franklin Institute. This report was published in September 1857. See William Hamilton, "Report on J. H. Cooper's Hand Printing Machine," *Journal of the Franklin Institute* (September, 1857), 3d Series, 34(3):213–214. The report is dated July 9, 1857, in its published version: "John H. Cooper to Committee on Science & Art, Philadelphia, April 8, 1857," Franklin Institute Archives.

22. *Ibid.*, p. 214.

23. "Description by the Inventor," *Journal of the Franklin Institute* (September, 1857), 3d Series, 34(3):214–215.

24. The Cooper Patent Model is in the Division of Mechanisms, National Museum of American History, Smithsonian Institution, catalogue no. 251,211. See also Royal Typewriter Company, *The Evolution of the Typewriter,* p. 11.

25. "Description by the Inventor," *Journal of the Franklin Institute* (September, 1857), 3d Series, 34(3):214–215. "Cooper's Hand Printing Machine," *Scientific American* (December 6, 1856), 12(13):100.

26. "Description by the Inventor," *Journal of the Franklin Institute* (September 1857), 3d Series, 34(3):214.

27. Michael H. Adler, *The Writing Machine: A History of the Typewriter,* p. 22. Hansen was born on September 5, 1835, in Hunseby, Denmark. He was apprenticed to a painter and studied theology before completing his studies in 1864 in Copenhagen. Following his examinations, he was appointed head of the Royal Deaf and Dumb Institute.

28. Michael H. Adler, *The Writing Machine, A History of the Typewriter,* p. 155. Adler pictures several different versions of the Hansen Writing Ball, as the machine is known to typewriter collectors. See also *The Evolution of the Typewriter,* p. 16. The modifications were primarily in the form of holding and moving the paper rather than in the type mechanism. The various forms of Hansen's machine included a Morse code version for telegraph use, and a form which created impressions on a metal sheet which was then run through a second machine which printed the impressed material by electrically reading the imprinted sheet. Most machines, however, were simply mechanical and featured either a cylindrical platen or a flat paper holder. Earlier models began with a flat carriage to hold the paper and an intermediate model used a curved frame to which the paper was clipped. All the modifications were evidently the work of the Jurgens firm which was the primary manufacturer of the machine for some forty years. Jurgens incorporated in 1887 as C. P. Jurgensens Mekaniske Establissement.

29. Adler, *The Writing Machine,* pp. 154–156.

30. *Ibid.*, pp. 150–157.

31. I have personally examined one of Hansen's Writing Balls in the collection of Uwe Breker, of Cologne, West Germany. Uwe Breker is one of the leading European typewriter collectors.

32. "John Pratt, Center, Alabama, Inventor of Typewriter," *B. P. W. State Bulletin* (March–April 1932), 5(4[issue no. 25]):8–10. Official Organ of the Alabama Feder-

ation of Business and Professional Women's Clubs. Data gathered and presented by Lola Taylor.

33. Thomas E. Crowson, *Sandlapper: The Magazine of South Carolina* (March 1971), 4(3):50–51.

34. Joe Azbell, "It Required Only A Short Time To Write This Story Because John J. Pratt Had Writers' Cramp In 1850s," *Montgomery Advertiser & Alabama Journal* (July 10, 1949), p. 14-B.

35. Marie Bankhead Owen, "Scholarship Fund Is Started by Women In Honor of John Pratt, Noted Inventor," *Montgomery (Alabama) Advertiser*, Monday, October 23, 1922. "John Neely, the "printer's devil" who furnished to Mr. Pratt his type and whittled his blocks, declares today (1922) that it was in the late 50s . . ." "Mrs. Marie Smyer, a niece of Mr. Pratt, . . . declares that Mr. Pratt conceived the idea and began work on his typewriter about 1857 or 1859 . . ." Pratt himself stated that the idea came in 1863, but this statement seems to refer to the completed machine.

36. Affidavit of John Mathis Neely, August 19, 1922, printed verbatim in Mrs. Frank Ross Steward, *History of Cherokee County, 1836–1956*, 2 vols. (Birmingham, Ala.: Birmingham Printing Company, 1959), pp. 282–283.

37. Pratt received his provisional British patent in 1864, a year before Hansen started work on his machine. It is reasonable to assume that Hansen may have searched the British technical literature and found Pratt's machine, which was of the radial strike design, as was Hansen's. Denmark had no patent laws in the 1860s, so Pratt's invention could not be protected there, although it could be and was protected in England. There is no solid evidence to suggest that Hansen knew of Pratt's work, but the circumstantial evidence, primarily the timing of Hansen's work, so shortly after Pratt's publicity in England, strongly suggests that he did know of Pratt's work and did build on it.

38. *The Evolution of the Typewriter*, p. 26.

39. "John Pratt, Center, Alabama, Inventor of Typewriter," p. 9.

40. *Evolution of the Typewriter*, p. 26; and *Report of the Commissioner of Patents for the Year 1868: Vol. 1: House of Representatives Ex. Doc. 52* (Washington, D.C.: GPO, 1869), pp. 212, 339, and 1,038; patent no. 81,000, August 11, 1868, "Mechanical Typographer." Pratt is listed as a resident of Grenville, Alabama at the time this patent was issued.

41. *Evolution of the Typewriter*, p. 26; G. Tilghman Richards, *The History and Development of Typewriters*; and Wilfred A. Beeching, *Century of the Typewriter*, pp. 23–25.

42. "Type Writing Machine," *Scientific American* (July 6, 1867), 17(1):3.

43. Barbara Craig, "'Lib' Movement Goes Way Back," *Alabama Sunday Magazine*, Sunday, September 2, 1973, pp. 3 and 10.

44. James Densmore to John Pratt, July 4, 1875 (on Densmore, Yost & Co. Stationary). Note from Hill Ferguson, November 15, 1957, in the Alabama Department of Archives and History, Montgomery. This letter is reported to be in the possession of "Chas. F. White, Librarian of Birmingham Bar Association, . . . a cousin of his family."

45. "John Pratt, Center, Alabama, Inventor of Typewriter," p. 9.

46. "By instrument dated December 6, 1879 (recorded May 16, 1890), John Pratt

transfers all his rights in said invention, to Jas. B. Hammond, Inventor of type writing machines, for $700, and other consideration. (A royalty of 50 cents per machine is mentioned.)" Note from Hill Ferguson, November 15, 1957.

47. Joe Azbell, "It Required Only A Short Time To Write This Story."

48. *Birmingham News*, June 18, 1935. A filler quoting the *Gadsden Times* of "fifty years ago this week," and "John Pratt, Center, Alabama." While Pratt never realized the full financial potential of his invention, he certainly seems to have been treated fairly and honestly by James Hammond. His ideas found their way into hundreds of thousands of Hammond typewriters and he served the firm both as a technical advisor and as a representative at expositions.

49. "John Pratt, Center, Alabama, Inventor of Typewriter." United States patent no. 267,367, Typewriter, November 14, 1882; United States patent no. 470,704, Typewriting Machine, March 15, 1892; and United States patent no. 477,224, Typewriting Machine, June 21, 1892.

50. It is not clear if Hammond was in business as early as 1879, but he clearly was considering such a venture by that time.

51. "Type Writing Machine," *Scientific American*, (July 6, 1867), N.S., 17(1):3.

52. Wilfred A. Beeching, *Century of the Typewriter*, pp. 18-21.

53. This designation appeared on the typewriter itself but was discontinued about 1877.

54. Richard Current, *The Typewriter and the Men Who Made It*. I have relied heavily on Current for a summary of the invention and early financing period, which is outside the scope of this work. The article Sholes read was "Type Writing Machine.," *Scientific American* (July 6, 1867), N.S., 17(1):3. Sholes was the collector of the port of Milwaukee. As a part-time inventor, Sholes developed a machine for addressing newspapers in 1860 and a page-numbering machine with successive improvements in 1864, 1866, and 1867.

55. Charles F. Kleinsteuber operated a small jobbing shop in Milwaukee that included a brass foundry in addition to general machine tools for model making.

56. Current, *The Typewriter*.

57. A Milwaukee-made typewriter may be found in the collections of the Buffalo and Erie County Historical Society, Accession Number 58-240. It was originally sold to the Dawes Brothers, lawyers in Fox Lake, Wisconsin. Julius H. Dawes donated the machine to the B&ECHS in 1889. He was, at one time, a member of the B&ECHS Board. This machine clearly dates after November 8, 1872, when James Densmore wrote complaining about the new lettering. I examined the machine in May 1983 and took numerous color slides, which are in the research files of the History Section, Milwaukee Public Museum.

58. Current, *The Typewriter*, p. 40.

59. Yost was part owner of the Corry Machine Company of Corry, Pennsylvania, which made agricultural machinery. Densmore, who was also a lawyer, did legal work for the constantly troubled Corry Machine Company. Densmore and Yost had earlier patented an oil tank car.

60. Current, *The Typewriter*, p. 67, citing the *Milwaukee Sentinel*.

61. Alden Hatch, *Remington Arms in American History*, pp. 167–176.

62. Hatch, *Remington Arms*, p. 137.

63. There is no good economic or technological history of the Remington com-

panies. The data here were drawn from various collectors' books and company histories. It is not clear if this dual strategy was a planned strategy or just happened the way it happened without any formal planning or direction, probably the latter. See Hatch, *Remington Arms;* Charles Lee Karr and Caroll Robbins, *Remington Handguns,* pp. 1–8.

64. For some fifteen years, government contracts came from Puerto Rico, Cuba, Chile, Colombia, Honduras, and China among other countries. See Karr and Robbins, *Remington Handguns,* p. 6. Surplus Civil War arms were sold to France in vast quantities. My interpretation is based on a conversation with Howard Madaus, February 6, 1984. Mr. Madaus is the Assistant Curator of History–Military, at the Milwaukee Public Museum and the curator of the Nunnemacher Arms Collection, one of the largest (about 3,500 arms) and best in public hands. He is particularly interested in military conversions of flint lock arms to percussion arms in the mid nineteenth century, but also has an exceedingly broad knowledge of American arms and manufacturers. He believes that if the Remingtons were paid for their shipments to France during the Franco-Prussian War, then they were indeed in strong financial shape in the early to mid 1870s. They evidently bought great quantities of surplus arms from the federal government after the Civil War, refitted them, and sold them abroad. Their famous "Rolling Block" rifle also enjoyed a wide sale. These foreign government arms sales carried the fledgling Remington Brothers Agricultural Works and the Remington (Empire) Sewing Machine Company. When the Remington Brothers Agricultural Works were formally incorporated on January 1, 1865, all its stock was held by the Remingtons. See Hatch, *Remington Arms,* p. 134.

The Remington Empire Sewing Machine Company was incorporated on October 18, 1870, with $200,000 capital. Little is known of these parts of the Remington operation. The sewing machine board of directors squabbled over patents in February and March 1871, while they followed the progress of "the model machine . . . to manufacture by." There was some concern with the manufacturing operation. On January 10, 1872, Article V of the bylaws was changed to read "The General Manager, Superintendent, and Inspector, and Foreman of the Manufacturing Departments; also . . . shall be appointed by the Board of Trustees." In other words, the board was extending its control directly onto the shop floor in the sewing machine department. On April 22, 1873, the firm changed from the Remington Empire Sewing Machine Company to the Remington Sewing Machine Company, evidently a sales organization under the direction of W. H. Hooper, who promptly spent the sewing machine company into debt. This contributed significantly to the bankruptcy of the Remingtons in 1886. See Hatch, *Remington Arms:* "Production of sewing machines evidently started in 1870, probably before the Remington Empire Sewing Machine Company incorporated, assuming if patent licensing data is reliable" (pp.168–169).

See "The Remington Empire Sewing Machine Co., Minutes" (Book 1), manuscript at the Remington Gun Museum, Remington Arms Company, Ilion, N.Y. Curator Lawrence Goodstall kindly took me through the Remington factory to an old storeroom in a far corner of the plant to see some uncatalogued historical material. We found this single volume, with many pages missing, and he kindly allowed me to use it. See *The Remington,* (Ilion, N.Y.: Remington Sewing Machine Company, July 1, 1874). A broadside in the files of the Division of Domestic Life, Museum of American History, Smithsonian Institution. See figure 4.6.

Sewing machines licensed by Remington Empire Sewing Machine Company.

Year	*Number*
1870	3,560
1871	2,965
1872	4,982
1873	9,183
1874	17,608
1875	25,110
1876	12,716

Grace Rogers Cooper, *The Sewing Machine, Its Invention and Development,* p. 40. Cooper's data is taken from Frederick G. Bourne, "American Sewing Machines," in Chauncey Mitchell Depew, ed., *One Hundred Years of American Commerce* (New York: D. O. Haines, 1895), 2:530. Depew lists machines licensed annually by the combination, not actual production data from particular firms. Sewing machine production may have continued as late as 1894, but probably ceased in 1888 when E. Remington & Sons were liquidated. Cooper, p. 73, lists the following firms and dates:

Manufacturer or Company	*First Made*	*Discontinued*
Remington Empire Sewing Machine Company	1870	1872
E. Remington & Sons, Philadelphia, Pa.	1873	1875
E. Remington & Sons, Ilion, N.Y.	1875	1888
Remington Sewing Machine Agency, Ilion, N.Y.	1888	ca: 1894

The records of the R. G. Dun & Co. Collection in the Baker Library at the Harvard University Graduate School of Business Administration provide some slightly different dates, but the general outline is the same.

65. R. G. Dun & Co. Collection, Baker Library, Harvard University Graduate School of Business Administration, vol. 373, "Sept. 21/75," Herkimer County, N. Y.; vol. 8, no. 12, "E. Remington & Sons", p. 1597. "A successful concern doing a large business chiefly with foreign governments." "Aug. 21/80" "settled w/Egyptians for $325,000 for arms plus 80K on 100K of gun making machinery." See also R. G. Dun & Co. Collection, Baker Library, Harvard University Graduate School of Business Administration, vol. 115, "Dec. 20, 1884" Herkimer County, N. Y., vol. 3, ". . . are said to be doing a very fair bus[iness] & are manufacturing per day of the Lea Magazine guns for the Chinese Government . . ."

66. R. G. Dun & Co. Collection, Baker Library, Harvard University Graduate School of Business Administration, vol. 114, "Sept. 1/85," Herkimer County, N. Y., 2:699.

67. R. G. Dun & Co. Collection, Baker Library, Harvard University Graduate

School of Business Administration, vol. 113, "Oct. 19/77," Herkimer County, N. Y., 2/1:596. "No Change" through "Jan. 31/83."

68. R. G. Dun & Co. Collection, Baker Library, Harvard University Graduate School of Business Administration, vol. 115, Herkimer County, N.Y., 3:300–302.

69. Hatch, *Remington Arms*, p. 175.

Firm	*Starting Date*	*Ending Date*
Densmore & Yost	March 1? 1873	to spring 1874
Densmore, Yost & Co.	spring 1874	to December 28? 1874
Locke, Yost & Bates	December 29,? 1874	to 1878
Fairbanks & Company	1878	to July 31, 1882
Wyckoff, Seamans & Benedict	August 1, 1882	to March 1886

This table of sales agencies is compiled from several sources, primarily Hatch, *Remington Arms*, and Current, *The Typewriter*.

70. Hatch, *Remington Arms*, p. 175, states the sale date as April 1886, but R. G. Dun & Co. Collection, Baker Library, Harvard University Graduate School of Business Administration, vol. 115, Herkimer County, N.Y., 3:301, "Mar. 27/86" "E. Remington & Sons" implies that the sale occurred in March. "The terms of the recent sale of the Type-Writer bus(iness) is not satisfactory to some . . ." The officers were William O. Wyckoff, President, of New York; M. W. Carter, Vice President, of Chicago; Clarence W. Seamans, Secretary, of New York; and Henry H. Benedict, Treasurer, of New York.

71. R. G. Dun & Co. Collection, Baker Library, Harvard University Graduate School of Business Administration, vol. 112, Herkimer County, N.Y., p. 337, "Standard Typewriter Manufacturing Co. Ilion."

72. *Ibid.* After they purchased the typewriter manufacturing operation, Wyckoff, Seamans & Benedict began a more aggressive advertising campaign. Part of that campaign included the publication of articles concerning the typewriter, its uses, and its manufacture. Prior to March 1886, advertising through articles was not a part of typewriter promotion. Consequently, little data concerning the earliest phase of typewriter manufacturing survives.

73. R. G. Dun & Co. Collection, Baker Library, Harvard University Graduate School of Business Administration, vol. 114, Herkimer County, N.Y., p. 736, "E. Remington & Sons, Ilion."

74. R. G. Dun & Co. Collections, Baker Library, Harvard University Graduate School of Business Administration, vol. 115, Herkimer County, N.Y., 3:302, "E. Remington & Sons, Ilion," "The receivers are running the foundry at present for the purpose of supplying the necessary castings to be used in their operations & at the same time furnishing supplies to the Standard Typewriter Manufacturing Co."

75. R. G. Dun & Co. Collection, Baker Library, Harvard University Graduate School of Business Administration, vol. 114, Herkimer County, N.Y., p. 736, "E. Remington & Sons, Ilion." "Their Sewing Machine Factory Burned today." "Dec. 14, 1882." "The building destroyed was used for making wood work of the sewing machine, typewriter, etc. Machinery was new. Loss abt. $20,000. It is doubtful if they (will) rebuild it."

The typebar lever remained wooden through the Remington Model No. 6, which ceased production about 1907 when the Models 10 and 11 were introduced. These new models were visible writers and had typebar levers of stamped steel with strengthening depressions stamped in them.

76. George Carl Mares, *The History of The Typewriter: Being an Illustrated Account of the Origin, Rise and Development of the Writing Machine,* p. 45. Mares states that "the Remington Armoury made three model machines which passed muster, and were regarded as satisfactory."

77. Current, *The Typewriter,* p. 67, quoting the *Milwaukee Sentinel.*

78. Current, *The Typewriter,* p. 71; R. G. Dun & Co. Collection, Baker Library, Harvard University Graduate School of Business Administration, vol. 115, Herkimer County, N.Y., 3:300. "Dec. 20, 1884" "E. Remington & Sons, Ilion." "They are also making 40–50 sewing machines per day & 20 [to] 25 type writers for which they receive cash as fast as manufactured."

79. Current, *The Typewriter,* p. 85.

80. R. G. Dun & Co. Collection, Baker Library, Harvard University Graduate School of Business Administration, vol. 113, Herkimer County, N.Y., 2/1:630, "December 14, 1880."

81. R. G. Dun & Co. Collection, Baker Library, Harvard University Graduate School of Business Administration, vol. 114, Herkimer County, N.Y., 2:699, "Sept. 18/85."

82. R. G. Dun & Co. Collection, Baker Library, Harvard University Graduate School of Business Administration, vol. 113, Herkimer County, N.Y., 2/1:427, "Remington (Empire) Sewing Machine Co., Ilion," "Dec. 9/76." Also, Jan. 72, "Just started."

83. William P. Tolly, *Smith-Corona Typewriters and H. W. Smith,* p. 14. Some background data on the firms under study are in order.

THE AMERICAN WRITING MACHINE COMPANY, HARTFORD, CONNECTICUT

In 1879 George Washington Newton Yost, one of the original promoters of the Sholes & Glidden (later Remington) Typewriter, began development work on a new writing machine, the Caligraph ("Beautiful Writer"). It incorporated many of the features of the Sholes & Glidden machine as well as new ideas developed by Walter J. Barron and Franz X. Wagner. Yost first opened a factory at 213 W. 31st Street in New York where the development work was carried out and the first machines were made. Later, after September 1881, Yost moved the factory to Corry, Penn., where, by 1884, he had some 125 people employed turning out 75 machines per week. By May 29, 1885, the factory had been moved again to Hartford, Conn., where, in 1888, the firm was prospering and where it remained. Eventually, the American Writing Machine Company joined the Union Typewriter Company, the "typewriter trust." In 1898, the firm discontinued its first models and brought out several new models called the "New Century Caligraph No. 5 & 6." However, by about 1910 the firm ceased production of the Caligraph and became the used typewriter rebuilder and dealer for the Union Typewriter Company.

Very little is known about the manufacture of the Caligraph. Nothing is known of the efforts in New York City or in Corry, Penn. Only a *Scientific American* article gives any data about the Hartford operation. Sources of information on the Caligraph and the American Writing Machine Company include: trade catalogue at Strong Museum and typewriter journal ads listing the company as a sales agent for used typewriters; Richard Current, *The Typewriter and the Men Who Made It,* especially ch. 10, "The Remingtons Are Mad," pp. 98–111; Milwaukee Public Museum trade catalogue H 43316/26801; Benjamin Whitman, *History of Erie County Pennsylvania,* (n.p., Warner Beers, 1884), "G. W. N. Yost," p. 1006. A later history of Erie County contains a number of errors and is not a reliable source—John Miller, *History of Erie County, Pa.* (1909), p. 578; R. G. Dun & Co. Collection, Baker Library, Harvard University Graduate School of Business Administration, 18:28; American Writing Machine Co., "Caligraph Writing Machine," *Scientific American* (March 6, 1886), N.S., 54(10):150; American Writing Machine Co., *The Caligraph Manufactured by the American Writing Machine Company,* a trade catalogue in the collections of the Milwaukee Public Museum, catalogue no. H 43769/26878.

THE HALL TYPE WRITER, SALEM, MASSACHUSETTS

The Hall Type Writer was invented by Thomas Hall of Brooklyn, N.Y., and patented in the United States in 1881. It was by far the most successful of the "index" or "indicator" machines and was sold in great quantities through the early 1890s by the Hall Type Writer Company of Salem, Mass. The Hall factory, located at 194–200 Derby Street, was "well stocked and well organized." It employed "various special machines and tools" to manufacture the Hall Type Writer, several of which were illustrated in the *Scientific American* in 1886. These illustrations provide the only data on the manufacturing techniques employed by Hall.

For information on the Hall Typewriter Company, see "The Hall Type Writer," *Scientific American* (July 10, 1886), N.S., 55(2):24–25; rpt. as "The Hall Type Writer," *Scientific American, Architects & Builders Edition* (August 1886), 2(2):30–31.

L. C. SMITH & BROS. TYPEWRITERS, SYRACUSE, NEW YORK

The L. C. Smith & Bros. Typewriter Company was incorporated on January 27, 1903 and capitalized at $5,000,000. This firm was composed of the four Smith brothers: Lyman Comelius, Wilbert L., Monroe C., and Hurlbut W., who had formerly organized the Smith Premier Typewriter Company, also of Syracuse, N.Y. The Smith Premier was originally an independent firm founded in 1887 as an outgrowth of the L. C. Smith Gun Company and the inventions of Alexander T. Brown. In 1893 the Smith Premier joined the trust known as the Union Typewriter Company, composed of Smith Premier, the American Writing Machine Company (the Caligraph), the Densmore Typewriter Company, the Yost Writing Machine Company, and the Remington Typewriter Company. Each firm in the trust continued to operate, manufacture, and advertise independently, despite their association and apparent control by Remington. In 1895 a new typewriter, the Wagner (soon known as the Underwood), appeared on the market featuring a "visible" or "writing-in-sight" mechanism. This mechanism allowed the typist to see what she was typing as she typed it. All the machines of the Union Typewriter Company were "blind" machines. The typist could not see what she was typing without lifting the carriage or rotating the platten.

The four Smith brothers recognized the importance of "visible" writing machines and believed that their firm, the Smith Premier Typewriter Company, should design and produce such a machine. The Union Typewriter Company evidently decided to reject the new "visible" machines and remained dedicated to the "blind" writers, thus prohibiting Smith Premier from innovating. Although two of the Smith brothers held offices in the Union Typewriter Company and all held offices in the Smith Premier Typewriter Company, they subsequently left both the trust and their own firm to organize the L. C. Smith & Bros. Typewriter Company in 1903. The first L. C. Smith & Bros. typewriter appeared on the market on November 24, 1904. It was a Model 2 machine with 84 characters. The L. C. Smith & Bros. typewriter was based on Carl Gabrielson's patent no. 753,972 of March 8, 1904. In addition, the new L. C. Smith & Bros. typewriter offered a number of innovative features besides its "visible" writing. Its segment shift—in which the relatively light type basket was lowered to produce capital letters, compared to raising the relatively heavy carriage—was its most notable feature. It also carried an inbuilt tabulator, a bi-chròmatic ribbon, stencil cutout, and interchangeable platens.

In 1926 L. C. Smith & Bros. Typewriter Company joined with another New York typewriter manufacturer, the Corona Typewriter Company of Groton, New York, which produced only portable machines, to form L. C. Smith & Corona Typewriters, Inc. Today this firm survives as an important division of SCM.

Sometime between August 1, 1915, and the corporate merger in 1926, the company published a trade catalogue entitled *A Visit to the Home of the L. C. Smith Typewriter*. "The purpose of this booklet is to take the reader on a trip through the factory of the L C Smith & Bros Typewriter Inc." It noted the company's " 'capacity for taking infinite pains,' " which had produced its "world wide reputation." At the time of the catalogue's publication, L. C. Smith & Bros. were manufacturing their No. 8 machine. The catalogue followed the flow of raw materials through the factory and illustrated virtually every major operation in some thirty photographs. These include views of the punch press operation, special machines, electroplating, automatic screw machines, casting, grinding and filing, baking ovens, ball bearing manufacture, six different assembly departments, alignment, and shipping.

Sources for the L. C. Smith & Bros. company biography include William P. Tolly, *Smith-Corona Typewriters and H. W. Smith*, pp. 12-14; and "A Condensed History of the Writing Machine, The Romance of Earlier Effort and the Realities of Present Day Accomplishments, Compiled in Celebration of the Semi-Centennial of the Founding of the Typewriter Industry, 1923," *Typewriter Topics* (October 1923), 55(2):126-128. There is very little known about the typewriter trust. It was evidently founded on March 1, 1893, and its primary function seems to have been keeping the price of office-grade typewriters at or above $100. It was a nebulous entity and deserves a detailed study in its own right. Some manufacturers went to great lengths to inform the public that they were not a part of the typewriter trust. Examples of the Franklin Typewriter Company's advertising in the Milwaukee Public Museum convey this clearly.

"A Condensed History of the Writing Machine," lists L. C. Smith as first vice president until 1896 and W. C. Smith as first vice president after L. C. Smith of the Union Typewriter Company. L. C. Smith was the president of the Smith Premier Typewriter Company from 1893–1903; W. L. Smith was vice president and factory manager; M. C. Smith was the secretary; and H. W. Smith was the treasurer. All quit to form the L. C. Smith & Bros. Typewriter Company in 1903. *The Evolution of the Typewriter*, pp. 40-41; *Typewriter Topics*, p. 128; *A Visit to the Home of the L. C. Smith Typewriter*, a trade catalogue is in the historical collections of SCM, and made available through the courtesy of Charles Molloy of SCM's Consumer Affairs Department, DeWitt, N.Y. A Xerox copy is in the research files of the Milwaukee Public Museum along with photographs and slides.

84. William P. Tolly, *Smith-Corona Typewriters and H. W. Smith*, p. 14.

85. E. Remington & Sons, *Directions for Setting Up, Working and Keeping in Order, The TYPE WRITER.*, ca. 1875. Photocopy in the research files of the Milwaukee Public Museum, original in a private collection. "If any type should get a trifle out of alignment, a gentle pressure against the inner end of the type-bar, one way or the other, as may be needed, will put all right again. If a type should get radically out of place, it can be adjusted by loosening the screw of its hanger-bearing, but this should not be attempted till one is fully familiar with the machine" (p. 6).

See also G. W. N. Yost and E. Remington & Sons, *The Perfected Type Writer, Number Two, Writes Capitals and Small Letters. Noiseless and Portable*, ca. 1877–1878, a trade catalogue in the files of the Division of Mechanisms, National Museum of American History, Smithsonian Institution.

86. American Writing Machine Company, *Directions and Suggestions for Using the "Caligraph" Manufactured by the American Writing Machine Company, (Revised Edition)*, (July 1, 1888), a trade catalogue reprinted by Dan Post, Post-Era Archives, Arcadia, Calif. (1981?), copy in the files of the Milwaukee Public Museum, catalogue no. H 39632/26229; and American Writing Machine Company, *Directions and Suggestions for Using the "Caligraph" Manufactured by the American Writing Machine Company, The Special Caligraph No. 3* (March, 1897?), a trade catalogue in the Milwaukee Public Museum, Carl P. Dietz Typewriter Collection, catalogue no. E 45707.L/12753.

87. *Ibid.*, pp. 5 & 6, illustrations in both catalogues.

88. American Writing Machine Company, *The Caligraph Manufactured by the American Writing Machine Company* (1888), pp. 3 and 4, a trade catalogue in the Milwaukee Public Museum, Carl P. Dietz Typewriter Collection, catalogue no. H 43769/26878.

ALIGNMENT

"All the letters and characters in the Caligraph are aligned to the small "n." When any of the type seem to be out of alignment, if the operator will strike such letters as seem to be too high or too low with the "n" each side of them, thus: nan, nBn, non, nMn and carefully observe the result, it will be seen which letters are out of alignment, and how much, and whether they strike too high or too low.

"To correct this, first with the tool provided for that purpose; (Adjusting Pin) (32), take up any lost motion there may be in the type bar pivot by turning the adjusting screw in the hanger (plate 1).

"To properly adjust the type bar, depress key attached to same, and turn adjusting screw in the hanger, until the bar remains in a horizontal position and will not drop; then holding the space key down, turn the adjusting screw backward slowly, until the type bar falls to its proper position.

"If this does not entirely correct the difficulty, loosen the large screw which holds the hanger of the faulty letter in the circle, and push the hanger in the desired direction, then put a light pressure of the screw on the hanger and test the letter between the two "n's." Repeat this operation until you have placed the letter in correct position, then set the screw home firmly.

TO PUT IN AND TAKE OUT TYPE

"To take out a type, turn it to the right; the opposite will tighten it. Should a type fall out, lift its type bar and put it in with the top of the letter towards you; hold a weight under the head of the type bar and drive it home with a piece of hard wood, bar of solder, or any heavy article softer than the type itself. Do not use iron for this. To align the type test it between the letter "n" (see article on alignment)."

See *Directions and Suggestions for Using the Caligraph Manufactured by the American Writing Machine Company (Revised Edition)*, p. 5; and *Directions and Suggestions for Using the Caligraph Manufactured by the American Writing Machine Company, The Special Caligraph No. 3*, p. 5. Illustrations in both catalogues.

89. Carl Mares, *The History of the Typewriter: Being an Illustrated Account of the Origin, Rise and Development of the Writing Machine*, pp. 75–77. *Typewriter Topics*, p. 67, states that the New Century Caligraph No. 6 was introduced in 1900. There were at least two models of the New Century Caligraph, a No. 5 and No. 6. Both models are represented in the Carl P. Dietz Typewriter Collection.

90. "The Hall Type Writer," *Scientific American* (July 10, 1886), N.S., 55(2): 24–25.

91. *Ibid.*, p. 24. Additional languages available on the Hall included "Greek, French, German, Spanish, Portuguese, Italian, Dutch, Norwegian, Russian, Swedish, etc."

92. Oliver Typewriter Company, *The Oliver Typewriters, The No. 3 Models* (ca. 1905), pp. 16–19, "A Visit to Our Factory." A trade catalogue in the files of the Division of Mechanisms, National Museum of American History, Smithsonian Institution.

93. American Writing Machine Company, *The Caligraph Manufactured by the American Writing Machine Company*. See also Romaine, p. 376.

94. "Producing a Wonderful Writing Machine," *Office Appliances* (December 1906), 5(1):20.

95. American Writing Machine Company, "Caligraph Writing Machines," *Scientific American* (March 6, 1886), N.S., 54(10):150.

96. "The Hall Type Writer," *Scientific American* (July 10, 1886), N.S., 55(2): 24–25.

97. Hounshell, *Mass Production*, p. 72. See figure 2.2, an illustration of the Wheeler & Wilson Manufacturing Company in 1879, from the *Scientific American* of May 3, 1879.

98. "The Hall Typewriter," pp. 24–25. The "Fig. 5" machine has only a single belt for power driving four belts to four separate but identical work stations. Each belt drives a shaft with an eccentric arm at each end. That arm is connected to the parallel movement device mounted on the machine and moves it as the belts turn. There is no other mechanism on this machine. Each work station could be started and stopped independent of the others by simply shifting its drive belt onto the idler pulley on the main power shaft. These idler pulleys are clearly illustrated in the drawing.

99. "Caligraph Writing Machines," p. 150: "Fig. 1. The Type Bars," illustrates the construction and assembly of the Caligraph typebars. The Caligraph differed sharply from the Remington and the American Writing Machine Company took advantage of that difference and advertised it widely. The typebars were first stamped from cold rolled steel and then folded lengthwise. A small piece of steel was then brazed into each end of the folded typebar, one piece to receive the type at the lower end of the bar and a second piece with conical points which protruded through holes in the sides of the typebar. The typebar hung in a U-shaped, typebar hanger, suspended by its conical points and secured with an adjustment screw.

The Milwaukee Public Museum specimens in the Carl P. Dietz Typewriter Collection consist of seven Caligraphs and one Frister & Rossman (German-made Caligraph) and all have the punched, formed, and "brazed" typebars. I have not seen a Caligraph with forged typebars. Milwaukee Public Museum specimens:

Machine	*Serial Number*	*Catalogue/Accession Number*
No. 1	3,369	E 34626/8616
No. 1	3,826	E 41773/12072
No. 2	3,392	H 34260/25372
No. 2	11,327	E 6944/2578
No. 2	14,638	H 43209/25775
No. 2	19,275	H 44000/26877
No. 3	5,450	E 41658/12045
Frister & Rossman	1,016	E 44044/12994

100. "The 'Oliver' Growth In Facts and Figures," *Office Appliances* (January 1908), pp. 33–48. An article removed from a trade journal in the Warsaw Collection, National Museum of American History, Smithsonian Institution.

101. *A Visit to the Home of the L. C. Smith Typewriter.*

102. *A Visit to the Home of the L. C. Smith Typewriter.* Although undated, this catalogue predates the 1926 merger of L. C. Smith & Bros. and the Corona Typewriter Company. It also postdates the introduction of the L. C. Smith & Bros. No. 8 machine on August 1, 1915. See "A Condensed History of the Writing Machine," *Typewriter Topics* (October 1923), 55(2):126–128. More precise dating is possible, since the catalogue states that "nearly 3/4 million" typewriters had been produced when the catalogue was published, thus putting the publication date closer to 1926 than 1915 according to "Serial Number Record, Office Typewriters" of L. C. Smith & Bros.

machines. SCM typescript (copy in MPM files) indicates that by the end of 1925 roughly 670,000 machines had been produced.

103. *The L. C. Smith & Bros. Typewriter, (Model No. 8)*, pp. 10 and 17. Milwaukee Public Museum, Carl P. Dietz Typewriter Collection, catalogue no. H 44182/26877. My estimates include only the ball bearings used on the typebars, not on the carriage, for example.

104. "The Remington Typewriter," *Scientific American* (December 15, 1888), N.S., 59(24):367 and 374–375; P. G. Hubert, Jr., "The Typewriter, Its Origins and Uses" (April 25, 1888), pp. 25–32. For a copy of Hubert, see the files of the Division of Mechanisms, National Museum of American History, Smithsonian Institution. In Hubert's article, the first illustration, entitled "Putting in Connecting-rods and Levers," shows a workman attaching the wire connecting-rods which run from the wooden type levers to the typebars which have already been mounted and aligned for the first time in the fixture described in the *Scientific American* article. The workman has the partially assembled typewriter mounted on a revolving table to facilitate the assembly by bringing each typebar into the same position. Hubert's second illustration shows a "Corner of the Aligning Room." Here, thirty-seven men are busy aligning the assembled typewriters. Aligning consists of adjusting each typebar so that all the letters share a common base line when typed. In other words, the letters must be straight with respect to each other. Each machine is again mounted on a revolving table. This is probably the second alignment after the initial alignment done during the assembly stage. Here, fully assembled machines are being aligned by typing. The paper is clearly visible and several men have lifted the carriages of their machines to examine the lines they have just typed. Hubert's third and final illustration shows a "Corner of the Adjusting Room." Adjusting occurred after alignment and consisted of a complete inspection and testing of the machine. Adjustment might include any aspect of the machine, the escapement, the shift mechanism, perhaps a minor alignment, a bit of oil if needed, etc. The ubiquitous revolving table is once again visible in front of each adjuster. It was a feature of every assembling, aligning, and adjusting department in virtually every typewriter factory.

The *Scientific American* article illustrated and described some aspects of the Remington operation under Wyckoff, Seamans & Benedict. The new works produced over 1,500 machines per month using specialized machinery and assembly techniques. "All parts require special machinery to secure uniformity and perfection in their construction. . . ." These illustrations included: "1. Engraving Original Type," "2. Case Making," "3. Annealing, tempering, and bluing," "Fig. 7. Type Arms and Type," "Fig. 8. Pivots of the Type Arms," "Fig. 5. Transverse Section of the Typewriter," and "Fig. 10. Centering the Type Arms."

One department was devoted to making the type. Manufacturing type was critical to the success of the typewriter, since it was the type itself which created the image on paper, the result of using the typewriter. An illustration entitled "1 Engraving Original Type" showed the initial step in manufacturing type. Here a workman, using a magnifying glass to see clearly, engraved the steel dies from which the type was made, a job that required "great skill and much patient labor." Later, the Remingtons, like other manufacturers, developed a pantograph to improve the quality of their type dies, but in the earliest years, type die making was a highly skilled job.

To make the type, blank type steel was first cut to size, then softened in the annealing furnace, shown in "Annealing, tempering, and bluing." The softened type was then forced into the handmade steel dies and assumed the shape of a particular letter. After this, it was hardened and tempered, "like wood or iron working tools," by heating to the appropriate temperature in a type hardening furnace and then being plunged into a liquid filled vat. Iron and steel parts that were not painted or nickel plated, were blued in a bluing furnace. Here, the parts were imbedded in sand which was heated while being constantly agitated to assure uniform heating. When the proper color was obtained, the parts were removed from the oven and separated from the sand by a sieve.

The Remingtons drop forged many parts, including their typebars, shown in "Fig. 7. Type Arms and Type." In addition, the pivots of their typebars were hardened and ground steel. These bearings ran in a typebar hanger which was formed into its final 'u' shape after having been punched from sheet steel.

105. "Producing a Wonderful Writing Machine," *Office Appliances* (December 1906), 5(1):20–24. The factory was largely the work of William K. Jenne, "the dean of typewriter makers," who was one of the original typewriter mechanics and manufacturers. Jenne had charge of the Remington sewing machine department in 1873 when Densmore and Yost brought Sholes' model to Ilion, and he evidently remained with typewriter manufacturing throughout his life. In 1906, the product of this new factory was the Remington No. 6 and the related special machines, Nos. 7, 8, and 9.

At the same time, relatively crude and simple machines were used in other operations, such as belt sanders in the casting preparation shop. Most parts were "drop forged, machined accurately to gauge in costly fixtures, ground, polished, annealed, plated, buffed, japanned, assembled, compounded, and finally inspected," before being stored "in an oiled condition" in the finished stock parts room. After each manufacturing step, an inspector checked the work and discarded any imperfect parts before they were further manufactured. Finished parts were stockpiled in the stock room, from which they were drawn for assembly.

106. "Producing a Wonderful Writing Machine."

107. "Caligraph Writing Machine," *Scientific American*.

108. *The Oliver Typewriters, The No. 3 Models.*

109. *A Visit to the Home of the L. C. Smith Typewriter.*

110. Bruce Bliven, *The Wonderful Writing Machine*, p. 179.

5. *The Development of Watch Manufacturing at the Waltham Watch Company, Waltham, Massachusetts, 1849–1910*

1. Leonard Waldo, "The Mechanical Art of Watchmaking in America." It is somewhat surprising to find that the watch manufacturing industry is perhaps the least studied of the American System industries. Robert Woodbury remarked as early as 1966 that "the work of the American Watch Company at Waltham, Massachusetts, deserves more attention; in fact, a monograph on their early anticipation of highly specialized, precision automated machine tools is sadly needed." Woodbury, review of "*A Short History of Machine Tools,* by L. T. C. Rolt," *Technology & Culture*. Most authors give the industry passing notice, realizing its importance and significance, but having no detailed information to offer. Thus, for example, Rolt notes that

"by this time (1855) watches were also being produced by the new methods by the American Watch Company at Waltham, Massachusetts." L. T. C. Rolt, *A Short History of Machine Tools*, p. 176.

2. Edward A. Marsh, *The Original American Watch Plant: It's* [*sic*] *Planting, Growth, Development and Fruit*, pp. 12–13.

3. From its earliest years the Waltham Watch Company hired women operatives. Throughout the industry, women accounted for perhaps a third of the work force. Edward A. Marsh, *"Workers Together:" A Story of Pleasant Conditions in an Exacting Industrial Establishment*. Elgin also had its miniature railroad tracks in front of automatic machines. See Roy Ehrhardt, *Elgin*, p. 105, one of a series of photographs taken in the Elgin factory, ca. 1930.

4. Charles W. Moore, *Timing a Century*, pp. 9 and 17, indicates the Springfield Armory as the source of Dennison's "inspiration." Dennison was known abroad as the "Father of Interchangeable Watchmaking." See Arthur Tremayne, *One Hundred Years After: Being a Little History of a Great Enterprise.*

5. The term *interchangeable* tends to imply the ability to assemble a mechanism such as a typewriter, watch, clock, or firearm, from a supply of parts chosen at random. In fact, every nineteenth-century manufacturer of complex mechanisms designed those mechanisms to be adjusted at the time of assembly. Thus the *interchangeable* parts were interchangeable, but only to the degree necessary, the degree stipulated by the design of the product.

6. Moore, *Timing a Century*, p. 87, table 11.

7. Waltham Watch Company, *Serial Numbers With Descriptions of Waltham Watch Movements.*

8. Dennison's letter is quoted in Leonard Waldo, "The Mechanical Art of Watchmaking in America," p. 187.

9. See John R. Harris, *Liverpool & Merseyside*, chapter on watchmaking; A. White, *The Chain Makers;* E. Surrey Dane, *Peter Stubs and the Lancashire Hand Tool Industry;* and "The Manufacture of Watches," in *Census of Manufacturing, 1880*, p. 61.

10. R. F. and R. W. Carrington, "Pierre Frédéric Ingold and the British Watch and Clockmaking Company"; *La Suisse Horlogerie et Revue Instrumentale de L'Horlogerie;* and *Quelques Notes sur Pierre-Frédéric Ingold et les travau de E. Haudenschild publié par la Société Suisse de Chronometrie à l'occasion de huitième Assemblée générale tenure à Berne, Au Burgerhaus le 4 Juin 1932.*

11. There were other lesser efforts, notably by Luther Goddard, who assembled English parts during the War of 1812 and shortly afterward, and by Jacob D. Custer of Norristown, Penn. Neither was significant.

12. American Watch Tool Company, *American Watch Tool Co. Waltham, Mass., Manufacturers of the Whitcomb Lathe*, p. 24. This early tool catalogue contains a short history of watch manufacturing in America. Its author believes the Pitkins made 1,000–2,000 watches. See also Chris H. Bailey, *Two Hundred Years of American Clocks and Watches*, pp. 193–194; Charles S. Crossman, "A Complete History of Watch and Clock Making in America, Number One," *Jewelers' Circular and Horological Review* (July 1886), 7(6):183–184; and Robert M. Wingate, "The Pitkin Brothers Revisited; Controversies and Comparisons of Their Hartford and New York Model Watches Revealed."

13. Henry Pitkin apparently designed the movement.

14. Edwin Battison explained several of the Pitkin watch design flaws to me.

15. Collectors should note that movements marked with Pitkins' name and New York are English imports sold by the Pitkins from their New York store but were not made in the United States. In this respect, they are like any other imported English movement with an American jeweler's name and they should not be confused with Hartford made watches.

16. An apocryphal story suggests that Dennison knew John R. Proud when Proud worked for the Pitkins in New York and that Dennison and Howard purchased the Pitkin machinery in the late 1840s and transported it to Boston where it became the basis of the Waltham Watch company. There is no evidence to support this assertion which is found in Wingate.

17. Charles S. Crossman, "A Complete History of Watch and Clock Making in America, Number Two," *Jewelers' Circular and Horological Review* (August 1886), 7(7):219.

18. This is a *greatly* simplified sketch of the early years of the Waltham Watch Company. For more detail, see Moore, *Timing a Century*.

19. Edward Howard to W. H. Keith, quoted in Crossman, "A Complete History of Watch and Clock Making in America, Number Three," *Jewelers' Circular and Horological Review* (September 1886), 7(8):259.

20. *Ibid.*, p. 259.

21. Edward A. Marsh, "History," p. 6. Waltham Watch Papers, Baker Library, Harvard University, Graduate School of Business Administration, MS. RC-2. This typed manuscript is entitled simply "History." From internal evidence, it is believed by Prof. Gitelman to have been compiled by Edward A. Marsh about 1921. It is complete with illustration notations and corrections from an early proofreading, but apparently it was never published. The most important published works on watch manufacturing are those of Edward A. Marsh, the master mechanic at Waltham who took the time to record much of the technical history.

22. Edward A. Marsh, "History of Early Watchmaking in America, 1889," p. 6. Waltham Watch Papers, Baker Library, Harvard University Graduate School of Business Administration, MS. RC-1. This manuscript was published as Edward A. Marsh, "The American Waltham Watch Company," ch. 50, in D. Hamilton Hurd, ed., *History of Middlesex County, Massachusetts*, (Philadelphia, Pa., 1890), 3:738–749.

23. Crossman, "A Complete History," p. 257.

24. Crossman, "A Complete History," p. 312.

25. The Springfield Tool Company was a small, short-lived machine tool company formed about 1855 to produce Chester Van Horne's new turning lathe.

26. Marsh, "History," p. 29.

27. Marsh, "History," p. 29. H. M. Gitelman, "The Labor Force at Waltham Watch During the Civil War Era," notes that the Connecticut River Valley was a source of early Waltham mechanics (pp. 216–217). Eugene S. Ferguson, "The Critical Period of American Technology," (ca. 1961), (see copy in the Eleutherian Mills Historical Library, Greenville, Wilmington, Del.), notes the "pivotal role" of machine shops and foundries in developing nineteenth-century technology. This is precisely what happened at Waltham. Here, the important research and development lab of its day, the machine shop, was brought into the factory and local foundries produced the necessary castings.

28. American Watch Tool Company, *American Watch Tool Company, Waltham, Mass., Manufacturers of the Whitcomb Lathe,* pp. 34–35.

29. Crossman, "A Complete History," p. 258, Dial feet are those small round pieces of brass soldered to the dial which protrude into holes in the pillar plate and thus secure the dial to the movement.

30. Edward A. Marsh, *The Original American Watch Plant,* p. 9. See also, James W. Gibbs, "Watch Hands: The History of J. H. Winn, Inc."

31. Edward A. Marsh, *The Evolution of Automatic Machinery: As Applied to the Manufacture of Watches at Waltham, Mass., by the American Waltham Watch Company,* p. 95. Hereafter cited as: Marsh, *Automatic Machinery.*

32. Marsh draws attention to Charles S. Moseley, although Charles Stark claimed to have been the "originator" and "pioneer manufacturer of the Spring Chuck Lathe." There is some support for Stark's claim, as he patented the concept on May 30, 1865. See United States Patent No. 47,997. Stark's idea involved a threaded spindle near the collet and not the sliding spindle concept. See Marsh, *Automatic Machinery,* pp. 14–18; and Stark Tool Company, *Stark Tool Company Manufacturers of Precision Bench Lathes and Fine Tools of Every Description,* back cover illustration caption. (MPM H 43766/26878.)

33. There is some question as to when the sliding spindle was first introduced at Waltham. It probably occurred sometime after 1865. Whenever it occurred, it was of great importance. This ambiguity about such a significant invention only emphasizes the need for more research.

34. Marsh, *Automatic Machinery,* ch. 1.

35. *Ibid.* See also "Appleton, Tracy & Co.'s American Watch Company, Waltham, Mass."

36. Edwin Battison reminds me that the subject of precision grinding at Waltham is extremely important, particularly in the production of precision machine tools. It deserves additional research.

37. Marsh, "History," p. 29.

38. "Appleton, Tracy & Co.'s American Watch Company, Waltham, Mass."

39. John R. Harris, an economic historian at the University of Birmingham, has written on the subject of skills and the Industrial Revolution. He argues that the British Industrial Revolution was based on the development of special skills in using coal fuel. The development of these new skills necessarily displaced old skills (such as charcoal iron smelting skills). The same is true of industrialization in America. The machines of the American industrial revolution simultaneously created and destroyed special skills. While these new skills differed sharply from the skills on which they were originally based, they demanded equal if not greater attention to detail. See, for example, John R. Harris, "Skills, Coal and British Industry in the Eighteenth Century"; *Industry and Technology in the Eighteenth Century: Britain and France;* and "Saint-Gobain and Ravenhead."

40. Moore, *Timing a Century,* p. 50. To a great degree, Waltham benefited from the inflation created by the United States government's method of financing the Civil War, the printing of *Greenbacks.* The resulting inflation catapulted Waltham into unimagined profitability in part because it could repay loans and debts in inflated (less valuable) currency. It is worth noting, however, that Waltham remained very profitable following the Civil War. See table 5.4.

41. So important was the Civil War not only in ensuring the financial success of Waltham but also in creating an image of Waltham watches that the company used the theme in advertisements in the early twentieth century. See, for example, a series of advertisements that ran in the *National Geographic* in 1923. In March 1923 one advertisement read "Granddad's WALTHAM ticked off the stirring minutes of 'Sixty-two'" (MPM H 39199.2/26005). In April another noted, "Nearly three-quarters of a century stretch between the great bulky Waltham of Civil War days and the light, thin, though equally sturdy model of today" (MPM H 39199.3/26005). Both were accompanied by nostalgic illustrations.

42. Crossman, "A Complete History," p. 313.

43. Thomas L. DeFazio, "The Nashua Venture and the American Watch Company."

44. Crossman, "A Complete History," p. 342.

45. Marsh, "History," pp. 40, 40 1/2, and 41.

46. The inside contract system may have been and probably was used earlier than 1863, but it was certainly in use by that time.

47. "The Factory of the American Watch Company," *Scientific American*, p. 226.

48. Edwin A. Battison, *From Muskets to Mass Production: The Men and the Times That Shaped American Manufacturing*. My understanding of this subject has been deepened through conversations with Howard Madaus, the Curator of Firearms at the Milwaukee Public Museum.

49. Smithsonian Institution, Museum of American History, catalogue no. 334,625. This watch is engraved "1852," suggesting that it may not have been finished until then. Some researchers believe the Marsh brothers were at work as early as 1848 or 1849.

50. Christy, Manson & Wood International, *Fine Watches and Clocks . . . Tuesday, June 28, 1983* . . . , "Lot 32—Two Unusual American Watch Movements . . . *American Watch Co., Waltham, Unnumbered* . . . a gilt full plate model 1857 18-size, lever movement signed MODEL *Appleton, Tracy & Co.* with compensation balance, jewels in settings, with signed white enamel dial," p. 14. This was standard practice in all watch factories. There are several examples in The Time Museum collection. See especially the Elgin models inventory nos. 315.5, .13, .17, .23, .30, .36, .37, .38, .41, .42, .44, .45, .46, .51, .52, .54, .55, .56, .58, .59, .65, and .77, and the Edward Howard prototype, inventory no. 1442, in my *The Time Museum Catalogue of American Watches*, (forthcoming, 1990).

51. "The Factory of the American Watch Company," *Scientific American*, p. 226.

52. *The Ordnance Manual for the Use of the Officers of the United States Army*, 2d ed., (Washington, D.C.: Gideon, 1850), Chapter 8, "Small Arms & Accoutrements," pp. 157–207, see especially, pp. 164–168, "List of Verifying Gauges," prepared by G. Talcott, Lt. Col. of Ordnance; and *The Ordnance Manual for the Use of the Officers of the United States Army*, 3d ed., (Philadelphia: J. B. Lippincott, 1862), Chapter 8, "Small Arms & Accoutrements," pp. 177–226, see especially pp. 215–221, "List of Verifying Gauges," prepared by Brig. Gen. J. W. Ripley, Chief of Ordnance.

53. "The Factory of the American Watch Company," *Scientific American*, p. 226.

54. The technology on which the new measuring techniques were built was the European pivot micrometer or *douzième* gauge, also known as a caliper gauge. Ferdinand Berthoud published a version of the *douzième* gauge as early as 1763 in

his *Essaie sur l'horologerie*. The idea is found in a number of other European horological treatises and expressed in a variety of tools. The pivot micrometer employs the simple geometric principle that a small movement of a lever pivoted near one end produces a proportionately larger movement at the other end. The ratio of those movements is directly proportional to the location of the pivot. The *douzième* gauge greatly multiplied the ratio and hence magnified the errors, the better to detect and read them. Gauges based on this geometric principle were well known to watchmakers and repairmen in the nineteenth century. By the end of the nineteenth century they were mass produced and commonly available through jobbers and regularly appeared in watch tool supply catalogues. Surprisingly, Marsh says nothing about gauging. See Theodore Crom, *Horological Shop Tools, 1700–1900*.

55. David A. Wells, "The American Manufacture of Watch Movements," p. 65.

56. "The Factory of the American Watch Company," *Scientific American*.

57. The Waltham Watch Company drawings at Smithsonian, collected by Edwin A. Battison, are uncatalogued.

58. "The Factory of the American Watch Company," *Scientific American*, p. 226.

59. It too employed the general principle of the *douzième* gauge, but in a different way. Its spindle slid in a vertical column and was held on the surface of the anvil with a spring. Attached to the spindle was a tiny chain, much like a fusee chain, that ran up the column, over a pulley, and down to a stationary pivot around which it was wrapped several times. The end of the pivot carried a watch hand that pointed to graduations on the outer edge of an adjustable dial. When the spindle was on the anvil, the dial was set at '0', then, any vertical movement in the spindle was transferred through the chain, pulley, and pivot to the hand, which indicated the change on the dial. To speed its operation, the upright gauge was equipped with a handle attached to the spindle. The handle allowed the operative to move the spindle with his or her left hand and manipulate the part to be gauged with the right hand. Like the "Fine Gage" and the "Balance Guage," the "Upright Gage" featured all the necessary features of mass production gauging. The spelling of the word "gauge" varies greatly in the nineteenth-century literature and on Waltham's drawings, appearing as "gage," "gauge," and "guage."

60. "Mechanical Appliances in America," extract, *London Times*, August 24, 1884; reprinted in American Watch Company, *Souvenir Catalog of the New Orleans Exposition, 1884–1885*, p. 31.

61. "The Watch as a Growth of Industry," *Appleton's Journal of Literature, Science, and Art*.

62. "The Factory of the American Watch Company," *Scientific American*, p. 226.

63. Surviving gauges from the Elgin National Watch Company indicate that it, like Waltham, commonly used the upright gauge in a wide variety of applications. In addition, Elgin adopted the idea of the *douzième* gauge and combined it with the go/nogo idea to produce highly accurate pivot gauges in the early twentieth century.

The upright gauge was copied and produced by several watch tool firms including Randall & Stickney, the American Watch Tool Company, and the Stark Tool Company. They were also featured in Waltham advertising in the early twentieth century along with another gauge, known as a needle gauge, which was used to gauge the diameter of jewel holes. Upright gauges were in use at Waltham when the factory ceased watch manufacturing in 1954 and as dial gauges they are widely used in all kinds of manufacturing today. See Donald R. Hoke, *The Time Museum Catalogue of American Watches* (forthcoming, 1990).

64. David A. Wells, "The American Manufacture of Watch Movements," p. 68.

65. H. C. Hovey, "The American Watch Works," *Scientific American*, p. 102.

66. To get an idea of the overall complexity of watch manufacturing, see Sersefi Vasilévich Tarasov, *Technology of Watch Production*. The book carries the note "Approved by the Administration of Educational Institutions of the *Ministry for Automation* and Instrument-Making of the USSR as a Textbook for Technical Colleges" (italics added).

67. The detailed process of screw production at Waltham can be found my *Time Museum Catalogue of American Watches*, (forthcoming, 1990). See the historical essay on watch manufacturing.

68. Edward A. Marsh, *Automatic Machinery*, p. 95; and David A. Wells, "The American Manufacture of Watch Movements," pp. 65–69, who cites "A correspondent of the *New York Times*," as furnishing the information.

69. "Letters Patent of the United States Patent, No. 21,824, and granted October 19, A. D. 1855, to George W. Daniels and Abraham Fuller." Cited in John Stark's patented "Turning Lathe," No. 47,997, May 30, 1865. George Daniels also holds patent no. 21,864 for a "Lathe Chuck," dated October 19, 1858. It appears that there is some confusion in Stark's patent about Daniel's date and number.

70. The process began with a small lathe fit with a double compound slide rest and a two or three spindle tumble tailstock. The steel wire was carried through the lathe spindle to the split chuck which held it during thread cutting. Earlier versions of this machine had a two bearing lathe with a hand tightened chuck, later versions featured the three bearing lathe with the sliding spindle and a foot pedal to operate the self-closing chuck.

The lathe was set in motion, and the tool held in one side of the double slide rest was used to turn the wire down to the diameter of the screw. Then the spindle in the tumble tailstock that held the screw cutting die was brought to the wire and turned on by hand, the operative (both men and women) using the wheel mounted on the end of the spindle and being careful not to break off the screw in the die. The second spindle of the tail stock held a stop that regulated the depth that the thread cutting die could be turned onto the wire. The die was then turned off by hand, the lathe set in motion, and the cut-off tool in the opposite side of the double slide rest was used to cut the screw almost free from its piece of wire. The operative then ran this partially finished screw into a "slotting plate," a piece of sheet metal with two rows of threaded holes, and the plate was twisted until the screw broke. This procedure was repeated until the slotting plate was filled.

The slotting plate was placed in a screw head slotting machine, essentially a milling machine with a milling cutter the diameter of the screw slot. The slotting plate was secured to the bed of the machine, which carried it past the milling cutter which cut the slots in an entire row of screws. The plate was then reversed and the second row of screws was slotted. A boy then removed the screws by hand and returned the slotting plate to the screw maker. This technique with slight variations was employed by other watch factories in the late 1860s and early 1870s. The output by a skilled operative and helper was 1,200 to 1,500 screws per day. See "Machine-Made Screws," *American Horological Journal*, pp. 251–254. This article describes the process of screw making at the United States Watch Company of Marion, New Jersey. There were minor variations in their technique compared with Waltham's. Elgin apparently used a similar technology. See Edward Eggleston, "Among the Elgin Watchmakers." This technology was made available to other manufacturers through the watch tool industry. The Sawyer Watch Tool Company of Fitchburg, Mass., offered it for sale in 1880, along with other tools including a "Pinion Cutting Engine," an "Automatic Leaf Polisher," and an "Automatic Staff and Pivot Turning Machine." See Sawyer Watch Tool Company, "Watch Making Machinery.," *American Machinist,* (August 7, 1880), pp. 4–5.

71. It was first used to produce small jewel screws (used to hold pivot or hole jewels in the plates), but about 1875 it was redesigned in a more substantial form to produce the other screws needed to assemble the watch. The Vander Woerd automatic screw machine is an especially interesting piece of machinery because it illustrates how the intellectual process of invention often progressed from hand work to machines that mimicked hand work, to machines that departed radically from hand work. Vander Woerd's machine is an example of the intermediate machine, which mechanized the earlier hand process but did not actually change the process. See Marsh, *Automatic Machinery*, p. 100. Note that Vander Woerd's machine predates the famous Spencer automatic screw machine. They were probably quite different, and a comparison of the two would be of interest.

72. Edwin A. Battison to Donald Hoke, May 30, 1986. I subsequently discussed the Vander Woerd screw machine with Battison, who is the director of the American Precision Museum in Windsor, Vermont, and a representative of the Waltham Screw Company.

73. The cams operated to open and close the chuck, advance the wire in the chuck, activate the clutch, move the two cutters into and away from the work piece, and move the tailstock (carrying the thread-cutting die) into and away from the work piece. A second camshaft activated an arm that carried the partially finished screw to a slotting operation and then ejected the finished screw.

It is unclear exactly how the products of these early automatic screw machines were checked. It is conceivable that Waltham employed some kind of statistical quality assurance by testing samples of screws, but whether these were random samples or not is unknown. Perhaps each machine's screws were tested hourly or daily. It seems obvious that each screw was not gauged or tested individually.

74. As early as 1884 and as late as 1919 it was featured in Waltham's advertising. Duane Church subsequently produced a simplified version of the Vander Woerd screw machine. Marsh, *Automatic Machinery*, p. 105.

75. U. S. Patent No. 329,182, E. A. Marsh, "Screw Machine," October 27, 1885.

76. A large drum carried four self-closing sliding spindles, each of which held a piece of wire. As the drum rotated to four successive work stations, each piece of wire was measured and chucked to size, turned to proper diameter, threaded, and removed to a threaded spindle in a second four-station drum. In this second drum, it was slotted and ejected. In essence, short sections of uncut wire were inserted into the machine, and finished (except for polishing) screws fell into a container, without the touch of human hands. The machine was entirely mechanical, being worked with belts, pulleys, cams, springs, ratchets, clutches, and levers.

The Marsh machine produced twelve plate screws per minute (one every five seconds). It was adapted for producing brass or gold balance screws, which are considerably softer than the steel plate screws and could be produced at the rate of twenty per minute or one every three seconds. When the wire from which the screws were cut was exhausted, the machine automatically shut down. One operative could run "six or more" of the Marsh Automatic Screw Machines, thus producing 50,000 to 60,000 screws per day. This compares with 1,200 to 1,500 screws per day using the earlier method that employed the three-bearing sliding spindle lathe and the screw slotting machine and with the Vander Woerd screw machine that produced 8,000 screws per day. See "Watchmaking in America," *The Illustrated London News* (June 19, 1875), pp. 591–592.

It should be noted that screw polishing followed screw production, and Waltham produced a series of machines to accomplish that task as well.

77. U. S. Patent No. 512,156, January 2, 1894.

78. "Duane Herbert Church," *American Machinist* (September 14, 1905), 28(2):359–360. The illustration of Duane Church found opposite p. 77 in Moore, *Timing a Century*, is the reverse of the same engraving used by Waltham in its advertising in the late teens and early twenties.

79. There are many variations of American watches. For example, early watches had balance wheels that oscillated only every 1/4 second, the "old style" or "slow train" movements as compared with the "quick train" movements that oscillated every 1/5 of a second. Some manufacturers used non-magnetic hairsprings and balance wheel materials. There is also a great deal of variation in plate design. See Donald R. Hoke, *The Time Museum Catalogue of American Watches*, (forthcoming 1990).

80. Olof Ohlson, *Helpful Information for Watchmakers*, pp. 7–12.

81. Just as there are variations in the general design of watches, there are variations in the types of balances and escapements.

82. Marsh, *Automatic Machinery*, p. 79.

83. Marsh, *Automatic Machinery*, p. 80.

84. Marsh, *Automatic Machinery*, pp. 81–82.

85. This dating is based on Marsh's identification of the machine and its inventor and the illustration of this machine in an 1874 publication. See figures 5.36 and 5.37.

86. Marsh, *Automatic Machinery*, pp. 82–83.

87. Marsh, *Automatic Machinery*, pp. 83–84. As with Woerd's machine, it is possible to date this machine to the 1883–84 period using Marsh's reference in 1896 of having invented the machine some eleven years ago, and its illustration in the 1884 *Scientific American* article. See figures 5.38 and 5.39.

88. See "The Watch as a Growth of Industry," *Appletons' Journal of Literature, Science, and Art*, p. 33; and H. C. Hovey, "The American Watch Works," *Scientific American*, p. 103.

89. H. G. Hovey, "The American Watch Works," *Scientific American*, p. 103.

90. The production of pallet and hole jewels is another interesting technology, for which there is no space in this study.

91. It is worth noting that this process was carried over from the craft methods of watchmaking, yet it remained standard practice at Waltham until the factory closed in 1954.

92. H. C. Hovey, "The American Watch Works," *Scientific American*, p. 103.

93. Marsh, *Automatic Machinery*, p. 133.

94. "The coiling of hairsprings seems to belong among the class of mechanical operations, or manipulations, which are not susceptible of marked improvement. There may be obtained a measure of superiority in the quality of the tools employed, but the processes of production admit of little variation." Marsh, *Automatic Machinery*, p. 135.

95. Marsh, *Automatic Machinery*, p. 142.

96. Marsh, *Automatic Machinery*, pp. 145–145. Avoirdupois is an archaic measure of weight "used for goods other than gems, precious metals, and drugs: 27 11/32 grains = 1 dram; 16 drams = 1 ounce; 16 ounces = 1 pound." *American College Dictionary*, (New York: Random House, 1966), p. 86.

97. H. C. Hovey, "The American Watch Works," *Scientific American*, pp. 103–104.

98. H. C. Hovey, "The American Watch Works," *Scientific American*, pp. 103–104.

99. H. C. Hovey, "The American Watch Works," *Scientific American*, p. 104.

100. Waltham produced a limited number of free sprung watches. These were especially high-grade watches, adjusted to such precision at the factory that they required no regulator levers. See figure 5.32.

101. The balance wheels on early low-grade watches were plain (uncut) steel and gold with flat hairsprings. Eventually, production got so cheap that all grades had cut bimetallic (temperature compensating) balances with adjusting screws and Breguet hair springs.

102. Waltham Watch Company, *Catalogue of Waltham Watch Material* (Waltham, Mass.: Waltham Watch Company, April 1909), rpt. (Yorktown Heights: Manfred Trauring, ca. 1970?). (MPM H 46162/27222).

103. *Ibid.*, p. ii. In 1888, after only twenty-four years in business, Waltham's largest competitor, the Elgin National Watch Company, listed only thirty-two different springs. Like Waltham, however, Elgin suggested that "in order that Springs of suitable strength may be selected, the Balance should be sent for springing." See *Net Price List of Materials Manufactured by the Elgin National Watch Co.* (Chicago: Knight & Leonard, April 1888), pp. 90 and 92.

104. "Duane Herbert Church," *American Machinist*, September 14, 1905, 28(2):359–360.

105. Edward A. Marsh, *The Original American Watch Plant, It's [sic] Planting, Growth, Development and Fruit*, pp. 19–20. Elsewhere, Marsh claims that the successful patent infringement suit against several smaller companies, notably the Aurora and Columbus Watch Companies, effectively drove them out of business.

106. United States Patent No. 360,234, March 29, 1887.

107. United States Patent No. 366,592. July 12, 1887.

108. "Appleton, Tracy & Co.'s American Watch Company, Waltham, Mass."

109. *Census of Manufacturing*, 1880, pp. 60–69.

110. H. C. Hovey, "The American Watch Works," *Scientific American*, p. 102.

111. Waltham Watch Company, *Serial Numbers With Descriptions of Waltham Watch Movements*, (1954), and *Census of Manufacturing*, 1880, p. 61.

112. Waltham had at least one of its "Crystal Plate" or "Stone" movements, orig-

inally produced in the late 1880s and 1890s, available for sale from stock as late as 1939. A. E. Mathews, "A Crystal Plate Waltham Watch," *Bulletin of the N.A.W.C.C., Inc.* (April 1971), 14(9[whole no. 151]):1,075–1,077.

113. See Elgin's advertisement for its Hulburd prestige watch in Ehrhardt, *Elgin*, p. 115.

114. *Census of Manufacturing*, 1880, p. 61.

115. H. C. Hovey, "The American Watch Works," *Scientific American*, p. 102; and Waltham Watch Company Papers, Baker Library, Harvard University Graduate School of Business Administration, MS. DG-1, Waltham Improvement Company, Cash A, p. 58, "pd. freight & expenses for watch blanks from Scoville Co. [$]6.03."

116. William F. Meggers, Jr., "A Tale of Two Watches, or Rectifying a Sixty-Eight-Year-Old Mistake," *Bulletin of the N.A.W.C.C., Inc.* (June 1983), 25(3[whole no. 224]):358–359, and "Update: A Tale of Two Watches," *Bulletin of the N.A.W.C.C., Inc.* (December 1983), 25(6[whole no. 227]):769. Meggers relates the delightful antiquarian's tale of finding a rare Illinois watch, made about 1914, with an incorrectly numbered train bridge. He subsequently found the watch with the correctly numbered train bridge in another collection and exchanged the parts which had been inadvertently exchanged in the factory. They interchanged completely.

117. Edward Howard, "American Watches and Clocks," 2:543.

118. Robbins & Appleton, *Net Price List: American Waltham Watch Co's Materials.*

119. Robbins & Appleton, *Net Price List: American Waltham Watch Co.'s Materials.*

120. "Appleton, Tracy & Co.'s American Watch Company, Waltham, Mass."

121. H. C. Hovey, "The American Watch Works," *Scientific American*, p. 103.

122. See the Carpenter's Department Books, Waltham Watch Company Papers, Baker Library, Harvard University Graduate School of Business Administration, MSS. B-1 to B-7; and David A. Wells, "The American Manufacture of Watch Movements," p. 66. "All the ingenious tools by which these remarkable results are obtained, were invented in this establishment, and constructed within its walls."

123. Edward A. Marsh, "Watch Manufacturing on the American System." Marsh was correct. But for space limitations, this essay could explore such Waltham machine tool developments as Duane Church's "Automatic Blank Feeding Device." See, for example, United States Patent of E. A. Marsh and Duane Church, Patent No. 366,592, July 12, 1887, "Automatic Blank Feeding Device."

124. R. E. Robbins, Treasurer, Waltham Watch Company, March 27, 1879, "Treasurer's Report, 1879, R.E. Robbins," Waltham Watch Papers, Baker Library, Harvard University Graduate School of Business Administration, MS. AD-1.

125. Edward A. Marsh, *The Original American Watch Plant*, p. 21.

126. Waltham Watch Papers, Baker Library, Harvard University Graduate School of Business Administration, see "Treasurer's Report, 1867, R. E. Robbins," MS. AD-1. Robbins continued, "it has been a question whether we should thus vigorously press the manufacture. The result of such a course . . . would be naturally to leave us temporarily without dividends . . . We nevertheless have not hesitated to follow this plan [increased mechanization]. It was the only way open to us by which to recede (keep down) the cost of manufacturing. (To diminish production was to in-

crease the cost of goods) the necessity of a reduction in our prices was apparent . . . In short, we have considered it much better to sacrifice our dividend than (to increase the cost of goods) or to check the development of our capacity to manufacture cheaply."

127. Waltham Watch Papers, Baker Library, Harvard University Graduate School of Business Administration, "Treasurer's Report, 1876, R. E. Robbins," MS. AD-1.

128. Waltham Watch Papers, Baker Library, Harvard University Graduate School of Business Administration, "Treasurer's Report, 1879, R. E. Robbins," MS. AD-1.

129. Waltham Watch Papers, Baker Library, Harvard University Graduate School of Business Administration, "Treasurer's Report, 1885, R. E. Robbins," MS. AD-2.

130. Moore, *Timing a Century*, p. 50.

131. Roy Ehrhardt, *Foreign and American Pocket Watch Identification and Price Guide, Book 3*, p. 19.

132. Moore, *Timing a Century*, "Table 11, Sales and Prices, 1890–1905, Waltham Watch Company," p. 87.

133. Edward A. Marsh, *The Original American Watch Plant*, p. 18.

134. Moore, *Timing a Century*, pp. 67–68.

135. Moore, *Timing a Century*, pp. 68–69, and Crossman, "A Complete History of Watch Making in America."

136. Donald R. Hoke, "The First Aurora Watch: A Preliminary Report on the Aurora Watch Company of Aurora, Illinois," *Bulletin of the N.A.W.C.C., Inc.* (August 1977), 19(4[whole no. 189]):331–350.

137. "Charles C. Hinckley," *Kane County History*, pp. 788–789.

138. *Catalogue of the American Watch Tool Company, Waltham, Mass.*, May 1884.

139. Edwin A. Battison, Addendum to "Introduction to *Technology of Watch Production*." S. V. Tarasov, *Technology of Watch Production*.

140. Leonard Waldo, "The Mechanical Art of Watchmaking in America," p. 190.

141. H. C. Hovey, "The American Watch Works," *Scientific American*, p. 102.

142. *Ibid.*, p. 102.

143. Crossman, "A Complete History," p. 313.

6. *The American System in Perspective: The Public and the Private Sectors*

1. Carroll W. Pursell, Jr., "*Yankee Enterprise, The Rise of the American System of Manufactures*, edited by Otto Mayr and Robert C. Post," *Technology & Culture* (January 1984), 25(1):150–151.

Bibliography

1. *Meanwhile, Over in the Private Sector . . .*

"The American System of Manufacturing," *Compressed Air Magazine* (September 1988), pp. 20–26.

Battison, Edwin A. "Eli Whitney and the Milling Machine." *Smithsonian Journal of History* (1966), 1:9–34.

Durfee, W. F. "The History and Modern Development of the Art of Interchangeable Construction in Mechanism." *ASME Transactions* (1893), 14:1225–1257.

Durfee, W. F. "The First Systematic Attempt at Interchangeability in Firearms." *Cassier's Magazine* (April 1894), 5(30):469–477.

"The Eagle Bicycle." *Scientific American* (January 11, 1896), 74(2): 17, 20.

Ferguson, Eugene S. "The Critical Period of American Technology." Paper, ca. 1961; copy in Eleutherian Mills Historical Library, Greenville, Wilmington, Del., 19807.

Ferguson, Eugene S. "The Americanness of American Technology."

Ferguson, Eugene S. "Enthusiasm and Objectivity in Technological Development." Paper read Dec. 29, 1970, at AAAS symposium, "Technology: Nuts and Bolts or Social Process."

Ferguson, Eugene S. "Toward A Discipline of The History of Technology." *Technology & Culture* (January 1974), 15(1):13–30.

Ferguson, Eugene S. "On the Origin and Development of American Mechanical "Know-How."" *Midcontinent American Studies Journal* (Fall 1962), 3(2):2–16.

Fries, Russell I. "British Response to the American System: Small Arms Industry after 1850." *Technology & Culture* (July 1975), 16(3):377–403.

Habakkuk, H. J. "The Economic Effects of Labor Scarcity," in Saul, ed., *Technological Change.*

Habakkuk, H. J. "Second Thoughts on American and British Technology in the Nineteenth Century." *Business Archives & History* (August 1963), 3(2):187–194.

Hacker, Barton C. and Sally L. Hacker. "Merritt Roe Smith, ed., *Military Enterprise and Technological Change; Perspectives on the American Experience.*" *Technology and Culture* (July 1987), 28(3):709–712.

Hoke, Donald. "Conference Report: A Symposium on the Rise of the American System of Manufactures." *Technology & Culture* (January 1980), 21(1):67–70.

Hounshell, David A. "From the American System to Mass Production." (Ph.D. dissertation, University of Delaware, 1977).

Hounshell, David A. *From the American System to Mass Production, 1800–1932.* Baltimore, Md.: Johns Hopkins University Press, 1984.

Hounshell, David A. "The Bicycle and the American System of Manufactures." May 1975.

Howard, Robert A. "Interchangeable Parts Reexamined: The Private Sector of the American Arms Industry on the Eve of the Civil War." *Technology & Culture* (October 1978), 19(4):633–649.

Noble, David F. "Command Performance: A Perspective on the Social and Economic Consequences of Military Enterprise." Ch. 8 in Merritt Roe Smith, ed., *Military Enterprise and Technological Change*, pp. 329–346.

Roe, Joseph Wickham. "Interchangeable Manufacture in American Industry."

Smith, Merritt Roe. "Army Ordnance and the 'American System' of Manufacturing, 1815–1861." Ch. 1 in Merritt Roe Smith, ed., *Military Enterprise and Technological Change*, pp. 63–64.

Smith, Merritt Roe, ed. *Military Enterprise and Technological Change: Perspectives on the American Experience*. Cambridge: MIT Press, 1985.

Smith, Merritt Roe. *Harpers Ferry Armory and the New Technology*. Ithaca, N.Y.: Cornell University Press, 1977.

Smith, Merritt Roe. "John H. Hall, Simeon North, and the Milling Machine: The Nature of Innovation among Antebellum Arms Makers." *Technology & Culture* (October 1973), 14(4).

Smith, Merritt Roe. "From Craftsman to Mechanic: The Harpers Ferry Experience, 1789–1854." In Ian M. G. Quimby and Polly Anne Earl, eds., *Technological Innovation in the Decorative Arts*. Charlottesville.: University Press of Virginia, 1974.

White, Jr., Lynn. "The Discipline of the History of Technology." *Journal of Engineering Education* (January 1964), 54 (10):349–351.

2. *Wooden Movement Clock Manufacturing in Connecticut, 1807–1850*

"Address of General Joseph R. Hawley." In J. J. Jennings, ed., *Centennial Celebration of the Incorporation of the Town of Bristol*, p. 72. Hartford: Case, Lockwood and Brainard, 1885.

Bailey, Chris H. "Notes on 'Torrington' Clocks." *The Bulletin of the N.A.W.C.C., Inc.* (December 1972), 15(7[whole no. 161]):803–833.

Bailey, Chris H. "George B. Seymour's Accounts." *The Cog Counter's Journal* (May 1974), no. 2, pp. 33–38. This account book is in the collections of the American Clock & Watch Museum in Bristol, Conn.

Bailey, Chris H. *Two Hundred Years of American Clocks and Watches*. Englewood Cliffs, N.J., Prentice Hall, 1975.

Bailey, Chris H. "Mr. Terry's Waterbury Competitors, the Leavenworths and Their Associates." *The Bulletin of the N.A.W.C.C., Inc.* (June 1979), 21(3[whole no. 200]):243–283.

Barr, Lockwood. *Eli Terry Pillar and Scroll Shelf Clocks*. Supplement to the *Bulletin of the N.A.W.C.C., Inc.* Columbia, Penn.: N.A.W.C.C., December 1952.

Barber, Laurence Luther. "The Clockmakers of Ashby, Massachusetts." *The Magazine Antiques*, (unknown date). Reprinted in *The Cog Counter's Journal* (August 1974), no. 3, pp. 21–23.

Bathe, Greville and Dorothy. *Oliver Evans: A Chronicle of Early American Engineering.* New York: Arno Press, 1972 [c. 1935].

Battison, Edwin A. "Screw-Thread Cutting by the Master-Screw Method Since 1480." *Bulletin 240: Contributions from the Museum of History and Technology, Paper 37,* pp. 105–120. Washington, D.C.: Smithsonian Institution, 1966.

Battison, Edwin A. *From Muskets to Mass Production: The Men and Times That Shaped American Manufacturing.* Windsor, Vt.: American Precision Museum, 1976.

Battison, Edwin A. and Patricia E. Kane. *The American Clock, 1725–1865: The Mabel Brady Garvan and Other Collections at Yale University.* Greenwich, Conn.: New York Graphic Society, 1973.

Brown, Abram English. "By Sun Dial and Noon Mark. Primitive Ways of Telling the Time Were Good Enough for Our Ancestors—Clocks Became a Necessity in Every House as Soon as Yankee Genius Made this Possible—Some Pioneer Manufacturers of Tall Clocks—Wonderful Growth of the Industry During the Present Century." In an unidentified newspaper, 1899. ":598 Photograph File Watches," Baker Library, Harvard University, Graduate School of Business Administration.

Burns, Bruce A. "Terry Standard Thirty-Hour Wood Movement Contemporaries." *The Bulletin of the N.A.W.C.C., Inc.* (December 1970), 14(7[whole no. 149]):781.

Camp, Hiram. "A Sketch of the Clockmaking Business (1792–1892)." Manuscript scrapbook in the Connecticut State Library entitled "Scrapbook of Newspaper Clippings" (1860–1890). Reprinted in part in Roberts, *Eli Terry,* p. 145.

Deyrup, Felicia J. *Arms Makers of the Connecticut Valley: A Regional Study of the Economic Development of the Small Arms Industry, 1798–1879,* vol. 33, Smith College Studies in History, Northampton, Mass., 1948.

Diehl, John A. "Luman Watson, Cincinnati Clockmaker." *The Magazine Antiques* (June 1968), 93(6):796–799. Reprinted in *The Cog Counter's Journal* (February 1978), no. 15, pp. 52–66.

Distin, William H. and Robert Bishop. *The American Clock: A Comprehensive Pictorial Survey, 1723–1900, With A Listing of 6,153 Clockmakers.* New York: Dutton, 1976.

Dyer, Walter A. *Early American Craftsmen.* New York: Century, 1915.

"Excerpts from an address by General Joseph R. Hawley. Bristol's Centennial Celebration, June 17, 1885, *Bristol Centennial Celebration* (Hartford, Conn., 1885)." Reprinted in Roberts, *Eli Terry,* p. 72.

Ferguson, Eugene S. *Oliver Evans.* Greenville, Del.: Hagley Museum, 1980.

Ferguson, Eugene S. "History and Historiography." In Otto Mayr and Robert C. Post, eds., *Yankee Enterprise: The Rise of the American System of Manufactures.* Washington, D.C.: Smithsonian Institution Press, 1981.

Francillon, Ward. "Some Wood Movement Alarms." *The Bulletin of the N.A.W.C.C., Inc.* (October 1970), 14(6[whole no. 148]):575–6.

Frank, Allan Dodds. "Genuine Phonies." *Fortune Magazine* (January 31, 1983), 131(3): 36–37.

"From the Archives—Connecticut State Library, Probate Files." *The Cog Counter's Journal* (November 1975), no. 8, pp. 3–4.

Green, Constance M. *Eli Whitney and the Birth of American Technology.* Boston,: Little, Brown, 1956.

Hall, Lawrence P. "Asaph Hall of Goshen and Clingon." *The Bulletin of the N.A.W.C.C., Inc.* (April 1980), 22(2[whole no. 205]):119–145.

Hindle, Brooke. *Technology in Early America: Needs and Opportunities for Study*. Chapel Hill: University of North Carolina Press, 1966.

Hindle, Brooke, ed., *America's Wooden Age: Aspects of Its Technology*. Tarrytown, N.Y.: Sleepy Hollow Restorations, 1975.

Hoopes, Penrose R. *Connecticut Clockmakers of the Eighteenth Century*. New York: Dodd, Mead, 1930.

Hoopes, Penrose R. *Shop Records of Daniel Burnap, Clockmaker*. Hartford: Connecticut Historical Society, 1958.

Hounshell, David A. "From the American System to Mass Production: The Development of Manufacturing Technology in the Untied States, 1850–1920." Ph.D. dissertation, University of Delaware, 1978.

Hounshell, David A. *From the American System to Mass Production: The Development of Manufacturing Technology in the United States, 1850–1920*. Baltimore, Md.: Johns Hopkins University Press, 1984.

Jerome, Chauncey. *History of the American Clock Business for the Past Sixty Years and Life of Chauncey Jerome Written by Himself*. New Haven, Conn.: F. C. Dayton, Jr., 1860.

Korten, Elmer C. "An Eli Terry Regulator." *The Bulletin of the N.A.W.C.C., Inc.* (August 1972), 15(5[whole no. 159]):436–453.

Morrison, Samuel Eliot. *Admiral of the Ocean Sea: A Life of Christopher Columbus*. Boston: Little, Brown, 1942.

Murphy, John Joseph. "Entrepreneurship and the Establishment of the American Clock Industry." Ph.D. dissertation, Yale University, 1961.

Murphy, John Joseph. "Entrepreneurship in the Establishment of the American Clock Industry." *Journal of Economic History* (June 1966), 24(2):169–186.

"Old Clock Shop To Be Exhibited In Smithsonian." *Torrington Connecticut Register*, November 16, 1959.

Palmer, Brooks. *The Book of American Clocks*. New York: MacMillan, 1950.

Palmer, Forrest. "Harwinton Clock Shop Earns Niche in U.S. History: Headed For Exhibit At Smithsonian." *Waterbury Connecticut Republican*, November 15, 1959.

Partridge, Albert L. "Wood Clocks: The Art or Mystery of their Manufacture." *The Magazine Antiques* (March 1946), pp. 179–181?. Reprinted in *The Cog Counter's Journal* (August 1975), no. 7, pp. 18–25.

Partridge, Albert L. "Wood Clocks, Connecticut Carries On." *The Magazine Antiques* (August 1946). Reprinted in *The Cog Counter's Journal* (August 1976), no. 11, pp. 43–51.

Partridge, Albert L. "Connecticut Enters the Eight-Day Field." *The Magazine Antiques* (June, 1952). Reprinted in *The Cog Counter's Journal* (February 1975), no. 5, pp. 29–33.

"Random notes from Samuel Terry accounts and loose papers. Courtesy American Clock & Watch Museum, Bristol, Conn." *The Cog Counter's Journal* (May 1974), no. 2, p. 4.

Richard, Paul. "National Gallery Names Chief Curator, Harvard's S. J. Freedberg Assumes Post Sept. 1." *Washington Post*, Tuesday, March 1, 1983, Style Section, pp. 1 and 3.

Roberts, Kenneth D. "Documented Listing of Connecticut Firms Manufacturing or Marketing Wooden Movement Shelf Clocks, 1816–1850." *The Bulletin of the N.A.W.C.C., Inc.* (February 1973), 15(8[whole no. 162]):993–996.

Roberts, Kenneth D. *Eli Terry and the Connecticut Shelf Clock*. Bristol, Conn.: Ken Roberts Publishing Co., 1973.

SAS Institute. *SAS User's Guide: Basics*. Cary, N.C.: SAS Institute, 1982.

Sloane, Virginia and Howard Sloane. "4,000 Clocks: The Story of Eli Terry and His Mysterious Financiers." *The Bulletin of the N.A.W.C.C., Inc.* (February 1980), 22(1[whole no. 204]):3–42.

Smith, Merritt Roe. *Harpers Ferry Armory and the New Technology*. Ithaca, N.Y.: Cornell University Press, 1977.

Taylor, Snowden. "Characteristics of Standard Terry-Type 30-Hour Wooden Movements as a Guide to Identification of Movement Makers." *The Bulletin of the N.A.W.C.C., Inc.* (October 1980), 22(5[whole no. 208, part 1]):442–530.

Taylor, Snowden. "Research Activities and News." *The Bulletin of the N.A.W.C.C., Inc.* (December 1982), 24(6[whole no. 221]):642.

Taylor, Snowden. "Daniel Pratt, Jr., Boardman and Wells, Levi Smith—The End of the Wood Clock Era." *The Bulletin of the N.A.W.C.C., Inc.* (February 1984), 26(1[whole no 228]):59–60.

Terry, Henry. "A Review of Dr. Alcott's History of Clock-Making: By A Clock-Maker." *Waterbury American*, June 10, 1853, 9(28[whole no. 444]). Reprinted in Roberts, *Eli Terry*, p. 35, table 5.

Unentitled note on Harrison in *The Cog Counter's Journal* (May and August 1977), no. 14, pp. 56.

Wood, Stacy B. C. and Stephen E. Kramer, III. *Clockmakers of Lancaster County and Their Clocks, 1750–1850*. New York: Van Nostrand Reinhold, 1977.

Wyke, John. *A Catalogue of Tools for Watch and Clock Makers by John Wyke of Liverpool*. Introduction and Technical Commentary by Alan Smith. Charlottesville: University Press of Virginia, 1978.

MANUSCRIPTS

Elisha Manross account book, Connecticut State Historical Society, Hartford, Conn., MS. 80834.

J. A. Wells, Esq. to Mr. (Daniel) Pratt, June 22, 1847. Reprinted in *The Cog Counter's Journal*, (February, 1975), no. 5, p. 3. ("The papers of Daniel Pratt, Jr., courtesy of the Reading Antiquarian Society, Reading, Mass.").

Smithsonian Institution, Museum of American History, Accession Files Numbers 224,779, 226,926, 260,025 and 306,558.

Hopkins & Alfred "Company Book," photocopy in American Clock & Watch Museum, Bristol, Conn.

Correspondence Files, Division of Mechanical and Civil Engineering, Museum of American History, Smithsonian Institution.

3. *Elisha K. Root and Axe Manufacturing at the Collins Company, 1830–1849*

Barnard, Henry. *Armsmear: The Home, The Arm, and The Armory of Samuel Colt: A Memorial*. New York, Alvord, printer, 1866.

Battison, Edwin A. "The Auburndale Watch Company." Smithsonian Institution Bulletin 218: *Contributions from the Museum of History and Technology, Paper #4*.

Bolles, Albert S. *Industrial History of the United States.* Norwich, Conn.: Henry Bill Publishing, 1881. "Axes and Saws," pp. 270–274.

Brinsmade, H. N., D. D. *An Address at the Funeral of Samuel W. Collins, May 2, 1871.* Newark, N.J.: Jennings and Hardham, Printers and Book Binders, Nos. 153 and 155 Market St., privately printed, 1872.

Burlingame, Roger. *Machines That Built America.* New York: Harcourt, Brace, 1953.

Edwards, William B. *The Story of Colt's Revolver: The Biography of Samuel Colt.* Harrisburg, Penn.: Stackpole, 1953.

"Elisha King Root." *National Cyclopedia of American Biography* 1922 ed. 18:313.

Ferguson, Eugene S., ed. *George Escol Sellers: Early Engineering Reminiscences, 1815–1840.* Washington, D.C.; Smithsonian Institution, 1965. U.S.N.M. Bulletin 238.

Gras, N. S. B. *Business History of the United States About 1650 to 1950s.* Ann Arbor, Mich.: Edwards Brothers, 1967.

Robert B. Gordon, "Material Evidence of the Development of Metalworking Technology at the Collins Axe Factory," *Industrial Archaeology,* (1983), 9(1):19–28.

Greeley, Horace et al. *The Great Industries of the United States.* Hartford, Conn.: J. B. Burr and Hyde, 1872, "Axes and Plows," pp. 122–147.

Hartford Daily Courant.

Hartford Daily Post.

Hartford Daily Times.

Hartford Evening Press.

International Library of Technology. Vol. 140, section 19, part 1, "Hand Forging."

Kauffman, Henry J. *American Axes: A Survey of their Development and Their Makers.* Battleboro, Vt.: Stephen Green Press, 1972.

Lundeberg, Philip K. *Samuel Colt's Submarine Battery—The Secret and the Enigma.* Smithsonian Studies in History and Technology, No. 29, Washington, D.C.: Smithsonian Institution Press, 1974.

Mitman, Carl W. "Elisha King Root." *Dictionary of American Biography,* 1935 ed., 16: 144–145.

"Our Visit to the Collinsville Ax Works." *Scientific American* (July 16, 1859), N.S., 1(3):36–37.

Post, Robert C. "'Liberalizers' versus 'Scientific Men' in the Antebellum Patent Office." *Technology and Culture* (January 1976), 17)1):24–54.

Roe, Joseph Wickham. *English and American Tool Builders.* New Haven, Conn.: Yale University Press, 1916.

Rohan, Jack. *Yankee Arms Maker: The Story of Sam Colt and His Six Shot Peacemaker.* New York: Harper, 1948.

Samuel Colt Presents: A Loan Exhibition of Presentation Percussion Colt Firearms. Hartford, Conn.: Wadsworth Athenaeum, 1962.

Smith, Merritt Roe. "John H. Hall, Simeon North, and the Milling Machine: The Nature of Innovation among Antebellum Arms Makers." *Technology and Culture* (October 1974), 14(4):573–591.

Stapleton, Darwin H. "Moncure Robinson: Railroad Engineer, 1828–1840." In Barbara Benson, ed., *Benjamin Henry Latrobe and Moncure Robinson.* Greenville, Del.: Eleutherian Mills-Hagley Foundation, 1975.

Uselding, Paul. "Elisha K. Root, Forging, and the 'American System.'" *Technology and Culture* (October 1974), 15(4):543–568.

Uselding, Paul. "Henry Burden and the Question of Anglo-American Technological Transfer in the Nineteenth Century." *Journal of Economic History* (June 1970), 30(2):312–337.

Woodbury, Robert S. *The History of the Grinding Machine*. Cambridge: MIT Press, 1959.

MANUSCRIPTS

Collins & Co. Connecticut Historical Society, Hartford, Connecticut. Especially: Collins, Samuel Watkinson, "The Collins Company" (unpublished corporate history).

4. *Typewriter Manufacturing, 1853–1924*

NOTES ON THE SOURCES

Because the scarcity of typewriter data noted by Terry Abraham (n. 1) is especially acute with respect to typewriter manufacturing, a short discussion of sources is in order. There were no manuscript collections available for this study and I was forced to rely on three sources of information: 1) surviving typewriters, 2) trade catalogues, and 3) magazine articles. The early Remingtons and Caligraphs in the Carl P. Dietz Typewriter Collection led me to the conclusion that adjustability was an integral part of typewriter design. The trade literature collections in the Milwaukee Public Museum and the Smithsonian Institution provided manufacturing data for several manufacturers since they were anxious to convince their customers of the quality of their machines. The method adopted by the major makers was to illustrate the factory and its operations. This form of promotion, illustrating the factory and the manufacturing process, extended into magazine articles as well, appearing in such diverse journals as the *Scientific American, Century Magazine, Office Outfitter,* and *Office Appliances*. These sources provide data on five firms over a thirty-eight year period, 1886–1924. The firms made the Remington, Caligraph, Hall, Oliver, and L. C. Smith & Bros. typewriters.

Abraham, Terry. "Charles Thurber: Typewriter Inventor." *Technology and Culture* (July 1980), 21(3):430–434.

Adler, Michael H. *The Writing Machine: A History of the Typewriter*. London: Allen and Unwin, 1973.

American Writing Machine Company. *The Caligraph; or, Perfect Writing Machine*. New York(?): American Writing Machine, ca. 1880. MPM H 45025/27054.

American Writing Machine Company. *The Caligraph Manufactured by the American Writing Machine Company*. Hartford: Star Print, 1888. MPM H 43769/26878.

American Writing Machine Company. "Caligraph Writing Machine." *Scientific American* (March 6, 1886), 54(10):150.

American Writing Machine Company. *Directions and Suggestions for Using the Caligraph Manufactured by the American Writing Machine Company*. Rev. ed. Hartford:(?), July 1, 1888. Rpt. Arcadia, Calif.: Post-Era Archives, ca. 1980.

American Writing Machine Company. *Directions and Suggestions for Using the "Caligraph" Manufactured by the American Writing Machine Company, The Special Caligraph No. 3*. Hartford: R. S. Peck, March 1897(?). MPM E 45707.L/12753.

American Writing Machine Company. *Ideal Caligraph*. 1884. MPM H 43316/26808.

Barbour, Harry E. "The Development of the Typewriter." In Robert Marion LaFollette, ed., *The Making of America: Science and Invention*. Vol. 7, Chicago: Making of America, ca. 1905.

Beeching, Wilfred A. *Century of the Typewriter*. London: William Heinemann, 1974.

Bliven, Jr., Bruce. *The Wonderful Writing Machine*. New York: Random House, 1954.

Brainard, George C. *A Page in the Colorful History of Our Modern Machine Age*. New York: Newcomen Society in North America, 1950.

Coleman, John S. *The Business Machine—With Mention of William Seward Burroughs, Joseph Boyer, and Others—Since 1880*. New York: Newcomen Society in North America, 1949.

A Condensed History of the Writing Machine. Reprinted from the October, 1923 issue of *Typewriter Topics*. New York: Business Equipment, 1924.

Cooper, Grace Rogers. *The Sewing Machine: Its Invention and Development*. Washington, D.C.: Smithsonian Institution Press, 1976.

Current, Richard N. *The Typewriter and the Men Who Made It*. Urbana, Ill.: University of Illinois Press, 1954.

Davies, Margery. "Woman's Place Is at the Typewriter: The Feminization of the Clerical Labor Force." *Radical America* (July–August 1974), 8(4):1–28.

The Edison Electric Pen and Duplicating Press. London: Edison Electric Pen and Writing Agency, 1881(?).

Engler, George Nichols. *The Typewriter Industry: The Impact of a Significant Technological Innovation*. 1969 Ph.D. dissertation, U.C.L.A. Ann Arbor, Mich.: University Microfilms, 1970.

Foulke, Arthur Toye. *Mr. Typewriter: A Biography of Christopher Latham Sholes*. Boston: Christopher Publishing House, 1961.

Frierson, Henry Cecil. *Text Book on Typewriter Repairing: The Only Written Authority on Typewriter Construction and Repairing*, Thorp I. Mortin, 1912.

Hall Type Writer Company. "The Hall Type Writer." *Scientific American* (July 10, 1886), N.S., 55(2):24–25. Reprinted as "The Hall Type Writer," *Scientific American, Architects and Builders Edition* (August 1886), 2(2):30–31.

Harrison, John. *A Manual of the Typewriter*. London: Isaac Pitman and Sons, 1888.

Hatch, Alden. *Remington Arms In American History*. Buffalo, N.Y.: Remington Arms Company, 1956 and 1972.

Herrl, George. *The Carl P. Dietz Collection of Typewriters*. Publications in History, No. 7. Milwaukee, Wis.: Board of Trustees, Milwaukee Public Museum, 1965.

History of the Typewriter. Detroit, Mich.: Metropolitan Typewriter, 1966.

History of the Typewriter. Reprinted from *Typewriter Topics*, 1923. New York: Geyer-McAlister, 1968(?).

Hoke, Donald R. "Mares, George Carl, *The History of the Typewriter, Successor to the Pen*; and Masi, Frank T., *The Typewriter Legend*." *Technology and Culture* (July 1987), 28(3):701–702.

Hubert, Jr., P. G. "The Typewriter, Its Growth and Uses." *Century Magazine*(?) (April 25, 1888), pp. 25–32.

Iles, George. *Leading American Inventors*. New York, Henry Holt, 1912.

Jenkins, H. C. "Cantor Lectures on Typewriting Machines." *Journal of the Society of Arts*. 1894.

Jones, C. Leroy (Rocky). *Typewriters Unlimited: History of the Typewriter*. Springfield, Mo.: Rocky's Technical Publications, 1956.

Karr, Charles Lee and Caroll Robbins. *Remington Handguns*. New York: Bonanza Books, 1960.

Lewis, Mel. "The Busy History of Early Office Equipment." *Art and Antiques* (May 26, 1973), 10(8):29–34.

Mares, George Carl. *The History of the Typewriter: Being an Illustrated Account of the Origin, Rise and Development of the Writing Machine*. London: Guilbert Pitman, 1909.

Mathews, W. S. B. *The Writing Machine*. Chicago?: Northwestern Christian Advocate, 1891.

McCarthy, James H., ed., *The American Digest of Business Machines*. Chicago: American Exchange Service, 1924.

Oden, C. V. *Evolution of the Typewriter*. London?: 1907.

Oliver Typewriter Company. *The Oliver Typewriters: The No. 3 Models*. Chicago: Oliver Typewriter, ca. 1905.

Post, Dan, ed. *Collector's Guide to Antique Typewriters*. Arcadia, Calif.: Post-Era Books, 1981.

"Producing a Wonderful Writing Machine." *Office Appliances* (December 1906), 5(1): 20–24.

Quaife, Milo M. *Henry W. Roby's Story of the Invention of the Typewriter*. Menasha, Wis.: Collegiate Press, 1925.

E. Remington and Sons. *Directions for Setting Up, Working and Keeping In Order, The Type Writer*. Ca. 1875.

Remington Rand, Inc. *Outline of Typewriter History*. Buffalo, N.Y.: Remington Rand, 1935(?).

"The Remington Typewriter." *Scientific American* (December 15, 1888), N.S., 59(24):367 and 374–375.

Richards, G. Tilghman. *The History and Development of Typewriters*. London: Her Majesty's Stationery Office, 1964.

Royal Typewriter Company, Inc. *The Evolution of the Typewriter*. New York: Royal Typewriter, 1921. MPM E 45707.PP/12753.

L. C. Smith and Bros. Typewriter Company. *A Visit to the Home of the L. C. Smith Typewriter*. Syracuse, N.Y.: L. C. Smith and Bros. Typewriter, ca. 1924.

The Story of the Typewriter, 1873 to 1923. Herkimer, N.Y.: Herkimer County Historical Society, 1923.

Tolly, William P. *Smith-Corona Typewriters and H. W. Smith*. Princeton, N.J.: Princeton University Press, for the Newcomen Society in North America, 1951. MPM H 43493/26819.

Varityper Corp. *The Story Behind President Wilson's Typewriter*. Newark, N.J.: Varityper, 1962.

Weller, Charles E. *The Early History of the Typewriter*. LaPorte, Ind.: Chase and Shepherd, Printers, 1918.

Yost, G. W. N. and E. Remington and Sons, *The Perfected Type-Writer, Number Two, Writes Capitals and Small Letters. Noiseless and Portable*. New York: D. H. Tuttle, ca. 1877–1878.

Zellers, John A. *The Typewriter: A Short History, On Its 75th Anniversary, 1873–1948*. New York: Newcomen Society of England, American Branch, 1948.

5. *The Development of Watch Manufacturing at the Waltham Watch Company, Waltham, Massachusetts, 1849–1910*

Abbott, Henry G. *The Watch Factories of America Past and Present.* Chicago: George K. Hazlitt, 1888.

Abbott, Henry G. *Watchmakers and Jewelers Practical Hand Book.* Chicago: George K. Hazlitt, 1892.

Abbott, Henry G. *A Pioneer: A History of the American Waltham Watch Company of Waltham, Mass.* Chicago: Hazlitt and Walker, 1904.

Abbott, Henry G. *Abbott's American Watchmaker and Jeweler.* Chicago: Hazlitt and Walker, ca. 1905.

Adt, Howard S. *Screw Thread Production to Close Limits.*. New York: Sterling Press, 1920.

Albert Bros. Wholesale Jewelers. Illustrated Catalogue 1903. Cincinnati, Ohio: 1903. MPM H 45783/27141.

Benj. Allen and Co. 1902–3 Tool and Material Catalogue. Chicago: 1902. MPM H 35211/25222.

American Institute: Annual Report (1863-64). "The American Watch Company," pp. 405–416.

American Waltham Watch Company. *The Equity Watch: The Square Deal in Watches.* Waltham, Mass.(?): American Waltham Watch Company, ca. 1915.

American Waltham Watch Company. *The Perfected American Watch.* Waltham, Mass.: American Waltham Watch Company, 1907.

American Waltham Watch Company. *Net Price List, American Waltham Watch Co.'s Materials.* New York, Robbins and Appleton, 1885; reprinted, Bristol, Conn.: Ken Roberts Publishing, 1972.

American Watch Company. "The Factory of the American Watch Company." *Scientific American* (April 11, 1863), N.S., 8(15):226.

American Watch Company. "I. The Watch as a Growth of Invention." *Appletons' Journal of Popular Literature, Science, and Art* (July 2, 1870), 9(66):2–5, and "II. The Watch as a Growth of Industry.," *Appleton's Journal of Literature, Science, and Art* (July 9, 1870), 9(67):29–36. Reprinted as *Watch-Making in America, Embodying The History of Watch-Making as an Invention, and The History of Watch-Making as an Industry.* Boston: Robbins, Appleton, ca. 1871.

American Watch Company(?). "Manufacture of the American Watch at Waltham." *American Horological Journal* (April 1870), 1(10):296–299; (May 1870), 1(11):327–330; (June 1870), 1(12):360–362.

American Watch Company. "Watchmaking in America." *The Illustrated London News* (June 19, 1875), pp. 591–592.

American Watch Company. *Souvenir Catalog of the New Orleans Exposition, 1884–1885.* Reprinted, Bristol, Conn.: Ken Roberts Publishing, 1972.

American Watch Company. *Illustrated Catalogue of Watches Manufactured by the American Watch Company, Waltham, Mass.* Boston: Kilburn-Mallory, ca. 1862–1863. Reprinted in Roy Ehrhardt. *American Pocket Watch Identification and Price Guide, 1977 Price Indicator.* Kansas City, Mo.: Heart of America Press, 1976.

American Watch Tool Company. *American Watch Tool Company, Waltham, Mass.: Manufacturers of the Whitcomb Lathe.* May 1884.

American Watch Tool Company. *Precision Machinery Manufactured by the American Watch Tool Co., Waltham, Mass., U.S.A.* ca. 1895–1900. Reprinted, Fitzwilliam, N.H.: Ken Roberts Publishing, January, 1980.

American Watch Tool Company Auction Catalogue. 1918. Original in Waltham Public Library, photocopy in Division of Mechanisms, Museum of American History, Smithsonian Institution.

"Appleton, Tracy and Co.'s American Watch Company, Waltham, Mass." *Ballou's Pictorial Drawing Room Companion* (October 2, 1858), 15(14[whole no. 230]):212–213.

Bailey, Chris H. *Two Hundred Years of American Clocks and Watches.* Englewood Cliffs, N.J.: Prentice-Hall, 1975.

Battison, Edwin A. *From Muskets to Mass Production: The Men and the Times That Shaped American Manufacturing.* Windsor, Vt.: American Precision Museum, 1976.

Battison, Edwin A. "The Auburndale Watch Company." *Bulletin 218: Contributions from the Museum of History and Technology.*

Beckman, Ed. *Cincinnati Silversmiths, Jewelers, Watch, and Clock Makers.* 1975.

Britten, F. J. *The Watch and Clockmakers' Handbook, Dictionary and Guide.* 13th ed. London: E. and F. N. Spon, 1922.

Carosso, Vincent P. "The Waltham Watch Company, a Case History," *Business History Society Bulletin* (December 1949), 23(4):165–187.

Carrington, R. F. and R. W. "Pierre Frédéric Ingold and the British Watch and Clockmaking Company." *Antiquarian Horology* (Spring 1978), 10(6):698–714.

Census of Manufacturing, 1880. "V. The Manufacture of Watches," pp. 60–69.

Chamberlain, Paul M. *It's About Time.* New York: Richard R. Smith, 1941.

Chapple, Bennet. "The Story of the Ingersoll Dollar Watch." *National Magazine* (September 1903), 18(6):. MPM H 44424.01/26877.

Circuit Court of the United States, Southern District of New York. In Equity. American Waltham Watch Company, Complaintent. Against Joseph H. Sandman, Defendant. Record. New York: Livingston, Middleditch, 1898.

Conley, Robert H. "Waltham: A City Becomes a Pocket Watch." *Bulletin of the N.A.W.C.C., Inc.* (October 1981), 23(5[whole no. 214]):481–483.

Crom, Theodore R. *Horological Shop Tools: 1700 to 1900.* Gainesville, Fla.: Storter, 1980.

Crom, Theodore R. *Horological Wheel Cutting Engines: 1700 to 1900.* Gainesville, Fla.: Theodore R. Crom, 1970.

Crossman, Charles S. "A Complete History of Watch and Clock Making in America." *Jewelers' Circular and Horological Review.*

Countryman, William A. "The Revolution in Watchmaking." In Robert Marion LaFollette, ed., *The Making of America.* Vol. 7, *Science and Invention.* Chicago: Making of America, ca. 1900.

Dahl, Randal E. "The American Watch Movement Manufacturing Industry." Ph.D. dissertation, Clark University, 1941.

Dane, E. Surrey. *Peter Stubs and the Lancashire Hand Tool Industry.* Altrincham: John Sherratt, 1973.

DeCarle, Donald. *The Watchmaker's Lathe and How to Use It.* London: Robert Hale, 1952.

DeCarle, Donald. *Watch and Clock Encyclopedia.* London: N.A.G. Press, 1978.

DeFazio, Thomas L. "The Nashua Venture and the American Watch Company." *Bulletin of the N.A.W.C.C., Inc.* (December 1975), 17(6[whole no. 179]):575–589.

Drost, William E. *Clocks and Watches of New Jersey*. Elizabeth, N.J.: Engineering Publishers, 1966.

Edwards, Bernard J. *Clock and Watch Advertising: A Grading Guide*. Privately published, 1976.

Eggleston, Edward. "Among the Elgin Watchmakers." *Scribner's Monthly* (April 1873), 5(6):785–791. Reprinted in the *Bulletin of the N.A.W.C.C., Inc.* (February 1977), 19(1[whole no. 186]):140–145.

Eguchi, Shigeru. *Japanese Made Pocket Watches, History Between 1890–1947 (Meiji, Taisho, Showa Era)*. Tokyo: Kaisei Shuppan, 1981.

Elgin Reminiscences: Making Watches by Machinery, 1869. Reprinted, Bristol, Conn.: Ken Roberts Publishing, 1972.

Ehrhardt, Roy. *Elgin National Watch Company Identification and Price Guide*. Kansas City, Mo.: Heart of America Press, 1976.

Ehrhardt, Roy. *Illinois [Watch Company] Identification and Price Guide*. Kansas City, Mo.: Heart of America Press, 1976.

Ehrhardt, Roy. *Production Figures with Grade and Serial Numbers: Hamilton Watch Co. Identification and Price Guide*. Kansas City, Mo.: Heart of America Press, 1976.

Ehrhardt, Roy. *Rockford Watch Company Grade and Serial Numbers with Production Figures*. Kansas City, Mo.: Heart of America Press, 1976.

Ehrhardt, Roy. *Trade Marks, Watch Cases, Pocket Watches, Precious Stones [and] Diamonds*. Kansas City, Mo.: Heart of America Press, 1976.

Ehrhardt, Roy. *Waltham Pocket Watch Identification and Price Guide, Serial Numbers With Description of Waltham Watch Movements*. Kansas City, Mo.: Heart of America Press, 1976.

Ehrhardt, Roy. *American Pocket Watch Identification and Price Guide, Price Indicators*. Kansas City, Mo.: Heart of America Press, 1975–1980.

Ehrhardt, Roy. *American Pocket Watch Encyclopedia and Price Guide*. Vol. 1. Kansas City, Mo.: Heart of America Press, 1982.

Elgin National Watch Company. *Net Price List of Materials Manufactured by the Elgin National Watch Company*. Chicago: Knight & Leonard, April, 1888. MPM H 45598/27141.

"Foreign Watch Trade Competition." *The Watchmaker, Jeweler, and Silversmith's Trade Journal* (March 5, 1878), p. 198.

Fuller, Eugene T. *The Priceless Possession of a Few*. Supplement No. 10 to the *Bulletin of the N.A.W.C.C., Inc.* Columbia, Penn.: N.A.W.C.C., February 1974. Subsequently privately reprinted by the author in a limited edition of 650. Columbia, Penn.: Mifflin Press, 1974.

Gibbs, James W. *The Dueber-Hampden Story*. Philadelphia: Adams Brown, 1954.

Gibbs, James W. "Watch Hands: The History of J. H. Winn, Inc.," *Bulletin of the N.A.W.C.C., Inc.* (April 1964), 11(3[whole no. 109]):224–227.

Gibbs, James W. "Hermann Von Der Heydt, Unsung Genius," *Bulletin of the N.A.W.C.C., Inc.* (April 1987), 19(2[whole no. 247]):83–88.

Giedion, Sigfried. *Space, Time, and Architecture*. 8th ed. Cambridge: Harvard University Press, 1949.

Gitelman, H. M. "The Labor Force at Waltham Watch During the Civil War Era." *Journal of Economic History* (June 1965), 25(2):214–243.

Goodrich, Ward. *The Watchmaker's Lathe*. Fox River Grove, Ill.: North American Watch Tool and Supply, 1974.

Gras, N. S. B. *Business History of the United States About 1650–1950s*. Ann Arbor, Mich.: Edwards Brothers, 1967. "Waltham Watch Company and the American Watch Industry," pp. 395–411.

Greeley, Horace et al. *The Great Industries of the United States*. Hartford, Conn.: J. B. Burr and Hyde, 1872. "Watches, and Machine Watch-Making," pp. 73–81.

Gruen Watch Co. *The Priceless Possession of a Few*. Cincinnati, Ohio(?): Gruen Watch, 1924.

Harris, John R. *Liverpool and Merseyside*. London: Frank Cass, 1969.

Harris, John. "Skills, Coal and British Industry in the Eighteenth Century." *History* (1976), pp. 167–182.

Harris, John. *Industry and Technology in the Eighteenth Century: Britain and France*. Published lecture, University of Birmingham, 1971.

Harris, John. "Saint-Gobain and Ravenhead." In *Great Britain and Her World*. Manchester: Manchester University Press, 1975, pp. 27–70.

Harrod, Michael C. *American Watchmaking: A Technical History of The American Watch Industry, 1850–1930*. Supplement No. 18 to the *Bulletin of the N.A.W.C.C., Inc.* Columbia, Penn.: N.A.W.C.C., Spring 1984.

Hoke, Donald R. *The Time Museum Catalogue of American Watches*. Rockford, Ill.: The Time Museum, forthcoming, 1990.

Hoke, Donald R. "The First Aurora Watch: A Preliminary Report on the Aurora Watch Company of Aurora, Illinois." *Bulletin of the N.A.W.C.C., Inc.* (August 1977), 19(4[whole no. 189]):331–350.

Hoke, Donald R. "Progress Report: The Aurora Watch Co." *Bulletin of the N.A.W.C.C., Inc.* (December 1978), 20(6[whole no. 197]):658–664.

Hoke, Donald R. "Some Incremental Rockford Watch Company Data." *Bulletin of the N.A.W.C.C., Inc.* (June 1981), 23(3[whole no. 212]):251–255.

Hoke, Donald R. "A Lever Setting Hampden Wristwatch," *Bulletin of the N.A.W.C.C., Inc.* (June 1983), 25(3[whole no. 224]):314–319.

Hoke, Donald R. "Aurora Watch Company: Serial Number and Grade Identification Data." *Bulletin of the N.A.W.C.C., Inc.* (December 1983), 25(6[whole no. 227]):724–733.

Hoke, Donald R. "An Unrecorded American Watch, The Rockford, 6 Size, Model 1." *Bulletin of the N.A.W.C.C., Inc.* (June 1984), 26(3[whole no. 230]):332–337.

Hoke, Donald R. "Another Piece to the Philadelphia Watch Company Puzzle." *Bulletin of the N.A.W.C.C., Inc.* (August 1984), 26(4[whole no. 252]):443–444.

Hoke, Donald R. "Philadelphia Watch Company—Addenda." *Bulletin of the N.A.W.C.C., Inc.* (February 1988), 30(1[whole no. 252]):52.

Hoke, Donald R. "Lever Setting Hampden Wrist Watches—Questions Answered." *Bulletin of the N.A.W.C.C., Inc.* (June 1988), 30(3[whole no. 254]):195–206.

Hoke, Donald R. "The American Watch Tool Industry—A Survey." Unpublished paper, 1974.

Hoke, Donald R. "Computerized Watch Research." *Bulletin of the N.A.W.C.C., Inc.* Forthcoming, 1989.

Hoke, Donald R. "The Aurora Watch Company, Skills, and the American System of Manufactures." SHOT Conference paper, Newark, N.J., October, 1979. N.A.W.C.C. Seminar paper, Rockford, Ill, October 1982.

"Horology. This City Its Home in America. D. D. Palmer's Horological School. Story of Its Successful Growth." *Waltham Tribune,* May 28, 1890.

Hounshell, David A. "Public Relations or Public Understanding? The American Industries Series in *Scientific American.*" *Technology and Culture* (October 1980), 21(4):589–593.

Hounshell, David A. *From the American System to Mass Production, 1800–1932. The Development of Manufacturing Technology in the United States.* Baltimore, Md.: Johns Hopkins University Press, 1984.

Hovey, H. C. "The American Watch Works." *Scientific American* (August 16, 1884), N.S., 51(7).

Howard, Edward. "American Watches and Clocks." In Chauncey M. DePew, *One Hundred Years of American Commerce,* vol. 2. New York: D.O. Haynes, 1895.

E. Howard Watch Company. *Howard Watches: Being a Concise Summary of Howard Watch Movements and Cases for the Information of the Howard Jeweler.* Waltham, Mass.: E. Howard Watch, ca. 1909. Reprinted 1969.

"Edward Howard's Non-Horological Ventures." *Bulletin of the N.A.W.C.C., Inc.* (October 1980), 22(5[whole no. 280, part 2]):554–556.

Hughes, Lawrence M. "Who Killed Waltham?" *Sales Management* (April 15, 1950).

Illinois Watch Company. *Illinois Watches and Their Makers.* Springfield, Ill.: Illinois Watch, ca. 1920. Lincoln Library, Springfield, Ill., Sangamo Valley Collection.

Ingersoll, Robt. H. and Bro. *Pointers About Ingersolls for Their 38;000,000 Friends,* ca. 1914.

Jackson, B. J. "Imitation American Key Wind Watches." *Bulletin of the N.A.W.C.C., Inc.* (August 1979), 21(4[whole no. 201]):426–434.

Keystone Watch Case Company. *Material Catalog: Products of the Keystone Watch Case Co., Material and Repair Department.* New York: Keystone Watch Case, January 1, 1919. MPM H 34284/25250.

Kemlo, F. *Kemlo's Watch-Repairer's Hand-Book, Being a Complete Guide to the Young Beginner In Taking Apart, Putting Together, and Thoroughly Cleaning the English Lever and Other Foreign Watches, and All American Watches.* Philadelphia: Henry Carey Baird, 1906.

Landes, David S. *Revolution in Time: Clocks and the Making of the Modern World.* Cambridge: Belknap Press, 1983.

Learned, William B. *Watchmakers' and Machinists' Hand Book.* Chicago: George K. Hazlitt, 1897.

Linberg, Steve. "The Production History of the Waltham Maximus." *Bulletin of the N.A.W.C.C., Inc.* (April 1985), 27(2[whole no. 235]):174–188.

Lund, John H. "Sizes of Watch Movements." *Bulletin of the N.A.W.C.C., Inc.* (June 1980), 22(3[whole no. 206]):264–265.

McCollough, Robert Irving. "Hamilton Watch Company, Lancaster, Penna., U.S.A. Some Notes on Its Founding and History, Part I, The Ezra F. Bowman Era." *Bulletin of the N.A.W.C.C., Inc.* (April 1965), 11(9[whole no. 115]):731–739, and "Hamilton Watch Company Lancaster, Penna., U.S.A. Some Notes on Its Founding and History, Part II, A Railroad Watch is Born." *Bulletin of the N.A.W.C.C., Inc.* (December 1965), 12(1[whole no. 119]):6–25.

"Machine-Made Screws." *American Horological Journal* (March (?), 1871), 1(9(?)):251–254.

Marsh, Edward A. "The American Waltham Watch Company." In D. Hamilton Hurd, ed., *History of Middlesex County, Massachusetts.* 3:738–749. Philadelphia: 1890.

Marsh, Edward A. *Information Concerning a Few Points in the Construction of a Pocket Watch: Also an Announcement of a Safeguard Against One of the Dangers to Which It Is Exposed.* Waltham, Mass.(?): American Waltham Watch Company, 1890. Rpt. Exeter, N.H.: Adams Brown, ca. 1970.

Marsh, Edward A. *The Evolution of Automatic Machinery: As Applied to the Manufacture of Watches at Waltham Mass., By the American Waltham Watch Company.* Chicago: Geo. K. Hazlitt, 1896.

Marsh, Edward A. "History of Early Watchmaking in America." 1889. This manuscript is the draft, with minor variations, for Marsh's article in Hamilton Hurd, *History of Middlesex County, Massachusetts.* Waltham Watch Papers, MS. RC-1, Baker Library, Harvard University School of Business Administration.

Marsh, Edward A. "Watch Manufacturing on the American System." In *New American Supplement to the New Werner Twentieth Century Edition of the Encyclopedia Britannica.* 29:496–498. Akron, Ohio: Werner Company, 1905.

Marsh, Edward A. *"Workers Together:" A Story of Pleasant Conditions in an Exacting Industrial Establishment.* Waltham, Mass.: Waltham Watch Company, 1916. MPM H 45301/26877.

Marsh, Edward A. *The Original American Watch Plant: It's [sic] Planting, Growth, Development and Fruit.* Waltham, Mass.: Waltham Watch, 1909. The Time Museum Collection.

Marsh, Edward A. "History." Waltham Watch Papers, Baker Library, Harvard University, Graduate School of Business Administration, MS. RC-2. This typed manuscript is entitled simply "History." From internal evidence, it is believed by Professor Gitelman to have been compiled by Edward A. Marsh about 1921.

Massachusetts Labor Bulletin. May 1903, "Trade and Technical Education in Massachusetts," no. 26, p. 73; "Waltham Horological School."

Mathews, Adin E. "Some Notes on the Osaka Watch Company." *Bulletin of the N.A.W.C.C., Inc.* (February 1974), 16(2[whole no. 168]):143–146.

Mathews, Adin E. "More on Osaka." *Bulletin of the N.A.W.C.C., Inc.* (June 1981), 23(3[whole no. 212]):274–275.

Meadsday, Walter. "The American Jeweled Watch Industry." Ph.D. dissertation, Massachusetts Institute of Technology, 1955.

Meggers, W. F., Jr. "Correcting Crossman." *Bulletin of the N.A.W.C.C., Inc.* (June 1985), 27(3[whole no. 236]):323–324.

Meggers, William and Roy Ehrhardt. *American Pocket Watch Encyclopedia and Price Guide,* Vol. 2, *American Pocket Watches, Illinois Watch Co.* Kansas City, Mo.: Heart of America Press, 1985.

Mitman, Carl W. "Watchmakers and Inventors." *Scientific Monthly* (July 1927), 25: 58–64.

Moore, Charles W. "Some Thoughts on the Early Labor Policy of the Waltham Watch Co." *Bulletin, Business History Society* (April 1939), 13(2):25–29.

Moore, Charles W. *Timing a Century: History of the Waltham Watch Company.* Cambridge: Harvard University Press, 1945.

Muir, William and Bernard Kraus. *Marion: A History of the United States Watch Company.* Columbia, Penn.: Mifflin Press, 1985.

The National Elgin Watch Company Illustrated Almanac. Rpt. Granada Hills, Calif.: Alvin A. Kleeb, 1972.

Niebling, Warren H. *History of the American Watch Case*. Philadelphia: Whitmore Publishing, 1971.

Niebling, Warren H. "Philadelphia Watch Company, 1868–1886, Eugene Paulus, President, Adolph C. Raufle, Treasurer." *Bulletin of the N.A.W.C.C., Inc.* (April 1978), 20(2[whole no. 193]):131–135.

Ohlson, Olof. *Helpful Information for Watchmakers*. Waltham, Mass.: American Waltham Watch Company, ca. 1900-1905. This trade catalogue was reprinted numerous times, the eleventh edition appeared in 1937. MPM H 45884/27141 and MPM H 45890/27141.

Penney, David. "The Name Engraved Upon a Watch Plate." *Antiquarian Horology* (March 1987), 26(5):491.

Piaget, Henry F. "*The Watch*." *Its Construction, Merits and Defects; How to Choose It and How to Use It*. New York: Published by the Author, 1868.

Piaget, Henry F. "*The Watch*." *Its Construction, Merits and Defects Explained and Compared. History of Watch Making by Both Systems*. New York: Published by the Author, A. N. Whithorne, 1877.

Pritchard, W. L. "American Watch Sizes." *Bulletin of the N.A.W.C.C., Inc.* (August 1975), 17(4[whole no. 177]):298–301.

Quelques Notes sur Pierre-Frédéric Ingold et les travau de E. Haudenschild publié par la Société Suisse de Chronometrie à l'occasion de so huitième Assemblée générale tenurèa Berne, au Burgerhaus le 4 juin 193?.

Rivett Lathe Manufacturing Company. *Rivett Precision Tools: Lathes, Milling Machines and Grinders a Specialty*. Breighton, Mass.: Rivett Lathe Manufacturing Company, ca. 1900. MPM H 35216/25222.

Robbins and Appleton. *American Waltham Watch Co's Materials*. New York(?): Robbins and Appleton, 1885.

Robbins and Appleton. *American Watch Co. Waltham, Mass., Incorporated 1854. New Orleans Exposition, 1884–1885*. New York: Eugene E. Adams, 1884.

Robbins and Appleton. "A CARD." *Harper's New Monthly Magazine* (July 1869), 39(230).

Robbins and Appleton. "To The Trade." *The Watchmaker and Jeweler* (September 1869), 1(1):15.

Rolt, L. T. C. *A Short History of Machine Tools*. Cambridge: MIT Press, 1965.

Sanderson, Edmund. *Waltham Industries: A Collection of Sketches of Early Firms and Founders*. Vt: Waltham Historical Society, 1957.

Saunier, Claudius. *Treatise on Modern Horology in Theory and Practice*. Julien Tripplin and Edward Rigg, trs. Newton Center, Mass.: Charles T. Branford, n.d.

Saunier, Claudius. *The Watchmaker's Hand-Book*. Julien Tripplin and Edward Rigg, trs. London: Crosby, Lockwood and Son, 1888.

Sawyer Watch Tool Company. "Watch Making Machinery." *American Machinist* (August 7, 1880), pp. 4–5.

Selchow, Frederick Mudge. "Belding Dart Bingham, 1812–1878 [and] The Nashua Watch Co., 1859–1862." *Bulletin of the N.A.W.C.C., Inc.* (December 1975), 17(6[whole no. 179]):539–574.

Shugart, Cooksey. *The Complete Guide to American Pocket Watches*. Cleveland, Tenn.: Overstreet Publications, 1981.

Smith, Alan. *The Lancashire Watch Company, Prescot, Lancashire, England, 1889–1910*. Fitzwilliam, N.H.: Ken Roberts Publishing, 1973.

Stark Tool Company. *Stark Tool Company: Manufacturers of Precision Bench Lathes and Fine Tools of Every Description*. ca. 1911. MPM H 43766/26878.

La Suisse Horlogerie et revue Instrumentale de l'horlogerie (December 1965), 80(?)(4):16–17.

Summar, Donald J. "The Osaka Watch Company—1889." *Bulletin of the N.A.W.C.C., Inc.* (December 1980), 22(6[whole no. 209]):670–671.

Tarasov, Sersefi Vasil'evich. *Technology of Watch Production*. Jerusalem: Israel Program for Scientific Translations, 1964. Translated from a 1956 Russian publication with an addendum by Edwin A. Battison, "Introduction to *Technology of Watch Production*."

Townsend, George E. *Almost Everything You Wanted To Know About American Watches and Didn't Know Who To Ask*. Vienna, Va.: George E. Townsend, 1970.

Townsend, George E. *The Watch That Made the Dollar Famous*. Arlington, Va.: George E. Townsend, Arva Printers, 1974.

Townsend, George E. *American Railroad Watches*. Alma, Mich.: George E. Townsend, 1977.

Townsend, George E. *E. Howard and Co. Watches, 1858–1903*. Kansas City, Mo.: Heart of America Press, 1982.

Treiman, Larry. "The Philadelphia Watch Company—Revisited." *Bulletin of the N.A.W.C.C., Inc.* (December 1978), 20(6[whole no. 197]):597–600.

Tremayne, Arthur. *One Hundred Years After: Being a Little History of a Great Enterprise*. Birmingham, England: Dennison Watch Case, 1912. The Time Museum Collection.

Tripplin, Julien. *Watch and Clock Making in 1889, Being an Account and Comparison of the Exhibits in the Horological Section of the French International Exhibition*. London: Crosby Lockwood and Son, 1890.

Turner, Anthony J. *Time Measuring Instruments*. Vol. 1, Part 3, Rockford, Ill.: The Time Museum, 1984.

Van Slyck, J. D. *New England Manufacturers and Manufactories*. "The American Watch Company," 1:23–25. Boston: Van Slyck, 1879.

Varkaris, Jane and Eugene Fuller. "H. R. Playtner and the Canadian Horological Institute." *Bulletin of the N.A.W.C.C., Inc.* (June 1987), 29(3[whole no. 248]):163–190.

Vigor, Bestfit Tools-Supplies-Equipment, Catalog BK-130. New York: B. Jadow and Sons, 1981.

Waldo, Leonard. "The Mechanical Art of Watchmaking in America." *Jewelers' Circular and Horological Review* (July 1886), 17(6):186–190. Read before the Society of Arts, London, May 19, 1886.

Waltham Watch Company. *Watchmakers' Handbook*. Waltham, Mass.: Waltham Watch, ca. 1930. MPM H 45855.01/27141.

Waltham Watch Company. *Serial Numbers With Descriptions of Waltham Watch Movements*. Waltham, Mass.: Waltham Watch, 1954.

Waltham Watch Company. *The Perfected American Watch*. Waltham, Mass.: Waltham Watch, 1907.

Waltham Watch Company. *Catalogue of Waltham Watch Material*. Waltham, Mass.:

Waltham Watch, April, 1909. Rpt. Yorktown Heights, N.Y.: Manfred Trauring, ca. 1970.

Waltham Watch Company. *Waltham and the European Made Watch*. Waltham, Mass.: Waltham Watch, 1919. MPM H 45886/27141.

"Watch Company of Massachusetts." *Reports of the Philadelphia International Exhibition of 1876*. London, 1877, 1:226.

Waterbury Watch Company. "Clock and Watch Land." *Century Magazine* (November 1881), 24(3):. Bound between no. 3 and no. 4. of volume 24.

Watson, James C. *American Watches: An Extract From the Report on Horology at the International Exhibition at Philadelphia, 1876*. New York: Robbins and Appleton, 1877. MPM H 34005/25250.

Wells, David A., ed. "The American Manufacture of Watch Movements." In *Annual of Scientific Discovery: or Year-Book of Facts in Science and Art for 1859*, pp. 65–69. Boston: Gould and Lincoln, 1859.

Wells, Louis Ray. *Industrial History of the United States*. New York: MacMillan, 1926.

The Whitcomb Family.

White, A. *The Chain Makers*. Christchurch: White, 1967.

Wingate, Robert M. "The Pitkin Brothers Revisited: Controversies and Comparisons of Their Hartford and New York Model Watches Revealed." *Bulletin of the N.A.W.C.C., Inc.* (August 1982), 24(4[whole no. 219]):381–391.

Woodbury, Robert S. "*A Short History of Machine Tools*, by L. T. C. Rolt." *Technology and Culture* (Spring 1966), 7(2):231–233.

Wyke, John. *A Catalogue of Tools for Watch and Clock Makers by John Wyke of Liverpool*. Introduction and Technical Commentary by Alan Smith. Charlottesville: University Press of Virginia, 1978.

Wyman, Walter F. *Export Merchandising*. New York: McGraw-Hill, 1922.

Ziebell, R. J. *Serial List from the Original Inventory of 1872 thru 1875, New York Watch Company, Springfield, Massachusetts*. Ipswich, Mass.: Old Post Office Clock Shop,

Index